The Peace Bridge Chronicles

Center Working Papers are works-in-progress or edited texts of interviews, lectures and symposia presented or supported by the Center for Studies in American Culture, a unit of the University at Buffalo, State University of New York.

The Peace Bridge Chronicles

Bruce Jackson

Center Working Papers

These articles first appeared in the weekly newspapers *Artvoice* and *Blue Dog*, and the online newsmagazine *Buffalo Report*.

The drawings on pages 12, 52, 161 and 431 by Tom Toles are reprinted with permission of the *Buffalo News*.

Center for Studies in American Culture
610 Capen Hall
University at Buffalo
Buffalo, New York 14209
csac@buffalo.edu
http://www.centerworkingpapers.com

ISBN 0-931627-04-4

For everyone on the right side in the Peace Bridge War.

In analyzing history do not be too profound, for often the causes are quite superficial.

Ralph Waldo Emerson, *Journals*

I have long been settled in my own opinion, that neither Philosophy, nor Religion, nor Morality, nor Wisdom, nor Interest, will ever govern nations or Parties, against their Vanity, their Pride, their Resentment or Revenge, or their Avarice or Ambition. Nothing but Force and Power and Strength can restrain them.

Samuel Adams, in a letter to Thomas Jefferson
October 9, 1787

Contents

The drawings on pages 12, 52, 161 and 431 are by Tom Toles

Abbreviations

BFEPBA	Buffalo and Fort Erie Public Bridge Authority
BNP	Buffalo Niagara Partnership (née Buffalo Chamber of Commerce)
DOT	Department of Transportation
EIS	Environmental Impact Statement
EPA	Environmental Protection Agency
FAQ	frequently asked questions
FOIA	Freedom of Information Act
FONSI	Finding of No Significant Impact
INS	Immigration and Naturalization Service
NAFTA	North American Free Trade Alliance
NEPA	National Environmental Policy Act
NFTA	Niagara Frontier Transportation Authority
NMG	New Millennium Group
NYSDOT	New York State Department of Transportation
PBA	Public Bridge Authority (short form of BFEPBA)
SEQRA	State Environmental Quality Review Act (New York)
UB	University at Buffalo

Foreword: DEADLINES

GETTING HERE

These articles all have to do with the political and social conflict that developed when the Buffalo and Fort Erie Public Bridge Authority (usually referred to as the PBA), a binational public benefit corporation, decided to increase the number of lanes on the Peace Bridge, the three-lane steel structure that connects Fort Erie, Ontario, and Buffalo, New York. They were responding to a volume of truck traffic that had been slowly increasing at the crossing through the 1970s and 1980s, and which increased dramatically with the implementation of the North American Free Trade Act in 1994. Their solution was to build a companion three-lane steel bridge, almost a direct copy of the 1927 bridge already in place.

In 1996, the city's only daily newspaper, the *Buffalo News* sponsored a bridge design competition, open to anyone. In 1998, the Buffalo Niagara Partnership (formerly the Chamber of Commerce) joined the Public Bridge Authority in sponsoring a design charette. Participants in the charette never were permitted to vote on the several alternatives to the PBA's plan and, at the end, the *News* and the Partnership both said that the public should support whatever choice the PBA made.

Opposition to the companion span developed steadily. A businessman and an architect, John Cullen and Clinton Brown, constituted themselves as "SuperSpan Upper Niagara PLC." They got people to sign petitions and make contributions that enabled them to take out ads arguing that instead of an anachronistic steel twin span, the PBA should instead build a six-lane bridge that would dramatize the border crossing.

Senator Daniel Patrick Moynihan, a passionate advocate of high quality public architecture, began saying that the PBA should plan imaginatively rather than copy dully, that it should use this opportunity to create a signature bridge, a bridge that would put an iconic stamp on the region.

Bruno Freschi, then dean of the University at Buffalo School of Architecture, in collaboration with the noted San Francisco bridge-maker T.Y. Lin, designed a bold and imaginative six-lane single-pylon, concrete, cable-stayed curved bridge to replace entirely the aging three-lane Peace Bridge.

Many people on the Buffalo side of the river thought the Freschi-Lin design a fine solution to the problem, preferable by far to the Public Bridge Authority's companion plan. Why use high-maintenance century-old technology, Freschi argued, and why build something ugly, when for the same amount of money or less you could build something beautiful? Senator Moynihan quickly became an advocate of Freschi's and Lin's signature bridge design. With that design, opponents of the twin span for the first time had a specific answer to the PBA's litany: "Our design works here. What do you have that's better?"

The problem was, the Buffalo and Fort Erie Public Bridge Authority wasn't the least bit interested in something better. It had made up its collective mind long before it even held public hearings to consider various bridge designs, long before the *Buffalo News* competition, long before the Partnership charette (which in retrospect turned out to have been a sham, "a charade, not a charette," as several architects later put it). The PBA would consider nothing other than its steel twin span. The Buffalo Niagara Partnership pushed hard for an immediate end to discussion and a quick start to construction. The *Buffalo News* agreed.

The PBA never wavered in its decision to copy the 1927 bridge until a New York judge and Buffalo's mayor and Common Council gave it no other choice.

In July 1997, Bruno Freschi urged me write an op-ed piece for the *News* about all of this. "They're always asking you to write opinion pieces for them. Write something about this. It might help."

He was right about my relationship with the *News*. The editorial page editor, Barbara Ireland, had several times asked me to write commentaries on public issues or cultural affairs. I usually did, and she always published them, as well as pieces I sent in on my own.

So I wrote a shorter version of the piece that stands first in this collection, "Talking Ugly." But things had changed at the *News*. Barbara Ireland had recently gone to the *New York Times* as op-ed editor, and Gerald Goldberg, her replacement at the *Buffalo News*, didn't know or care who I might be. Barbara had always responded in a day or two; Goldberg didn't respond at all. After few weeks I called and left a message, and then a few days later I called again. He said he'd looked at the piece that day but wouldn't use it because it was of no interest.

What Goldberg said next astonished me and is probably the reason I wound up writing the other 63 articles: "We had an editorial board decision and decided that the bridge is a dead issue. The *News* is not interested in the bridge question any more and we won't be running any more editorial comments on it." He wasn't unfriendly. He said that my writing showed promise and I shouldn't let their lack of interest in this piece deter me from sending something else in the future, if it was timely and of broad public interest. The only reason they were rejecting it was because it lacked timeliness and significance.

The largest construction project in this part of the country since the new campus of State University of New York at Buffalo 30 years earlier, an opportunity to have something grand instead of something boring, a chance to do something useful for a community that had been savaged by lousy public works decisions for decades, a project with huge environmental issues – and the only daily paper in town decides it's "a dead issue"!

I told everybody. My son Michael, a Buffalo attorney and musician, said, "Send it to Jamie Moses at *Artvoice*. He'll print it."

I tweaked the article a bit, made it a little longer, and mailed it. Moses published it about five days later, in the August 5 *Artvoice*.

There was an immediate reaction. Old friends and people I'd never met called or emailed me about the bridge issue. Some said, "Leave it alone, you can't change decisions like this once they've been made." Others said, "What can I do?" WNED, the city's public television station, asked me to do a pro-and-con about the signature versus twin span issue with Andrew Rudnick, president and CEO of the Buffalo Niagara Partnership, an aggressive steel twin-span advocate.

That, I thought, was the end of it, at least of my part of it – until the last week of January 1999, when Jamie Moses called and said, "Don't you think it's time you did another Peace Bridge article?"

I wrote "A Bridge Not Too Far," which appeared February 18. A few days later, Gerald Goldberg called and said that he'd read it, that the *News* thought the paper should encourage discussion about journalistic practice, so they would like to publish parts of the article. "Only," he said, "it has to be a little shorter."

"You want to cut out the parts criticizing the *News*," I said.

"Oh, no," he said. It was specifically the criticism of the paper they wanted to bring to a broader readership. It would be fine with him if I shortened the piece and they'd go with my cuts and wouldn't alter it without talking it over with me first.

That seemed fair and reasonable. Moses agreed to let the *News* reprint the piece. I think he was delighted and amused at my report of Goldberg's call.

It appeared as an opinion piece on the March 27, 1999, *Buffalo News* editorial page, with the title "There's still time to fix this, so let's build the right bridge." Goldberg, contrary to his promise, made two cuts without telling me, both in the same paragraph:

> Buffalo is a one newspaper town. Were the *Courier-Express* still alive we would have seen a lively journalistic debate about the merits of the various proposals and the Authority's refusal to consider them seriously. But we have no second daily paper, so there was no debate. "We had an editorial board decision and decided that the bridge is a dead issue," Jerry Goldberg, the *News* editorial page editor, told me. "The *News* is not interested in the bridge question any more and we won't be running any more editorial comments on it." The only journalistic challenge to the Authority has come from *Artvoice*, a weekly alternative paper.

I didn't know it then, but that encounter would prove typical of the coverage the *Buffalo News* gave the Peace Bridge War over much of the next four years: selective, distorted, at once self-serving and in the service of unnamed interests. All that understanding would come later. At that point, I was just appalled at the broken promise, at how the *News* had cut the core sentences about its own behavior and identification of what *Artvoice* was.

TARBABY

I meant to write that first article on the bridge that Bruno talked me into writing. And I meant to write the second one six months later, after Jamie Moses called. But I never planned to do the rest of them. They just happened. The Peace Bridge War was like Tarbaby: once I started slapping at it, there was no way I was able to shake loose.

The longer it went on, the more I learned. The steel twin span wasn't just bad design; it was also lousy planning. The project would disrupt traffic in Buffalo for most of a decade and the final truck flow would add a huge amount of air and noise pollution to a section of the city already suffering some of the region's highest rates of lung disorders. In the short run, there would be jobs created by the steel construction project – but most of them would go to distant manufacturers; a bridge incorporating modern design and technology – concrete cable-stayed – would bring jobs here. In the long run, the anachronistic design favored by the Public Bridge Authority and the Buffalo Niagara Partnership would add virtually nothing to Buffalo's economy, and it would impose continuing harm on Buffalo's infrastructure. The steel bridge would be vastly more expensive to maintain, which most people saw as a disadvantage but which a few saw as a way to make a lot of money for a long time, and those few seemed to be influencing the major decisions.

People had to know those things, they had to know that many of the claims by Bridge spokesmen were absurd or mendacious, and that the *Buffalo News* wasn't reporting all it might. More and more people got involved and they sent tips, questions, reports, leads. One email message began, "I work at the Partnership and I want you to know that a lot of us on the staff don't agree with what Andy Rudnick is saying and I can tell you that they never polled the membership about the bridge issue."

How can you ignore things like that?

THE ARROW OF TIME

Most of my career has been as an academic and most of my books are academic books. In the course of writing about the Peace Bridge War I came to appreciate the difference between journalism and scholarship as I never had before.

With journalism you hand in your copy some time between the event you're reporting on and your paper's deadline. If it's an ongoing story, you publish ongoing reports. There's an arrow of time: what you know and think at the end of a series of stories may be very different from what you knew and thought at the beginning of the series, and the writing reflects that. Every article along the way includes, overtly as part of the text or covertly as part of your understanding, what you learned last time and the time before.

With scholarship, you publish your work after everything is over. The last thing you find out or decide modulates every single page. If at the end of your research and thinking you have a key insight, you alter everything to incorporate that key insight. What the reader sees first – the introduction – is the section you write last of all. How could it be otherwise? You cannot introduce what does not yet exist. What you're reading at this moment is the 65th Peace Bridge Chronicle.

Years ago, when I was writing primarily for academic journals (which published articles a year or two after submission) and for literary periodicals such as *Atlantic* and *Harper's* (which published articles three or four months after submission), a friend who was a reporter on the *Charleston Gazette* said to me, "You have all the time you want to write what you think, and by the time it comes out you don't remember what it was any more. What I write tonight is on my doorstep tomorrow morning, and it's wrapping fish the next day."

At the time, I thought what she said applied only to cloture: she could be done with something and on to something new far faster than I. But it's more than that: in scholarship, you're responsible for all of your facts. In journalism, you're responsible for only that day's facts. If, in journalism, you meet on Thursday someone who gives you data showing a totally new aspect of what you had in the paper on Wednesday, that's just material for an article that will be in Friday's or Saturday's paper. If, in scholarship, you meet on Thursday someone who gives you data showing a totally new aspect of what you just sent to the publisher, you send the publisher a letter saying, "Hold it, gotta make some changes." And if the publisher says, "Too late," you take a lot of Maalox.

There's another major difference in the two modes of writing: the way each regards and utilizes what people say or the documents people or organizations create. In daily journalism, spoken or printed utterances are their own validation; they're what the story is. It's perfectly adequate to have an article about the mayor's press conference that does nothing more than quote or summarize what the mayor said, or to have an article about a report delivered to the press by a corporation or public agency that does nothing more than quote or summarize what the report seems to say.

In scholarship, utterances are only data; they're not just used, they're also examined. What the mayor said and what the report claimed are of no moment in themselves. They matter only insofar as they become part of the analysis or exploration or explanation of something larger, more complex, more ranging, deeper.

Neither mode is of itself better or worse than the other. They serve different ends and therefore ask different questions. Writing without a deadline for a weekly newspaper lies somewhere between the two: sometimes you're just getting the facts out, sometimes you're trying to figure out what the facts mean, and sometimes you're trying to do both at once.

As time went by, I learned a good deal more about the players and the games. Some things I said in the early articles I might not have said later on, or I might have phrased them differently. I might have asked different questions. People I thought were straight turned out to be lying without pause, people I thought mendacious turned out to be impeccably straightforward and honest. In scholarship, when you don't publish till you're satisfied with what you've got, there would have been time to sort such things out. But journalism is serial, it takes place, as I said earlier, only one way in time. You don't get to undo early errors; you just try to get it right this time.

One of the great pleasures of scholarly writing is, when you're done with it, you're done with it. You can move on to something else. One of the great pleasures of newspaper writing is, when you're done with today's version, tomorrow always gives you a chance to do it better, deeper, another way.

CHANGES

I didn't do any rewriting for this collection, though the temptation to gloss the earlier pieces with what I learned later was pretty strong. This is a book about a developing event, and part of the story is how thing moved in and out of focus. The myriad aspects of the story call to mind exuberant Whitman in the penultimate section of "Leaves of Grass":

> Do I contradict myself?
> Very well then I contradict myself
> (I am large, I contain multitudes).

There are repetitions here, and contradictions too.

Daily and weekly journalism are by their very nature repetitious and contradictory. People reading your article this week may not have seen the articles you wrote last week and the week before, so you have to identify the players and explain the actions if they are to know what you're talking about. Some behaviors are repeated and some issues keep coming up: just as a thief who is apprehended stealing again and again gets his name in the paper again and again, so too in these pages, such behaviors as the *Buffalo News*'s persistent practice of referring to the Buffalo and Fort Erie Public Bridge Authority by a deceptive name is noted again and again.

Contradictions happen because you learn things as you go, so the way some behavior appeared last week may not be how the same behavior appears this week. Daily or weekly newspapers rarely have space for reexamination of previous notions: mostly, you tell it the way it seems to be now.

There's one syntactical inconsistency I noticed when I reread all these pieces that I thought best left alone: sometimes I refer to the Public Bridge Authority in the third-person singular (it) and other times in the third-person plural (they). I think that's because sometimes I was thinking about a bunch of people and other times I was thinking about an organization. As a long-time English professor, I know that's probably not a very good rationale, so I'll take refuge in the fact that I live on the Canadian border and the Peace Bridge is a bi-national entity: Americans handle collective nouns as singulars ("The team is...."), Canadians handle them as plurals ("The team are....")

I did correct a few errors of fact, such as occasional incorrect first names for a Canadian federal official and a member of the Public Bridge Authority. In the *Artvoice* articles I often referred to the "Olmsted Conservancy"; the correct name of that organization is the "Olmsted Parks Conservancy." I decided that historical fidelity to the way things first appeared in *Artvoice* was less important than using the correct name here, so I added the missing medial. None of the changes is more substantial than that.

THE PEOPLE RESPONSIBLE

My byline appeared on these 64 articles but I could not have done any of them on my own. Scores of individuals provided ideas, questions, answers, leads, reports, suggestions, corrections, pictures, tapes, tips, charts, xeroxes, and memos. It would be impossible to name everyone who helped but I must mention one group and one individual in particular: the New Millennium Group and Jamie

Moses.

The New Millennium Group, young professionals trying to make Buffalo a viable city, took on the Peace Bridge as one of their major issues, and over the two most important years of the Peace Bridge War they consistently kept the issue in front of the public, provided detailed and accurate data to public agencies and the press, did everything that a fine advocacy group should do. Buffalo should be proud of them. I'm in their debt. We all are.

Jamie Moses, publisher and, until recently, editor of *Artvoice*, decided early on that this was a key issue for the city and he put the paper at the disposal of those of us who chose to write about it. The average op-ed piece in the *New York Times* or the *Buffalo News* is under 300 words. Eight of the pieces here are over 5000 words – one of them, the interview with Bob Knoer is over 7000, and one, "The Great Autumnal Q&A" is over 13,000. Jamie was equally generous with space for the excellent Peace Bridge articles by Tom Schofield and several other writers. Jamie's own article on the subject, "IT'S WAR!" published in the March 18, 1999 issue, made many Buffalonians aware of the battle for the first time. None of the other alternative newspapers in Buffalo covered the Peace Bridge War until much later. One of them published articles arguing that the whole fight was pointless and that people who weren't professional urban planners had no business interfering in decisions made by their betters. Until it was clear that public opinion has shifted strongly against the twin span and in favor of a signature span, the *Buffalo News* counseled the same. *Artvoice* provided the only public forum in which continuing informed discussion about the Bridge could occur, and that was critical to the outcome.

On the subject of the *Buffalo News*: I scold it frequently in these articles for shoddy news coverage of and irresponsible editorial commentary on the Peace Bridge affair. However derelict the *Buffalo News* was in its Peace Bridge news coverage and editorial commentary, it also provided space for the work of three observers of the Bridge scene who frequently contradicted the paper's news and editorial writers: columnists Doug Turner and Donn Esmonde and, especially, editorial cartoonist Tom Toles (who became editorial cartoonist for the *Washington Post* in July 2002).

I'm grateful to *Buffalo News* editor Margaret Sullivan for allowing us to use of four of Tom's cartoons in this book.

I'm especially grateful to Tom for letting us include his drawings

and, more importantly, for having done them and all the others on so many other topics for so many years, for using his great wit and talent to illuminate so many important issues.

Thanks also to Diane Christian, whose generous assistance all the way was indispensable. Diane read or listened to all of these articles before they went off, often helped in the information gathering and sorting along the way and, at the end, checked the proofs for errors and inconsistencies.

A huge number of people and groups were involved in stopping the twin span disaster and forcing the Public Bridge Authority to behave decently. Many, but nowhere near all, of those people are named in these 64 articles. They're all people who cared enough about the quality of life in their hometown to fight a long and difficult battle in the hopes that the kind of thinking that, in previous decades led to public works projects that destroyed parkland and gutted neighborhoods, cut the city off from its waterfront, destroyed downtown business life and even let a huge office building deny lower Main Street direct sunlight, wouldn't, this time, get to degrade the quality of urban life even more.

These are some of them: Guy Agostinelli, Bill Banas, Elissa Banas, Jeff Belt, Clint Brown, Sebastian Ciacio, Luis Clay, Bob Coles, Ed Cosgrove, Joe Crangle, Jack Cullen, Pam Earl, Donn Esmonde, Gene Fahey, Bruno Freschi, Tony Fryer, Andres Garcia, Kevin Gaughan, Sam Hoyt, Gail Johnstone, Jim Kane, Bob Knoer, Bob Kresse, Steve Lana, Mary Catherine Malley, Randy Marks, Vic Martucci, John Mander, Tony Masiello, Mark Mitskovski, Laura Monte, Jamie Moses, Eli Mundy, Pat Moynihan, Jim Pitts, Ross Robinson, Sister Denise Roche, Joe Ryan, Andrea Schillaci, Tom Schofield, Chuck Schumer, Judie Takas, Joe Tauriello, Rich Tobe, Tom Toles, Jeremy Toth, Doug Turner, Don Wharf, Ed Weeks, Fr. Jud Weiksnar, Lynn Williams, all the members of the New Millennium Group and the Olmsted Parks Conservancy, the board of the Episcopal Church Home, WNED, and all the other groups and individuals who continue to believe that a better Buffalo is worth the struggle and that the public has every right to decent air, usable parkland, and beautiful architecture.

They cared, they worked, and what they learned this time will profit us all in the public works wars yet to come. I dedicate *The Peace Bridge Chronicles* to all of them because these articles are every bit as much

PEACE BRIDGE AUTH.
MAYBE WE DON'T NEED A NEW BRIDGE AFTER ALL.
MAYBE, IF YOU INSIST ON PARTICIPATING IN THE DECISION-MAKING PROCESS, WE DON'T NEED ANY BRIDGE.
SO THERE!
TOLES
2000 THE BUFFALO NEWS
SPEAKING OF THINGS WE MAYBE DON'T NEED...

theirs as they are mine.

THE WAR GOES ON

I wish I could say there would be no more Peace Bridge Chronicles, that the Buffalo and Fort Erie Public Bridge Authority, the town of Fort Erie and the city of Buffalo had agreed upon a design, that the design was aesthetically pleasing, environmentally responsible, and politically and economically feasible.

But I can't. The design process is still going on, and it will continue for some time to come. Anti-terrorism legislation and conditions that may be imposed by the Department of Homeland Security have complicated issues that not long ago seemed simple, and re-opened issues that seemed to have been resolved. Individuals and corporations more interested in making money than serving the public continue to make all kinds of mischief, some of it public, some of it secret. After three years of almost total silence on the subject, Bruno Freschi and T.Y. Lin, whose imaginative design energized this whole process five years ago, seem about to enter active competition with the other designers.

So the Peace Bridge War continues, and I'll continue chronicling it as best I can. What follows is what I've been able to learn thus far. More to come.

B.J.
Buffalo, June 1, 2003

August 5,1988

1 TALKIN' UGLY

Some years back one of the Buffalo TV stations ran a public service feelgood campaign about the city. The theme was "Talkin' proud." There were all these logos and short commercials in which "talkin' proud" figured. I don't remember the commercials saying what Buffalonians should haven been talkin' proud about or even how one talked proud. The commercials just said that we were supposed to DO it, and they implied that talkin' proud alone could provide things to be proud of.

There was one commercial in particular that I hated. It showed a thirty-ish woman in a high-school twirler's outfit who bounced down the steps of the Albright-Knox with her body and head tilted far back like she had a spinal deformation or hip dislocation. She twirled her baton and the voiceover said, "Buffalo: talkin' proud!" A friend visiting from Paris saw the commercial. He said, "That is the most depressing commercial I've ever seen. And you say they show it all the time? God!"

I thought about conversation a few days ago while I was driving along 190 south and looking at the Peace Bridge. There's been a lot of sentimental drivel about the Peace Bridge going back and forth in the *Buffalo News* in the past few months and some people have talked and written about it as if it were a significant part of the Buffalo landscape. That it is, but it's an also an ugly part of the landscape.

Now we've got a chance to put up something really beautiful in its place but the Peace Bridge Authority wants to twin the old bridge, to have two ugly bridges where we now have only one ugly bridge, and that will be really depressing. It will be like that "Talkin' proud" commercial, one of those wasted moments when resources were available to do something wonderful but they were applied in the service of the dreary.

And every time you drive that drive, whether or not the ugliness rises into your conscious thoughts or not, ugliness will be coming into you. Every time. Trust me on this.

Here's something all poets, artists, musicians, backpackers, kings and rich people know, and you know it too: beauty is a good thing to have around because it makes you feel good. It's that simple and it's unvaryingly true: beautiful things make you feel good. A gorgeous sunset, your lover's or baby's smile, a hawk back-and-forthing on the afternoon wind for an hour without a single wing beat, a painting or photograph that gets it just right, a riff that soars beyond the moon, bread just out of the oven. God talks to us in beauty.

And the opposite is also true: ugly things make us feel lousy. People who are surrounded by ugliness all the time often act ugly when they don't even want to. The ugliness leaks into their pores. People trying to fight their way out of poverty know that all too well. People in war know that all too well. No way you can feel good when you're surrounded by ugly. That's what ugliness is all about. That's what ugliness does.

There are wonderful bridges out there in the world. The three I know best are the Brooklyn Bridge connecting Brooklyn and Manhattan, the Golden Gate connecting San Francisco and Marin County, and Pont Neuf connecting Paris droit and Paris gauche. All three make people feel good when they go across them, when they look at them, when they remember them. The pleasure never stales. I first walked on the Brooklyn Bridge when I was a little kid and I did it again a few years ago. There's more than half a century in that sentence. The Bridge is every bit as glorious as it ever was.

The great thing about great bridges is they work for everybody. Not just for rich folk who can buy paintings for their walls and theater and concert tickets. Not just for art critics and aestheticians, talking only to one another. Not just for politicians and crooks and people who care only about money to be made or power to be gotten.

The Freschi-Lin Bridge could work for us. And for our kids – these bridges last a long time, as the present ugly bridge demonstrates so well.

"I don't see what the fuss is about," a Canadian official said to me a few weeks ago. "A bridge is supposed to get people from one side to the other. Beauty has nothing to do with it. If it gets people from one side to the other it's done its job. That bridge in Detroit is ugly. It gets people from one side to the other and you don't hear anybody complaining about that bridge, do you?"

Do you think for a moment Canadian officials would be saying "beauty has nothing to do with it" if the city in the equation were Montreal or Toronto or Vancouver? Beauty has nothing to do with it for the Canadian official because Fort Erie is a zit on Canada's ass. Fort Erie is where trucks pay import duty and head north to dump their loads. It's gooch joints where American teenagers and businessmen get their laps damp without the guilt of penetration. Fort Erie is Ugly. Canada doesn't give a shit about Fort Erie. Nobody gives a shit about ugly.

Since taste, logic, common sense, budget, timelines and even the eminently sensible Pat Moynihan say "Go with the Freschi-Lin Bridge," I keep asking myself "Why doesn't the bridge Authority thank Freschi and Lin for their enormous gift to the region and start implementing their plan immediately?"

I'm a writer so in my professional work I regularly try to come up with plots that will make sense of things that don't make sense. If this group appointed to oversee bridge tolls and bridge maintenance is opting for ugliness, then maybe there are massive political debts being paid you and I know nothing about. Or major payoffs you and I know nothing about. Or major gangster interests you and I know nothing about. Or profound aesthetic cataracts you and I know nothing about.

Well, okay: if those things are there, we know nothing about them and if they're anything like other things of their type, we never will, so there's no point our fretting about politics or payoffs or gangsters or aesthetic cataracts. We should stick with what we know.

What we know is we can have beauty or we can have ugly. There's no choice – no rational, legal, decent, aesthetic choice, that is.

If the people making the decision decide that "beauty has nothing to do with it," we'll get another chance to get it right 80 or 90 years from now, when the descendant of the current Bridge Authority thinks about tuning or redoing the job. You and I will have spent what's left of our lives, and our children will have spent their entire lives, seeing again and again the results of a sensibility that says "Beauty has nothing to do with it."

2 A Bridge Not Too Far

The Done Deal

It was supposed to be a done deal. The twin span bridge from Buffalo to Fort Erie proposed by the Buffalo and Fort Erie Public Bridge Authority and endorsed by the Greater Buffalo Partnership (the fancy name for what used to call itself the Chamber of Commerce) was inevitable. The alternative proposal, the graceful signature bridge designed by UB Architecture Dean Bruno Freschi and San Francisco engineer T.Y. Lin, was dead.

No matter that the Freschi-Lin bridge would have been fully functional years earlier, cost less, be kinder to the environment, consume less public land and help reclaim some of the Olmsted park land now consumed by the inefficient toll plaza the Authority's plan would not only keep but enlarge. No matter that it would have given Buffalo an identity recognized around the world. No matter that it would have provided the city something that is really beautiful.

The Bridge Authority and the Greater Buffalo Partnership had their reasons for their stodgy design and they weren't budging. We'll never know the real reasons why they came up with and got so defensive about that ugly and expensive twin span. We'll never know if the determining factor was ego, money, bad taste, or all three. All we know is they had their twin span idea early on and they never seriously considered any alterative.

That became clear at a public meeting of the International Joint Commission held in Amherst on January 28. Commissioner Susan Bayh asked the Bridge Authority "if you ever considered a single-span, single-pylon bridge?" The answers were all evasive. One focused on the problem of having a bridge close to a water intake, which it turned out had nothing to do with the Freschi-Lin design. Another argued that the current bridge couldn't be taken down because it was a historic structure. The current bridge has never been declared a historic structure by any agency of any government; it's old and ugly, it's not historic. The bridge manager argued that the cables on the bridge might cause problems for migrating birds. No ornithologist has suggested that migrating birds are so stupid they can't fly around a bridge. The PBA never gave Commissioner Bayh a straight answer to any of her questions.

PAT TO THE RESCUE

The PBA plan began to unravel when the construction bids came in last month One of their major justifications for rejecting the Freschi-Lin signature bridge was their twin span was far cheaper. Their numbers, we now know, were fictional. On January 5, an editorial in the *Buffalo News* reported that the bids put the project "$10 million to $15 million over budget." It was far worse: a news article in the same edition of the *News* said the low and high bids were $15 to $45 million over the projected budget, differences of 38% and 69%. The *News* editorial justified the Authority's gross error in budgeting: there was, the paper said, "a national abundance of bridge projects that erodes competition among bridge builders."

That seemed so goofy I did a Lexis search. No other newspaper in the United States has reported increasing bridge costs because of "a national abundance of bridge projects." I couldn't even find the abundant bridges.

The PBA probably would have had its way had it not been for the persistence of Senator Daniel Patrick Moynihan. He'd been a supporter of the signature bridge plan all along, but except for Buffalo Common Council member James Pitts, he was out there by himself. Then Charles Schumer replaced Alfonse D'Amato in the Senate and shortly thereafter he and Moynihan coauthored letters to the American Canadian Joint Authority and the US Coast Guard asking them to block construction of the twin span.

They asked Admiral James M. Loy, Commander of the Coast Guard, to "deny the permit for the Peace Bride Capacity Expansion Project as submitted by the Buffalo and Fort Erie Public Bridge Authority on January 26, 1998. We make this request in light of recent developments that cast serious doubts as to the economic and environmental viability of the Authority's proposal to construct a new twin span and rehabilitate the current Peace Bridge structure." They noted that the "attempt to segment the bridge construction from its related plaza and approach road improvements represents a purposeful effort to evade the scrutiny of alternatives that would otherwise be required."

(The Bridge Authority has consistently insisted that the bridge and toll plaza projects be separate. Freschi's original design for the signature bridge was up river from the present site, a design that would have made for a shorter bridge and would have returned to the public some of the Olmsted park area that the current plaza occupies. When he submitted his plan to the Authority he was encouraged to draw it so it would land at the current plaza. He said that was poor planning and poor design, but the Authority insisted.)

In their letter the next day to the International Joint Commission, Moynihan and Schumer repeated the problems they had reported to Admiral Loy and also said that "alternative proposals exist that have not been fully evaluated."

TALK IS MONEY

The Authority has consistently argued that any inquiry into the grounds for their decision – the money, the water flow, whatever – would extend the time for the new bridge, so the public would be hurt. To every attempt to question any of their assumptions they say, delay will increase costs. They argued the Moynihan-Schumer letters on the grounds that looking at the impact of the two-phase construction project would cost more money. The gambit is, even if they did a rotten job we shouldn't question it because pausing to fix their errors will increase costs.

The *Buffalo News* bought that cockamamie logic. In a December 28 editorial urging Attorney General Spitzer to keep Dennis Vacco appointee Brian Lipke (president and CEO of Gibraltar Steel and a big Republican party contributor) on the PBA, they concluded that we shouldn't "reopen the design debate" because that would confound the timetable and increase costs. "What is clear," the *News* said, "is that the Authority cannot let the unexpected extra costs of bridge construction erode its stated commitment to building a 'gateway' US plaza that will embody the community's hopes for symbolism and greatness. Buffalo deserves, and demands, that much at least."

No: Buffalo deserves a good deal more, both from its bridge Authority and from its only daily newspaper.

Attorney General Spitzer did not take the *News*' advice: he replaced Brian Lipke with former Buffalo Common Council member at large Barbra Kavanaugh, and announced that he and Kavanaugh were strong supporters of the signature bridge and the efforts of Senators Moynihan and Schumer. There was an immediate Republican response: State Department of Transportation Commissioner Joseph Boardman pulled Robert Russell off the Authority and gave his seat to Brian Lipke. Lipke won't chair the PBA any more – it's time for that position to cycle to one of the five Canadian members – but he'll still be on the PBA to argue for the twin span, to which he is passionately devoted.

A One Paper Town and the Silence of the Pols

This is a major construction project, one that will matter in public life here for a hundred years, yet the mayor has kept silent, as have all of the common council members except James Pitts, all of the state legislators, and all of the region's congressmen.

What's really odd about all this is, hardly anybody except Brian Lipke defends the twin span. You talk to the mayor, the council members, all the others, they say "Oh, it's terrible, it's just awful." You talk to executives at the *News* and they say the same thing. It's like they're talking about the weather: something that just happens, something we relate to passively because there's nothing to be done about it.

Being a one newspaper town hurt us badly in this sorry process. Had the *Courier-Express* still been alive we would have seen a lively debate in the press about the merits of the various proposals and the PBA's refusal to consider them seriously. Some of those silent pols would have been challenged and they would have had to speak out. But there is no second daily paper and the only one we have decided the issue just wasn't important enough for editorial examination. "We had an editorial board decision and decided that the bridge is a dead issue," the *News* editorial page editor told me late last July. "The *News* is not interested in the bridge question any more and we won't be running any more editorial comments on it." Without major competition and with nothing to gain, the *News* just backed off and the politicians hid.

I've heard two hypotheses for the silence of the pols. One is that rich developers paid everybody off. I think that's just conspiracy theory. Developers don't care which plan gets adopted – they just care that SOME plan gets adopted. The construction industry is going to make its money whatever design is selected. The other hypothesis is more likely: the pols all felt that the Authority was so entrenched in its position, the Canadians so opposed to Buffalo getting a signature bridge, that there was no way they could win.

Politicians rarely join battles they don't think they can win.

Word is out that the enthusiastic pursuit of the signature bridge by our two senators and the state's attorney general has the editorial board at the *News* rethinking its hands-off decision of last July. One editor there said, "Maybe there's time to do something after all." Well, that's news.

THE BIG LETDOWN

There's been all this drivel the past few months about how Bill Clinton let us down by unzipping his fly for Monica in the oval office and then trying to pretend it hadn't happened. Bah! That did me no harm that I know of. I'm far more concerned by how we were let down by the NFTA, the chairman of which joined the PBA board and just went along with everything Brian Lipke said. I'm concerned by how we were let down by every one of our local, state and national representatives except Moynihan, Schumer, Spitzer and Pitts. I'm concerned by how we were let down by the *Buffalo News*, which early on argued against the twin span and then when it seemed that the Authority wasn't going to budge just folded and said to hell with it.

They all stood silent and let this idiocy happen. Bill Clinton's turpitude was nothing compared to the way those elected and appointed and hired officials let you and me down.

HERE'S WHAT TO DO

Cities don't get many chances for greatness and Buffalo has shot itself in the foot before: Route 190 isolated most of us from the waterfront and Route 198 bisected one of Frederick Law Olmsted's greatest urban parks. The convention center chokes downtown streets, a huge office building squats over lower Main Street, and the trolley system strangles downtown business. Building the new UB campus on a suburban Amherst swamp rather than one of the three available Buffalo sites shifted a huge middle-class wage-earner base out of the city. It's time we got one right.

After my previous article on the Peace Bridge follies in these pages several people asked me what they might do. At the time, I didn't have much to suggest because the project seemed dead. Now I don't think it is. Moynihan, Schumer and Spitzer are right: we've got to fight this, and we can.

Call the mayor and tell him you're sick of his inaction and his silence and you'll remember it next election. Call your Common Council representative, your congressman's office and your state representatives' offices. Call the governor's office. Senator Moynihan has asked Governor Pataki to join the fight for the signature bridge, but he hasn't responded. Pataki is stoking up for a presidential campaign and he might not want to go into that tagged as the governor who looked at beauty and endorsed ugly.

Let them all know that you feel you feel cheated by their failure to act now and that you'll remember it next November and the November after and the November after. Tell them you'll remember what they did for you or to you every time you drive along the Buffalo waterfront and see either the pathetically ugly and hugely expensive twin span or the gorgeous soaring signature bridge that is still a very real possibility. Call Moynihan, Schumer, Spitzer and Pitts and say thank you for fighting for us and please keep it up.

All those calls will take you maybe 30 minutes. If enough of us do it and the pols take us seriously, we'll get paid back every time we cross the Freschi-Lin bridge and see it curving into space before us, and every time we drive into or out of the city. It is still possible to avoid the truly ugly.

3 SINKING THE TWIN SPAN

Last week the Buffalo and Fort Erie Public Bridge Authority ran a full-page ad in the *Buffalo News* justifying its refusal to consider any Peace Bridge plan other than its own. They've also been running multiple radio ads. The ad campaign remind me of Mary McCarthy's famous evaluation of Lillian Hellman: "Every word she writes is a lie, including the 'ands' and the 'thes.'"

Contrary to the ad's claims, the current bridge is not a historical treasure, it is decrepit and needs massive repairs, the only deadlines missed by taking the time to do this right are deadlines the Authority itself imposed, and the Authority's twin span will cost more, take longer to complete, and will generate far worse traffic congestion than the six-lane concrete signature bridge proposed by Bruno Freschi and T.Y. Lin. Even the end of the ad – "Peace Bridge Authority" in large letters – misleads: that's not the Authority's legal name.

Attorney General Elliot Spitzer says the Authority never held the required public hearings. Why didn't a board member point that out before we got into the present mess? Because in March 1997 Attorney General Dennis Vacco substituted Brian Lipke, president and CEO of Gibraltar Steel, for the Board's traditional watchdog of the public's interest, the head of the Attorney General's Buffalo office.

Brian Lipke loves the twin span design. Steel is his business and he has no interest in the Freschi-Lin concrete signature bridge. Neither is he interested in aesthetics: "The most beautiful bridge in the world," he said in 1997, "is one that works and pays for itself." Lipke owns more than half his family's $250 million in Gibraltar common stock, a company his father bought in 1972 for $1 million. He is a major contributor to and fundraiser for the Republican Party, which is why he got the seat on the PBA board and why, after Spitzer replaced him, the Pataki administration gave him the seat ordinarily held by the State's Department of Transportation. That's two people's representatives Brian Lipke got to displace because of his clout with high-placed Republicans.

Both of New York's US senators, its Attorney General, the Erie

County Executive, nearly all members of the Buffalo Common Council, the Baird Foundation, the Landmark Society, many other officials and organizations, and most of the public favor the Freschi--Lin signature bridge. The only elected officials of note who support the twin span are Mayor Anthony Masiello and Congressman John LaFalce. Neither has offered any substantial reason for his position, but Masiello's choice is easy to figure out: he's been in bed with Pataki since before the last election and he's now dancing to tunes you and I can't hear. I can't understand LaFalce's position on this, since on most public interest issues he's been a pretty sensible guy.

Masiello and LaFalce are now in an ever-thinning minority, so we needn't fret about them. The really puzzling question is why Brian Lipke and his supporters cling to and vigorously defend a design hardly anyone wants or likes. Some people say it's all about money, but that's too simple. Most of the people who are going to make money on this will make money whatever bridge design is selected. Not all, but most.

The Authority's entrenchment began to make sense to me last week when I heard another PBA board member angrily tell a TV reporter that if people wanted a signature bridge they should have come up with the idea three or four years ago, that the board had invested a lot of time and energy working on the twin span and it wasn't right to just move on to something new now. No, it just wasn't right. Not after they'd done all that work.

Sound familiar? It's the logic of poor poker players seeing bets when other players obviously hold better cards, and it's the logic of the US government staying in Vietnam long after it should have gotten out: "We've invested so much, we can't back out now." It is the fallacy of sunk costs, familiar to anyone who has taken Economics 101. "Sunk costs" are what you spent in the past; the fallacy is when you let those old expenditures control what you do next.

Would Brian Lipke modernize a decrepit steel plant if projections showed it would be a loser anyway? Why can't he and his supporters say, "Our staff produced a plan we liked, a better plan has emerged, let's go with that"? You and I shouldn't have to pay for and live with an expensive, ugly and anachronistic twin span only because the Authority can't or won't apply to public architecture the same good sense Brian Lipke applies every day to matters of cold hard steel. It's time for them to cut their losses and tend to our future.

4 "Don't You Build No Ugly Bridge": This Week's Meeting of the SuperSpan-Signature Bridge Task Force

Mister Mayor

Mayor Masiello's spokesman, Peter K. Cutler, said Mayor Masiello couldn't attend Monday's meeting of the city's SuperSpan Signature Bridge Task Force because he had a long-scheduled vacation in Florida and his wife would have punished him if he took time out for city business. People smiled politely; it is possible but not likely that someone in the Common Council conference room bought the excuse. We all knew that there are a good number of planes to and from Florida vacation spots every day and that he could probably have flown up for this meeting and been back with the family before dinner. But this was not just a meeting with Senator Charles Schumer, the Task Force, a good number of other area political figures and concerned citizens; it was also a meeting at which someone would almost certainly have asked, "Where and with whom do you stand, Tony?"

Cutler said the mayor wanted all to know that he was not, as has been charged by some, the only elected Buffalo official in favor of the twin span (although he phoned *Artvoice* on March 16 to go on record as being in favor of the twin span). He wasn't now in favor of twin span OR signature bridge, what he was in favor of was "what's best for the community." In addition, "The mayor is a supporter of the best possible bridge between the two countries." People smiled politely: only a churl could be opposed to that kind of international goodwill.

Then Cutler described a meeting between Mayor Masiello and the executive committee of the Buffalo and Fort Erie Public Bridge Authority at which the mayor seems to have proposed a more complex companion bridge, sort of a signature bridge next to the current bridge. Kind of a combination of the two plans. The smiles disappeared and people looked down and at the walls and at pieces of paper, anything except making eye contact with someone else thinking the same thing: is he nuts?

Much later in the meeting, Corporation Counsel Michael B. Risman said that the mayor supported the Common Council's unanimous resolution to file a lawsuit if the Coast Guard doesn't deny the PBA its construction permits. I think, coming after the earlier report of a mayor engaged in deep waffling, Risman's statement stunned many people in the room. This seemed to be a real stand on the issue. There was an immediate burst of enthusiastic applause.

I have the feeling that Masiello is starting to see the handwriting on the wall and is looking for a safe way to avoid being the city's only elected politician on the wrong side of what has become a very popular public issue.

REPORTS FROM ALL OVER

Early in the meeting, committee chairman and Common Council President James Pitts invited representatives of the various groups in attendance to make statements. First all the Common Council members in the room got to speak. Three Councilpersons are on the committee – Pitts, Barbara Miller-Williams and Robert Quintana – but by the time the meeting really got going there were about ten of them there.

The Council statement I remember most was David Franczyk saying he now realized this bridge issue was of international importance so we should "think big" and start looking for an international bridge designer to take over the job. He said he didn't want to disparage the work done by Bruno Freschi (he mispronounced Freschi, as did every single person who uttered it during the entire meeting: the first syllable rhymes with 'desk' not with 'mesh'), but, well, it was an international project and "What's the name of that Spanish guy?" I hope someone tells him that the T.Y. Lin of Freschi-Lin is one of the worlds most honored and experienced bridge engineers.

After they were done, Jeff Belt of New Millennium, presented his group's analysis of the economic and social effects of various plaza configurations and construction plans. It was smart, lucid, and informed. I thought that if the PBA had ever heard it and taken it seriously there was no way it could have insisted on its current bridge and plaza designs. Then I found out Mr. Belt had made a presentation and it had made no impact whatsoever.

He was followed by Dr. Sebastian Ciancio, chair of Periodontology at UB, who reported that he had circulated a petition among 155 people: 150 were in favor of the signature bridge, three were in favor of the twin span, and two said "What bridge?" He urged the politicians present to note the seriousness of the information. I called him the next day to make sure I had the numbers right. He said, "You did as of yesterday. But I got another 26 today. And there was one for the twin span and one who didn't care."

TWO LETTERS

Attorney Kevin Gaughan, a member of the Task Force, asked for a show of hands of people who favored the signature bridge. The only hand I didn't see go up (tv cameras and their operators blocked part of my view) belonged to Natalie Harder, representative of the Greater Niagara Partnership (GNP), which alone among community organizations has been out there opposing the signature bridge. Earlier, Ms. Harder distributed a three-page letter from GNP president and CEO Andrew Rudnick addressed to "Dear Peace Bridge Stakeholder."

(I must digress from this report to tell you this: every time I read or hear that city-planner cliche "stakeholder" I see either a grinning butcher holding up a huge bloody slab of beef, or a nerdy guy holding the tip of a sharply pointed piece of wood to the sleeping vampire's breast while his highly educated employer raises a mallet with which he will momentarily drive the stake home. Neither of those individuals, I presume, is one of the stakeholders Mr. Rudnick had in mind.)

The GNP had investigated the matter thoroughly, Rudnick wrote, so there was no reason for anyone to look into it further; the twin span was the only way to go. I don't remember anyone at the meeting referring to the letter, the GNP, or President and CEO Andrew Rudnick.

The other letter handed out, much shorter, much more eloquent, and more concerned with the community's needs than with ratifying old decisions, was Senator Daniel Patrick Moynihan's, who couldn't attend because of a prior commitment in Europe. Moynihan's letter began, "The grand American architect and scholar Daniel Burnham – whose gift to Buffalo was the magnificent Ellicott Square Building – challenged us to 'make no little plans. They have no magic to stir men's blood.'

"The Buffalo and Fort Erie Public Bridge Authority proposal to construct a 'companion span' alongside its existing bridge is indeed a little plan. Should it ever be built, it would stand for decades to come as a bitter reminder of what might have been done in its place."

What I've always liked about Pat Moynihan is the way he immediately goes beyond the trivial and epidermal to the heart of the matter, and how eloquently he speaks and writes. We're going to miss him when he leaves the Senate in two years.

He concluded: "A new bridge over the Niagara could become for Western New York what the Golden Gate has become for San Francisco's Bay Area. A new gateway, a defining moment of entry. To reach Buffalo or Fort Erie over the Authority's dull trestle when one could otherwise soar across on a bold new single-span would be rough justice for anyone who had the opportunity to imagine what could have been. Ada Louise Huxtable said, 'Cities and men get what they deserve.' Surely Buffalo deserves better."

THE JUNIOR SENATOR FROM NEW YORK

During the last election some people in this part of the state worried that Charles Schumer was too much of a Brooklyn boy to take us seriously. If his involvement in the bridge issue is any indication, that worry can be put to rest. He not only takes us seriously, but he does his homework and he listens to what people have to say. Schumer said, "The potential of having a worldwide symbol here is so exciting and could have so many other external benefits that to not even explore it would be a major crime, I think, in terms of the future of Western New York, something I care very much about."

MR. PITTS' BUMPER STICKER

James Pitts is a canny politician, he knows you don't have all this energy mobilized and do nothing with it. He told the group that he proposed (which translated as "here's what we're going to do") that the Task Force and the mayor's office need to meet with the Peace Bridge Authority to work out a way to develop an alternative plan. He said he would ask all the major and minor players to sign the invitation letter: the Task Force, Senators Moynihan and Schumer, the Western New York delegation in the State Assembly and Senate, Mayor Masiello, Energize Buffalo, New Millennium, and others. The purpose of the meeting is "so that we can all encourage the Peace Bridge Authority to take this moment of history and take it seriously."

In one brilliant stroke he mobilizes and coordinates a huge range of community, individual and political interests, and presents them to the PBA not in terms of finding blame but rather in terms of fixing a problem.

Pitts ended the meeting with this story:

> I received a bumper sticker last week that said, 'Don't you build no ugly bridge.' I happened to be sitting with my children in Anderson's on Delaware avenue over the weekend, and I believe from their accent there were some Canadians that were there. They were sitting at the table behind us. One, I guess he was the father, he said, 'Well, have you noticed that bumper sticker. It says 'Don't you build no ugly bridge.' And they started to deal with the syntax. And they said, 'Don't you think there should be an exclamation point behind 'Don't you build?' And he says 'No ugly bridge' means – and they were playing with it, so we finally turned around and said, 'We want a signature bridge. We want a beautiful bridge.' And that is American lingo to say, 'Don't you build no ugly bridge.'
>
> "So the issue is this: we have an opportunity to gather this moment, we have tremendous support. I haven't seen a groundswell like this in a long time. What we will do is we will get that letter together, all of us will sign it, we will appeal to the Peace Bridge Authority to sit down and stop their recalcitrance and work with us and this community to build a symbol that's going to be indeed great. With that, I'll move to adjourn.

THE SECRET ARMY

The second person to arrive for the meeting told me that the day before he'd attended a rally where people had made speeches and had charts and diagrams. He talked about how important this was. After a while I said, "For which bridge?" "The signature bridge, of course," he said. He's a retired high school art teacher. Another man came into the room, maybe 25 or 30 years old. He went on about the foolishness and ugliness of the twin span. I asked what organization he was with. He wasn't with any organization, he said, he was here just because he was interested. He said he lived in Buffalo and the bridge mattered to him. He didn't seem to be a politics freak, he wasn't part of an organization with a complex agenda of which this was just a part, he wasn't someone with a special interest of any kind. He was just a guy who lives here who is insulted that a closeted group answerable to no one wants to force him to live with ugly when there is no need for it. He said that.

There was a lot of speechifying over the next two hours, some of it expectable, some of it absurd, some of it very good stuff indeed, but what that man said more than anything else represented what was going on in Buffalo. I think he, and thousands of people like him, are an army the Public Bridge Authority had no idea was there, and they're the reason the Public Bridge Authority will eventually have to take its public responsibility seriously.

April 15, 1999

5 The Other Side of the Bridge

Stalemate at the Border

The Canadian government Wednesday issued construction permits so the Buffalo and Fort Erie Public Bridge Administration could go ahead with its plans to build a companion span for the decrepit Peace Bridge now in place. That's more symbolic than functional because everything is on hold until the US Coast Guard decides whether or not it will issue the parallel US permits. If the Coast Guard declines, it will be to force the environmental impact study the PBA tried to sidestep by separating the bridge and plaza construction projects. If the Coast Guard issues the permits, there may be lawsuits and other legal action from citizens' groups and even the city of Buffalo.

US senators Moynihan and Schumer, New York Attorney General Spitzer, the Buffalo Common Council, and most Western New York delegates to the state legislature and assembly have asked the Coast Guard to refuse to issue the construction permits. The Common Council voted to interfere with construction should those permits be issued, and Mayor Masiello, after months of fence-straddling and waffling, decided to endorse its actions. In the last few months, what had been scattered mumblings of dissatisfaction with the twin span design commissioned by the PBA has developed into a firestorm of public activity on this side of the border.

But not on the Canadian side. Ask people in Fort Erie whether they prefer a gorgeous signature bridge or a duplication of the current bridge and almost everyone says they prefer the signature bridge, of course, who wouldn't? That's where it stops. There is no agitation, no apparent public concern, no questioning of the five Fort Erie and Port Colborne residents who are Canada's appointees to the ten-person PBA. Most Canadians I've talked to seem to have the idea that a pretty bridge would be nice, but the issue didn't really concern them because it was a governmental matter.

The Canadian Position

That is the position the Canadian government has taken as well. According to Consul-General Mark Romoff, the "stakeholders" in the bridge question only marginally include people who live close to the border. The people who really matter are the people scattered across eastern Canada and the United States who make, distribute, buy and sell the hundred-million dollars' worth of goods that cross the Peace Bridge every day. For those distant buyers and sellers, the primary concern is velocity: they want a bridge that will let them move their goods as quickly as possible. It's not that they're opposed to aesthetic questions; rather it's that they just don't care about them because they live and work nowhere near where our aesthetic issues come into play. There is no difference to them if the bridge that provides the traffic lanes they need is gorgeous or ugly.

The official Canadian position, in that regard, is identical to the position advocated by Andrew Rudnick, president and CEO of the Greater Buffalo Partnership: the important stakeholders are everyone but the people who live here and have to look at the thing. Money is all that matters.

John D. Maloney, Canadian member of parliament from Erie-Lincoln, issued a statement on April 7 announcing that even arguing over Peace Bridge design was unacceptable. Not that one side or the other was wrong, but wrangling itself was improper. "The original decision must be respected by all concerned," he wrote. "I am deeply disturbed by the last minute intervention and intimidation for no compelling, supportable or valid reason by political elements in Washington, Albany, and Western New York." I assume by "political elements" he means everyone who has expressed an interest in the affair.

"Let no one lose sight that this is not just a Buffalo concern. It is a national and international issue that impacts industry and commerce throughout Western New York and the entire eastern seaboard; Fort Erie, the Niagara Region and the Province of Ontario and, most importantly, the jobs that go with them. It impacts the lifestyle of a forgotten element in the equation, the good people of both our countries who regularly traverse the border." (Oh that prose. If he uttered those words on that wonderful cable channel that shows sessions of British Parliament you'd hear half the chamber going "Harrumph, harrumph, harrumph.")

Then Maloney leaves the bombast in favor of a curious threat: "If the project is killed by parochial US political interests, let those responsible be prepared for the fallout which will surely follow."

What fallout could there possibly be? Mark Romoff suggested it: "There doesn't have to be a bridge at all," he said. "Buffalo and Fort Erie are lucky that they have a bridge and there is all this traffic. But it's possible that no new construction could take place. Or that a new bridge could go up somewhere else."

Not likely. If the Canadian interest is in getting a bigger bridge operational more quickly, then it makes little sense to start from scratch in a totally new location when much of the work – whichever design is adopted – has already been done here. If time really is essential, then even if we go on hold for the environmental impact study or for various details to be ironed out, we're still far ahead of a completely new design somewhere else. Countries don't have principles, as the diplomats say, they just have interests. It's probably in the interest of neither country to start afresh.

Much of the talk about what a signature span will do for Buffalo focuses on local pride and tourist traffic. The Canadian officials, most members of the PBA, and people like Rudnick aren't interested in local pride or tourist traffic at all. They're interested in commercial traffic. But some people question how much good Buffalo and Fort Erie actually receive from this huge caravan of semis roaring thru the area en route to and from places like Ottawa and Pittsburgh. Do they buy enough gas here for us to accept their fumes and pounding of our roads and consumption of our riverfront parkland? Maybe, maybe not. What if this were just a passenger passover? If the trucks found another way to go to and fro, would Buffalo be better or worse off? When Canadians threaten to take the trucks away there are people who say, "Yeah? So?"

I think we can ignore Maloney's saber-rattling, but I don't think we can ignore the real difference in the way Canadians and Buffalonians regard this project. The gap in opinion is far wider than the physical distance that separates us. The Canadians are neither fools nor Philistines. After all, most of our own elected officials didn't get interested in this affair until there had been a huge amount of grassroots agitation pushing them in that direction. Until a few months ago, the *Buffalo News* had all but abandoned the signature bridge. Things turned around because Senator Moynihan and Buffalo Common Council President James Pitts continued hammering at the issue, and because the alternative press kept running articles about it, the New Millennium action group kept increasing pressure on politicians to reexamine the bridge question, and Democrats Schumer and Spitzer replaced Republicans D'Amato and Vacco.

Nothing like that has happened on the Canadian side, so there has been no reason for Canadian officials, let alone Canadian members of the PBA, to rethink their position, or to redefine who the "stakeholders" are.

TASTE AND TIME

Consul-General Romoff said he thought only a fool would prefer the ugly to the beautiful. He said if there were alternate plans on the table that could reasonably be compared, then there might have been a different outcome. But, he said, there weren't any alternative plans that were more than preliminary ideas. By the time Freschi-Lin was presented 18 months ago, the PBA's plan had been worked out in great detail, and Freschi-Lin was little more than a great design. "There was only one fully substantiated plan on the table," he said. The numbers, he said, just weren't there for anything else. What we have, he said, is a conflict between "pragmatic" interest on one side and "emotional" interests on the other. I think "pragmatic" means "we've got goods to move across the border" and "emotional" means " we don't want to look at an ugly bridge every day."

I asked him if Canada would built this ugly bridge in Montreal or Toronto or Vancouver. He paused a moment and said, "Probably." I bet they'd also probably take a lot more time considering the alternatives once the alternatives were offered. The simple fact is, no one in official Canada gives much of a hoot about Fort Erie aesthetics, let alone Buffalo aesthetics. If you doubt that, take a look at downtown Fort Erie: customs brokers, gooch joints, Chinese restaurants, and one of the biggest hard liquor stores and signature sweatshirt duty-free joints on the Canadian-American border.

Romoff doesn't believe that a signature bridge can be done in a reasonable amount of time, nor does he buy Schumer's and Moynihan's insistence that shifting to a signature bridge will cost no more than a year. He says that the permissions and clearances from a dozen or so Canadian and American agencies that have already been obtained by the PBA will take much longer than that, and that the process of seeking those clearances can't even begin until the proponents of a signature bridge deliver something far more specific than a beautiful design.

TRUTH AND CONSEQUENCES

What's the truth of all this? If the Freschi-Lin numbers are soft, how long would it take for them to get hard? (I'd prefer "vague" and "specific" myself, but "soft" and "hard" are the adjectives everybody involved in this seems to prefer.) We know now that PBA numbers were off the mark, but how much does any of that matter anyway? Does a difference of ten or twenty million dollars count for much in a project of this size involving a structure meant to last a century? How long would it really take to get what Romoff says is needed to make a fair judgment, which is plans of equal detail? Is there a way to expedite the clearances, as Schumer and Moynihan insist, or is the bureaucracy to methodical, as Romoff believes? There's no way for you and me to know. We only get reports from the front lines.

Romoff argues that people are going off in all different directions and no agency or no official seems capable of bringing everyone to the same table to have a conversation that lets all interests have equal and fair voice. He may be right. Surely the PBA hasn't provided that service. Its solution is to tell us all to shut up now because it opened its doors to full public participation five years ago and we didn't come up with a signature bridge idea then. I haven't been able to find any evidence of any such invitation, of any openness to the public at the beginning. Lately, the PBA tactics seem more bullying than reasoning. A week ago the five American members, in an attempt to forestall the likely lawsuits, suggested a 45-day timeout while some outside evaluator was asked to check the options. The five Canadian members refused en bloc. No more conversation about anything, they said. And the attempt by the American five to avoid a time-consuming lawsuit has nothing to do with their position on the bridge itself: it's still nine to one, with Barbra Kavanaugh the only member of the PBA who would vote for a signature bridge.

Is there a way to mediate the Canadian and American economic interests and the Buffalo and Fort Erie quality of life interests? Need those issues be totally separate and need the sides to be at war with one another? Isn't there a way both can be represented in the conversation and both honored in the resolution?

A Modest Proposal

Next month, the terms of three of the five Canadian members on the PBA run out: Dr. Patricia K. Teal (eye doctor and surgeon, Fort Erie), John A. Lopinski (CPA from Port Colborne and the present chair of the PBA), Roderick H. McDowell (attorney, Fort Erie). The terms of the other two run out in November (Deanna DiMartile, Fort Erie) and Peter Caperchione (Port Colborne), both retired.

Canadians keep saying that they're insulted because people on our side of the border aren't spending adequate time considering their feelings and opinions. They're right; we haven't and we should. But people on our side think the Canadian members of the PBA don't give a hoot about what we think or care about, they're interested only in ratifying their own actions and choices over the past five years. What will Canada do about that? Give them another six years to continue doing it or replace some of them with people who are really interested in talking to Americans about a problem that matters to people other than masters of commerce?

How about appointing to that Authority a few people who are not so glued to the sunk costs of the past five years they can't take a fresh look at a problem that is causing a worsening wound between people who should and can be far better friends? And while they're getting up to speed, how about giving Freschi and Lin an invitation to harden their numbers to everyone's satisfaction?

We've got a troubled bridge over ordinary waters. Maybe it's time for a little de-escalation of rhetoric on both sides and a return to the civil assumption that neither side really wants to hurt or ignore or trivialize the other. It just looks that way.

6 Bruno Freschi: "I'm Talking Pragmatism Here"

If one person is responsible for the continuing controversy over whether to build a companion span for the 1927 Peace Bridge or to put up an entirely new signature bridge, it is Bruno Freschi, dean of the University at Buffalo School of Architecture and Planning. Freschi is an internationally-known architect who holds the Order of Canada, that nation's highest civilian award. He is a Fellow of the Royal Architectural Institute of Canada, he was elected to the Royal Academy of Art, and he is an Associate Member of the American Institute of Architects.

Dean Freschi wasn't the first person to suggest that the companion or twinning idea of the Public Bridge Administration was a bad idea and a missed opportunity for the region, nor was he the only person to offer an alternative design, but his design is the only significantly developed alternative and it is the one that has fired the public's imagination and garnered support from nearly all elected officials with responsibilities to this part of New York State.

All five Canadian representatives on the Public Bridge Authority have insisted that the issue get no further consideration and that construction on the companion bridge start immediately. The Canadian government has taken the position that the bridge exists solely for the purpose of moving trucks and tourists, so aesthetic issues are irrelevant. We may be heading into an international diplomatic argument as intractable as the political one now going on between the PBA and the residents of the Buffalo area.

In all of this, Bruno Freschi has been cited, quoted, paraphrased, talked about, attacked, referred to – but never interviewed. *Artvoice* thought it would be useful to hear his responses to the questions that have been raised. What follows are extracts from a conversation we had last Sunday afternoon. My questions are in italics.

You threw the apple of discord on the table when you came up with the design you worked out with T.Y. Lin. Why did you do it?

I was working on the waterfront for the city, helping to get the waterfront plan back on track. We took the big picture, we looked at the whole waterfront. In the course of that I was fascinated by this issue of bridges because there's a little bridge proposed in the inner harbor as well. This is '95, '96. I thought the claim of public consultation on this Peace Bridge was strange, so I refused get involved because I just saw it as a charade. I wasn't involved, but I had been working on some concepts. I had a small studio in the School and I had some students working with me because it's a fascinating idea.

Then I was in Korea and that's what really tipped the scale. I was speaking to the Kyonggi University faculty and after my talk a faculty member came up to me and said "I saw in your c.v. that you live in Buffalo, New York." I said "Yes." "And Buffalo, New York, has a PEACE bridge." I said, "Yes," totally unaware of what it meant until that moment: *Peace Bridge*! What a phenomenal metaphor. I thought about how the bridge was built to commemorate a century of peace between our two countries. I had never related the name on the bridge to Peace.

I took the idea that we'd been percolating for about a year and we evolved this bridge. We had a magnificent structural idea and a magnificent kinesthetic idea of driving in a curve. The concept made intuitive sense, it made symbolic sense, and all the ideas came together. One of the things that pushed me over the edge of going public with it was that no matter what scheme anybody came up with, it was going to be a massive decade-long disruption of traffic. We looked for an alternative to that problem.

Then the notions came into the newspaper of a twin, to keep the old bridge. That didn't make much sense. I knew a lot about why the old bridge was built and how the mistake over the canal was made, and how it was arbitrarily patched together. Then the designer of the second twin span, having realized that the existing span was kind of an ugly bridge, suggested that they could change the Parker truss on the old bridge and make it look like the new twin, which was put there initially to represent the first bridge. Things were getting absurd.

I had the good fortune of sharing the idea with Stan Lipsey and in turn with Stan Lipsey and Senator Moynihan, and both felt as I did that public icons and the public work of the city are important symbols and it all ought to be looked at again. At that point, we weren't concerned about time or money or any of that.

How did T.Y. Lin get involved?

This whole thing had its genesis in the School of Architecture and Planning at UB. T.Y. Lin is one of the world's foremost engineers, one of the most experienced bridge builders in the world. I know him from a little bit of work on the West Coast. He happened to have been giving a lecture in the School of Engineering and the School of Architecture and Planning and I was able to do the traditional napkin sketch for him.

T. Y. got very excited. More than I anticipated. He took away my napkin. Later, he called me from San Francisco, and said "When can you come?" Then I was out there for a conference, I walked across the street to his office and sat down with his team. They'd been working. And they proved the engineering was even BETTER than even I thought. We were right. They convinced me to tip the pylon one way from another way. We worked out the spacing of cables, we talked about the issue of weather and how to avoid birds flying into cables – all the ordinary bridge design issues.

All that stuff had been played out in a hundred different bridge projects. There are bridges in the Arctic where ice was a far bigger problem than Buffalo, New York. then we talked about constructability, which is perhaps forgotten but most important point of all. He explained that and how this could be more efficient. This is why HE was excited.

I came back and said, "We gotta do it." And so I did it.

I believed that in a thing this big, this important, if there was political will, timing was irrelevant. Then when we did the research, it turned out we were also faster and cheaper!

Your opponents say that your numbers are just pulled out of the air, while they've spent five years developing their numbers.

No, our numbers are not pulled out of the air.

I involved the firm of Cannon, which I'm associated with (I'm not a partner in Cannon, I'm not in any way a corporate officer, I have a functional title) because I can work with them, I can consult with them on projects all over the world. I went to Cannon because I know they have a good engineering office, and they have an extraordinary construction cost analysis staff. I've worked with them on large-scale projects and I trust their numbers. They have very competent project management and understand scheduling and the whole process. They did a thorough analysis.

I also went to T.Y. Lin and said, "T. Y., tell us what it costs and tell us how long everything's going to take – construction documents, bidding time, construction time, and what kinds of questions we should have about the Environmental Protection Act and its reviews, the Coast Guard, et cetera." They researched it from current bridge projects.

Then a local major contractor volunteered to do yet another arm's-length cost analysis and to add his understanding of this area's weather and the conditions of the Niagara River. He looked at other comparable bridge constructions project in the country. His work included a thorough estimate of the demolition costs of the existing bridge.

We have those numbers. Those are the numbers that we reported. This alterative is indeed cheaper. In other words, this bridge will cost less and be completed sooner.

I'm talking six lanes in one step. Not three and three in two steps which takes three or four years longer.

One of your critics said that you gave the Bridge Authority a good design, but not a real plan, whereas the twin span designers had a real plan. He said there was no way to evaluate your plan because it was apples and oranges.

This a concept design by a competent team. Behind this concept there's a lot of talent that honestly believes it's a legitimate alternative. Certainly, further engineering details should be done. We have not had three to five years to do this. All the effort by all members of the team has been volunteered.

The twin span design has been detailed to a far greater extent. However, one must address the premise of maintaining the old bridge and the subsequent design concepts and no solution to the plaza problems. Nobody knows what fixing the old bridge will cost. It would appear that it's going to be ten years until you see six lanes and a new plaza, maybe more.

They also say you would have to get new clearances and you couldn't do that until you had detailed drawings so it would put the project back by five years.

That assumption was based on their designs being accurate, and their costing being accurate, which of course since has been proven wrong.

And it was based on the questionable policy decision to keep the old bridge. So who's going to take longer?

We are a group of people who really know this complex process, who've done it, and we went to all the agencies on the American side. The Canadian side had apparently all been approved. The American side was the question mark. Our reviews included meetings with the Coast Guard, the Army Corps of Engineers, the EPA. We have included in our schedule a worst-case projection of time for review: eighteen months. And keep in mind that these reviews are processed simultaneously with the preparation of the engineering details. It's not like the project is on hold while you wait for them.

Would your bridge be cheaper to maintain?

Yes.

Why and how?

Firstly, the pylon. The pylon can be concrete or steel. Steel today can be coated and permanently protected as opposed to just paint. There are alternatives in the coating that protect steel from salt and the weather and all that. But we've actually proposed a concrete pylon with space inside for an elevator.

Second, the new cabling is permanently protected and wrapped. They used it on the bridge in Tampa. I've gone and touched and felt those cables and I've talked with the guys who built it. They don't need the sandblasting and painting that older bridges need. How many times have you driven across the Peace Bridge and not seen them doing that kind of work?

They're painting it all the time. But the Tampa bridge exists in an ambience very different from here.

Salt! Seawater! Worse! Worse than here!

Never mind. So what about ice?

Heating coils. All you have to do is prevent the formation of ice and you don't need a lot of heat for that. This isn't the first bridge in the north, there are worse places than Buffalo, New York. It's a concern on a steel bridge: a steel superstructure, just like cable, will hold ice.

The old bridges don't have coils, they're a problem; the new bridges have this technology; they're not a problem.

What about inspecting the cables? Your pylon is so high. How can anybody check it out and make sure it's holding up properly?

You use the modern technology of stress gauges. You can measure and keep track of the stresses. You can get up the cables, they have rigs that do that, but since there's so much less need for it, it's not a permanent structural thing that you need to build. None of this is a big deal. There isn't a bridge in the world that hasn't provided for these routine inspections.

What about the birds? Opponents of your design keep bringing up bird migrations.

One: there is not a river in the world that is not a migratory path. They thought it was a discovery that this was a migratory path. Name a river that isn't. Birds use rivers. That's the way they fly.

Two: Most rivers have bridges on them.

Three: Definitive research has proven that birds fly into unlit objects, like dark buildings at night. You will notice in that poster of the bridge in Tampa that it's brightly lit and it looks gorgeous. If you built this bridge you'd WANT to light it. Birds, believe it or not, are just like us: they can see a lit object. Sure, there's going to be a dumb bird that flies into a cable. Not often.

Would you keep the same plaza they're using now, and if not, why didn't your design reflect that?

I would not keep the plaza that is there now. We proposed in the original design a bridge which delivered to a new plaza just downstream, or north of the existing plaza. The new plaza would be at the point where Niagara Street divides into an access road to the Thruway and, more importantly, becomes a primary gateway to downtown Buffalo. This area and its adjacent escarpment is empty. We have proposed a plaza to be constructed in the air and on the ground over the Thruway and railway below. The design of this plaza could deliver traffic to the Thruway, and most importantly, a Niagara street gateway to Buffalo.

A block that's empty now?

Yes. It's just a leftover triangle. To properly address this gateway, a block of houses and small commercial buildings should be removed to provide an appropriate entrance to the city. Remember that Niagara Street takes you right to Niagara Square and City Hall. What we have now, is a spaghetti roadway system that has destroyed Front Park. The park is dead, it's inaccessible, it's a green patch in the middle of nowhere. Our design would restore the Olmsted Front Park as a park for the West Side.

So why didn't we see this plaza in your design?

It was in the original proposal, but we were asked to consider the alternative of connecting to the existing plaza because the process which we agreed to participate in had removed the plaza from consideration. And believing in the process, as a good Canadian architect, I went along with it. You can do either one, but if you ask me now what my choice is, it's a new plaza at a place that serves the traffic and the city far better.

In one construction phase and in the shortest time, one could have a signature bridge of six lanes, a new American plaza, an integration of the bridge on the Canadian side – all with minimal disruption of traffic. Further, this design would support the green parkway concepts for the Niagara River. In one day, you open the new bridge and plaza and close the existing bridge. After that, demolish the existing bridge and plaza and restore Front Park.

Won't demolishing the old add a huge amount to the cost?

Our analysis shows that it will be between $7 and $9 million.

You referred to yourself as "a good Canadian architect." How come there's such a distance between what you seem to find important and what the five Canadian members of the Public Bridge Authority and the Canadian foreign office find important? They seem to be taking a cold business-oriented approach to this whole issue and the Americans, who are usually accused of that, are taking an aesthetic and green space approach. Does that make any sense to you?

It makes sense, but I'd use different adjectives. I think that the

Canadians really believed the Authority. Canadians tend to trust their government, they tend to trust their elected authority. Canadians aren't less democratic, but they have less process. I like it, it's kind of nice that way. I believe that they were sold a concept and they've invested heavily in that concept and don't want to change direction now and are upset with the timing of all this. I don't blame them.

The only issue here is, why the decisions were made. I'm not here to condemn anybody. I make good ones and bad ones every day, just like everybody else. So we should be big enough to come forward and admit there were some bad decisions. To quote David Crombie, there's a "hunger for this symbolic issue." What's important isn't to justify a decision made three or five years ago, it's to make the right decision now.

What's happened is a moment of learning. That doesn't mean you accuse somebody of being stupid or backward. It's a discovery, it should be a really great discovery, it is a visionary discovery by the community. The people want this. That's what leadership must recognize. The only person who reacted at first was Senator Moynihan. Now there's Senator Schumer, Attorney General Spitzer, the county executive, the Common Council, the mayor, and all the others. Gradually, I think, we Canadians will also appreciate the faster and less expensive signature proposal.

But if the Canadians, are primarily interested in this project only to increase truck and automobile traffic capacity, why should they incur the extra costs required to change plans?

It's the opposite: we would have more traffic faster for less money. More traffic. Less cost. Sooner. Forget the symbolic quality. I'm talking pragmatism here. That's what we are all talking about.

It seems to me that if Canada's main interest is truck throughput –

It should be. I agree with them.

– then your proposal is not in conflict.

That's my point. The issue has been misunderstood. They want to say "It's not about aesthetics!" They're right: it's about faster cheaper. Wake up! You're not going to get it faster and cheaper the way they're going now.

Here are the salient facts:

One: We have a single pier in the water. Big – but one. They want ten.

Two: This signature bridge can be constructed more quickly than building a companion bridge and rehabilitating the old bridge.

Three: There could and can be a new gateway plaza and a restored green Front Park.

Four: This can be achieved at less cost in shorter time with the least disruption of traffic.

That's just true.

I'm saying this: We'll give you a spectacular bridge, six lanes, and a new plaza in 2004. Cut the ribbon! It's all there! I'm serious.

April 29, 1999

7 The Nail and the Teeter-Totter: Ending the Peace Bridge Stalemate

It's no longer a local issue and our newest secret is out. The *Washington Post* ran a not very accurate or insightful article on it last Thursday, the *New York Times* ran it as a front page story, and *Time* magazine columnist Stephen Handelman is analyzing it in depth. We are no longer known only as the town of world-class snow and chicken wings, but, along with our friends in Fort Erie, we're the place that can't agree on how to get from one side of the river to the other.

The Nail

Everybody knows that old saw "For want of a nail the shoe was lost, for want of the shoe the horse was lost...." and on and on through the lost soldier, the lost battle, and finally the lost war. The moral is "Little things mean a lot" or "Pay attention to the details" or "What you don't notice can kill you," or something else equally obvious and no doubt true.

But rarely in real life are we able to find exactly the thing that set in motion the sequence that only in retrospect reaches a conclusion that seem inevitable and unchangeable.

Bruno Freschi thinks the nail in the Peace Bridge affair was when someone on or working for the Buffalo and Fort Erie Public Bridge Authority decided that the 1926 bridge Frank Baird built was of historical value and that if they tore it down we would scream and yell at them, that we would do exactly what we're doing now because they're NOT tearing it down. When I first heard him say that I thought he was being unnecessarily generous, but now I think he may very well be right. "Somebody gave them bad information years ago," Freschi said, "and it all follows from that."

I don't know when it happened. I don't know if it was a member of the Authority or a staff member (a lot of the really steamy rhetoric coming out of the PBA lately has come from staff, not the ten appointed members). I don't know if it even rose to the point of conversation or if it was just a tacit assumption made by everyone sitting around the table back when the first decisions were being made.

I do know that thirty years ago, long before NAFTA, the New York Department of Transportation predicted that the current bridge would not continue to be adequate. DOT pointed to four possible solutions: increase the manned toll, immigration and customs services on both sides of the border or increase the number of lanes in one of three ways: widen the current bridge, build a parallel bridge, or build an entirely new bridge.

Staffing of the booths is still a problem, and just last week Customs Commissioner Raymond Kelly said he would assign 25 new part-time inspectors to the area's four bridges this summer. That will help, but it won't solve the problem in the short or long run. Few of those part-time employees will be coming to the Peace Bridge and, more important, the likely increase in traffic here is such that in a few years the bridge will have major traffic jams even without customs and INS inspections or toll booths.

THE 1994 STRUCTURAL DESIGN CONFERENCE

People opposed to the signature span say its proponents should have come forward five years ago, and they frequently remind us that the PBA considered eleven alternatives to its current companion span design. But proponents of a signature span, had any been around five years ago, couldn't have come forward because they were already locked out of the conversation: the historical assumption had been made and the PBA was considering only widening or twinning.

This is documented in the *Peace Bridge Capacity Expansion: Draft Environmental Assessment,* a report prepared by the Public Bridge Authority in September 1996. Chapter three, "Bridge Crossing Alternatives," reports on a structural design conference held in September 1994. "The Conference was attended by the Bridge Authority and representatives of the four consulting engineering companies." Nobody from the general public, no designers with other ideas, nobody but the PBA, its staff, and the engineering firms it was paying to provide advice. We needn't assume malign intent here: in those days, everybody assumed that the PBA represented the public, that the five Americans and five Canadians were looking out for all of our interests.

Here is the key paragraph, the one that established the policy decision we're all fighting about now:

> The Structural Conference concluded that capacity expansion of the existing bridge would be required within the very near future. Therefore, the group of consulting engineers collectively prepared and identified preliminary alternatives for expanding the bridge's capacity. The alternatives included widening the existing structure, or alternatively constructing a parallel structure either to the north or south of the existing bridge, that would be visually compatible with the existing bridge.

There was no consideration at all of entirely replacing the old bridge. The consulting engineers were never asked or allowed to deal with that possibility. (You can read their full report in another PBA document: *The Peace Bridge: Structural Conference, Report of Findings*, October 1994.)

The 1996 consultants examined eleven possible ways to build a companion bridge. One of their hypothetical bridges was cable-stayed, the kind of structure proposed by Bruno Freschi and T.Y. Lin, but theirs was a cumbersome affair with three huge towers constructed of materials now obsolete that needed a good deal of maintenance and frequent replacement. The consultants faulted their own design on two primary counts: expensive maintenance (not an issue if it were built of the materials used by modern bridge builders) and ugly in juxtaposition to the current bridge (which is indeed true, and why everyone thought Mayor Masiello goofy when he proposed something like this a month ago). They also cited as a possible problem interference with bird traffic. They may be the source of that cockamamie canard.

TURTLING UP

So the ten members of the PBA, appointed to handle mundane affairs like tolls, repaving, and rust on the truss (the Americans serving without pay since Mario Cuomo was governor), found themselves enmeshed in a project of huge public significance. For a while it seemed simple enough: their assumption that the Baird bridge had historical value meant a replacement bridge was out of the question, and widening was impractical because it would block traffic in one or more lanes of the current bridge for years. A companion span seemed to make a great deal of sense; it may even have seemed the only way to go.

They then spent a good deal of money getting someone to design a companion span that wouldn't look absurd next to the Baird bridge or cause problems for myopic birds.

It all fell apart as people in the area became aware of the design, started talking about the absurdity of it, challenged the initial assumptions, and came up with new proposals. Locked into its companion bridge idea, the PBA separated the bridge and plaza projects, a move no one has ever adequately defended. Freschi and others argue that you can't build a bridge without knowing where it's going to land, that the plaza has to be part of the bridge design. "Form follows function," one prominent businessman puts it. "I keep telling them, 'It's the plaza, stupid.' Nobody listens."

The attacks started and the PBA went into defensive mode. They stopped listening and started fighting. They spent tens of thousands of dollars buying full-page ads in the *Buffalo News* and several more thousands buying spot commercials on local radio stations. One official said, "All of us sudden everybody was yelling at them so they turtled up."

Then, three weeks ago, there seemed the possibility of a break: the five American members of the PBA were willing to let an outside team consider whether or not there were good reasons to reopen the conversation. If nothing else, such a study would say to the public, "We're willing to listen." But the five Canadian members of the PBA were not willing to go along. The vote was an even five to five, which meant nothing would change. As far as the Canadians were concerned, it was over, finished, done with, set in concrete. As far as the Americans were concerned, the Canadians had decided to be deaf.

WHAT THE CANADIANS DON'T KNOW

But it doesn't work like that. People don't arbitrarily decide to be bullies or blind and deaf to information. It's easy to demonize people by leveling such charges, but rarely do such charges get us anywhere. We're better off asking questions that help us understand what has happened than we are blaming everyone in sight.

So why are the Canadians apparently so much less interested in the Freschi-Lin signature design than people this side of the border? So far as I've been able to find out, it's because they've gotten most of their information from PBA staff-written press releases and the *Buffalo News*. They don't know that there is a viable alternative. They never read Bruno Freschi's responses to the charges against the signature bridge because the *Buffalo News* never interviewed Bruno Freschi. They never had a chance to learn how Boston solved an almost identical problem because the *Buffalo News* never ran a story about the Boston experience. And, most important of all, no public official they know well enough to trust has told them that there is every likelihood that the Freschi-Lin bridge will bring them what they want – velocity and volume of traffic – sooner and more cheaply than the companion design.

Even when the *Buffalo News* does cover the story it's difficult to trust the information. Last Sunday, for example, the *News* ran "In Fort Erie, twin span support is solid," a man-in-the-street article by Patrick Lakamp. The longest quotations were from a man identified only as "Fort Erie lawyer John Teal." Teal got to kvetch without challenge about all the things presumably wrong with the signature bridge proposal. "Where were these people seven years ago?" he said. "I see no reason to debate this. It starts to strain the bounds of logic. This is just politics on the American side. It's a shame that would get in the way of this project."

No: what's *really* a shame is that *Buffalo News* reporter Lakamp didn't include in his article the key fact he knew about John Teal's opinions: John Teal, Lakamp told me, is the brother of Peace Bridge Authority board member Dr. Patricia Teal. John Teal is not a man-in-the-street. He's a spokesman for the Canadian half of the PBA and the *Buffalo News* knew it.

There's something truly ironic about this I think no one has mentioned before: the people who will get the most benefit out of a gorgeous signature bridge won't be the residents of Buffalo, it will be the residents of Fort Erie. Once you get off the three or four blocks of joints near the power lines, Fort Erie is a lovely town. It has quiet residential streets with very few of the abandoned buildings so common in much of Buffalo. If you drive along its waterfront you go along a carefully-maintained green space running all the way from Old Fort Erie to Niagara Falls. You'll be able to see the new bridge from all the houses along that route, and from all the picnic tables and walking areas, all the boat areas, all the bike paths. The Canadians have kept their side of the river accessible and beautiful, and the signature bridge will immediately become part of that precious landscape.

But on the American side: how do you get to the river? Not how do you cross it, but how do you get to it? Most of the river is blocked by the Thruway, by factories, by warehouses. People on this side of the river will get to see the new bridge as they drive by, but they'll have none of the easy access afforded the Canadians.

Nothing in the Canadian or Buffalo press has mentioned that in addition to a bridge providing six functional lanes and two fully functional plazas faster and cheaper, the Canadians will reap most of the visual benefits of the entire project.

So we can hardly blame the residents of Fort Erie for their failure to understand how things look from over here. In Buffalo, we've had an alternative press that has refused to let the issue go, a group of articulate concerned young professionals who have refused to let the issue go, and a growing number of local, state, and national politicians who have refused to let the issue go. They all have worked very hard to inform the public of the facts, of the real choices available. There has been no such effort and no such resource on the Canadian side of the river.

A QUESTION OF LEADERSHIP

But that's not how things have to remain. It was a long time before we learned the salient facts and once we did, public opinion began to develop and change. Knowledge is power.

Canadian Consul-General Mark Romoff and other Canadian officials have frequently complained that American advocates of a signature span have failed to communicate with Canadians, that Americans have treated this as if it were an issue affecting the southern border only. Canadians are quite properly sensitive about things like that. I assume one of the reasons Fort Erie Mayor Wayne Redekop issued his statement a week ago about maybe forgetting the whole thing is he was annoyed when he learned from the newspapers that American politicians were planning on crossing the bridge to talk things over with him.

But something is missing: we *have* communicated with the Canadians, and we've done it for a long time. Canada has five federally-appointed members on the PBA and they have heard from significant numbers of Americans about how things appear from this side of the river.

Senators Moynihan and Schumer, New York Attorney General Spitzer, and all the other federal, state and local officials got involved in this for one simple reason: because the ten members of the PBA did not seem responsive to the rest of us or aware of the opportunity at hand. It was never a matter of our officials bullying the PBA, it was never a matter of us ignoring the Canadian people or their government.

Mark Romoff said last month that what was needed in this sad affair was leadership. "Where's the leadership, tell me that?" he said. I think he meant that good leadership might resolve the mess to everyone's satisfaction and benefit. He's right: the leaders have to come out and lead.

But the required leadership cannot come from the American side alone. It's already there on the American side: the two US senators, the young professionals, the legislators. Even Mayor Masiello is starting to make sense. But it's like playing on the teeter-totter by yourself: you can have the best intentions in the world but without a partner who's close to equal you're stuck on the ground or you're up in the air, neither a good place to be. A good place to be is both, which is what the signature bridge design can provide, if only everyone will stop defending the past and start thinking about the future.

The first poem in John Berryman's Pulitzer Prize-winning *Dream Songs* (1964) begins,

> Huffy Henry hid the day,
> unappeasable Henry sulked.
> I see his point, – a trying to put things over.
> It was the thought that they thought
> they could *do* it made Henry wicked & away.
> But he should have come out and talked.

Indeed. And likewise our Canadian friends: it's time to do more than sulk, it's time to act like equal partners, it's time to come out and talk. Consul-General Mark Romoff, Ambassador Raymond Chrétien, Transport Minister David M. Collenette, Prime Minister Jean Chrétien: it's time to come out and talk. The cars and trucks go both ways across the bridge, the river, and the border; it's time for conversation to do the same.

4/29/99

8 THE PEACE BRIDGE WAR: SHIFTING CONSTITUENCIES, THE NAUGHTY *NEWS*, AND THE CRUISE MISSILE OPTION

Three big Peace Bridge stories broke in the past week: the Coast Guard approved the permit request from the Buffalo and Fort Erie Public Bridge Authority, Congressman John LaFalce has been lobbying the Coast Guard on behalf of twin bridge proponents, and Governor George Pataki announced that he and the five American members of the PBA were now strong supporters of the signature bridge. The *Buffalo News* gave major play to the first, all but ignored the second, and fragmented and hid the third, which was the most important of the three.

BURIED DYNAMITE

Governor Pataki's surprise endorsement of the signature span and his report of companion endorsement by the five American members of the Public Bridge Administration appeared in the *Buffalo News* on Friday, April 30. It was broken into three parts tucked into a Douglas Turner article on page A-8 titled "Coast Guard dropped options to twin span as too sketchy." Turner's political pieces are usually given far better placement and nothing in the title or any subhead would have given a reader any idea of the dynamite buried there. You don't get to Pataki's surprise until you're five paragraphs into the Coast Guard story:

> Gov. Pataki criticized the authority proposal as "just a functional bridge." He told a Georgetown University audience that a signature bridge is a better investment.

This was, so far as I know, the first time Governor Pataki has publicly expressed any opinion at all about the signature bridge. The *News* article returned, without comment, to the Coast Guard story, then, five paragraphs further on and as a total non sequitur, told us that:

> Pataki said the twin span has "no architectural or cultural value at all" and gave his strongest endorsement yet to the signature bridge.
>
> "The five US authority commissioners," Pataki said, "have concluded let's build a signature bridge, something that adds some character to the entrance to Western New York and the northern regions of America.
>
> "Quite simply I think that this is a great idea. I think that when you are going to make this investment that is going to last not just for decades, but for generations, let's do it right. The Americans have taken that position. The Canadians have not. So there we have an impasse.
>
> "'Maybe when the situation in the Balkans calms down, we can send (US Ambassador) Richard Holbrooke up to Canada to see if he can broker an agreement."

Astonishing stuff: not only is the Governor is in favor of a signature bridge, but so are all five American delegates to the Public Bridge Authority. (On Monday of this week three of those American board members once again endorsed the twin span, contradicting the governor responsible for their appointments.)

After two more paragraphs about the Coast Guard, the article again returned to the Governor:

> Pataki on Thursday visited Sen. Daniel Patrick Moynihan, D-N.Y. at his Washington apartment. Moynihan, whose September, 1997, speech in behalf of a "great bridge" started the controversy over bridge design, is recovering from back surgery.

The remaining five paragraphs say nothing more about Pataki or his unexpected endorsement of the signature bridge at a university four hundred miles from home, or his visit with the senator the *News* credits with starting it all.

Anyone familiar with the issues, politics, and players in the Peace Bridge saga knows that the Governor's strong endorsement of the signature bridge and his report that the American board members agreed with him, is major news. Why did the *Buffalo News* fragment the story and bury it in an article innocuously parked on page A-8?

Why were there no followup articles or commentaries on Saturday, Sunday or Monday? Why did the *News* run as its front page Peace Bridge story that day a bunch of anecdotes of far less importance?

THE PAGE ONE UNIMPORTANT STORY

That anecdotal page one story in Friday's *News* was about reactions to Wednesday's Coast Guard announcement that it had given the PBA a permit to begin construction on the companion span even though there were as yet no clear plans for a plaza and the PBA had not done the environmental impact statement usually required in such complex projects. Most of the article was comments from various Buffalo and Fort Erie political figures.

The only American in the article who applauded the Coast Guard decision was Andrew J. Rudnick, president and CEO of the Greater Buffalo Partnership (né Chamber of Commerce) who said, "Every day that our community continues the divisive rhetoric and threats that recently have surrounded the Peace Bridge expansion issue is another day that our community suffers." Profound thoughts eloquently expressed. Rudnick said nothing about how long we'd suffer if we got the ugly twin span and had to look at it year after year.

What's with him and the Greater Buffalo Partnership? Every business owner I've talked to says Let's build the signature span, it's better for the region in all conceivable ways. Who is Rudnick really speaking for? Are the moguls (who have been silent on this major public policy issue) pushing the Partnership in some other direction for reasons Rudnick won't or isn't allowed to talk about? Is there some powerful drummer who plays only for him and whom only he hears? I doubt that Rudnick is following orders from his long-time patron Bob Wilmers, who I'm told has worked very hard on the city's behalf. I can't imagine Bob Wilmers pushing Andy Rudnick to endorse an ugly bridge for Buffalo. It's equally unlikely that Wilmers' old friend and *Buffalo News* owner Warren Buffett gives a hoot what kind of bridge goes up here. So who is pushing Rudnick's buttons?

THREE CANADIANS, ENCORE

The story also said that the Canadian Minister of Transport had reappointed three twin-span hardliners to new five-year terms on the Bridge Authority: John Lopinski (current PBA chairman), Roderick

H. "Roddy" McDowell, and Dr. Patricia K. Neal.

I didn't know anything about the PBA's Roddy McDowell until I read David Chen's article on our Peace Bridge fiasco in the April 27 *New York Times.* Chen quoted McDowell as saying we shouldn't have a signature bridge because "Cinderella's castle won't work here." I had to read that line twice. I can't understand why the Canadian Minister of Transport would reappoint to so important a position someone who can't differentiate between a castle and a shoe.

(McDowell's malapropism didn't appear in the national edition of the NY Times, which is the one sold and delivered in Buffalo, but it did appear, along with a good deal of other additional material, in the edition distributed in New York City. That version was also more prominently-positioned: our version was on page C18, after the stock market numbers and below the fold; the city edition version was the first three columns across the top left of page B1. Major placement for a major story.

Since Lopinski, McDowell, and Neal continue to reject all requests to consider alternatives – not just to *build* an alternative, but also to *talk* about the possibility – we should read these reappointments as an indication that the Canadian government is telling us to fuck off. I've heard that whatever personal opinions the three harbor, they've all been told by Ottawa to be recalcitrant and to take the heat for the national government, which is absolutely opposed to a signature span between Buffalo and Fort Erie and which does not want to come out and say so. More on this below.

THE EDITORIAL WAFFLE

There was more journalistic weirdness in the following day's *Buffalo News* (Saturday May 1). The lead editorial, echoing last week's *Artvoice,* said Canadians and Americans should begin talking rather than posturing. The editorial at first seemed to argue for the signature span, or at least give it serious consideration, but then it went on to suggest that proponents of the signature span should do nothing that would upset anyone anywhere. They certainly shouldn't file lawsuits that might slow down the construction process and get people on the other side angry. A beautiful bridge would be nice, the *News* seemed to say, but if the Canadians won't agree to it, signature bridge proponents should roll over and take wherever the Canadians are shoving wherever they're shoving it.

Keep in mind that last August, *News* editorial page editor Gerald

I. Goldberg said "We had an editorial board meeting and decided that the bridge is a dead issue." I can't remember a single editorial in the *News* advocating the signature bridge. Immediately after signature bridge advocate Elliot Spitzer replaced Dennis Vacco as New York attorney general, the *News* ran a lead editorial urging him to keep steel-magnate, major Republican party contributor, and twin-span advocate Brian Lipke on the PBA. Instead of urging Spitzer to give that seat to a deputy attorney general who was sworn to represent the peoples' interests, as had been the tradition before Dennis Vacco, there was the *News* urging him to give the job to the owner of a steel mill.

The only contributor to the *News* editorial page who has consistently pointed to the PBA's obstinacy and wrong-headedness is political cartoonist Tom Toles. When Warren Buffett and the *News* drove the *Courier-Express* out of business in 1982, thereby making Buffalo a one-paper town, the loss I regretted most was Tom Toles, who was then drawing for the *Courier-Express.* Happily, the *News* dumped its dull editorial cartoonist and hired Toles, which some of us thought meant the *News*, now that it had total journalistic power, was going to take a far more acute look at social issues in the community. That never happened, but, to its credit, the *News* kept Toles and gave him freedom to draw, even though his drawings were frequently in clear opposition to the editorials on the same page. The old *Buffalo Evening News* won two Pulitzer Prizes: Bruce M. Shanks for editorial cartooning in 1958, and Edgar May for local reporting in 1961. The only Pulitzer Prize ever won by the *Buffalo News* went to Tom Toles in 1990 for his editorial cartoons.

A CONGRESSMAN AND HIS CONSTITUENTS

Proponents of the signature bridge were disappointed but not really surprised about the Coast Guard approval of the PBA permit request. Some people think that the Coast Guard made a partial concession to Senators Moynihan and Schumer by sitting on the announcement for a month after it was ready, thereby giving signature bridge supporters like the vigorous New Millennium Group more time to develop opposition to the twin span.

The *Buffalo News* said on Thursday that Congressman John LaFalce had been informed of the Coast Guard ruling and he hoped for a cooling-off period. The bridge lands in his district (which runs from Buffalo's west side all the way over to Rochester), so it's reasonable he'd be informed of the decision and that he'd have an opinion about it. But the *News* article said nothing about LaFalce's direct involvement in the process or that he had been working in direct opposition to his two Washington colleagues, Senators Moynihan and Schumer, as well as Attorney General Spitzer, the Common Council, the mayor, this region's delegates to Albany, and what seems to be the great majority of voters in this part of his district. LaFalce had written Secretary of Transportation Rodney Slater in March about the Peace Bridge, but unlike Senators Moynihan and Schumer, LaFalce wouldn't make his letter public. He also refused to have press conferences in which he discussed the issue or in which he gave reasons for opposing the two senators from his own state and party.

Word is that LaFalce very much wants to be named US ambassador to Canada in two years, should the Democrats get the White House again, so he's doing everything he can to keep cozy with his friends across the border. They want a dull bridge here, he'll help them get a dull bridge here. They want no complications here, he'll do what he can to kill complications here. That's what a Congressman does, represent his constituents, right? I guess the problem sometimes is remembering who one's constituents are.

Gossip like that shouldn't go unchecked, so I called his office early last Thursday morning and said *Artvoice* wanted to interview him about his position on the Peace Bridge and the report of him secretly lobbying against the two senators and almost everyone else this side of the river. The woman who answered the phone said I'd have to talk to a Mary Brennan-Taylor, who wasn't in yet but who would return my call shortly.

She didn't.

I was in the Federal building in the early afternoon, so I stopped by Congressman LaFalce's office, identified myself and said I'd like to make an appointment to interview him. The receptionist was snippy. "I talked to you this morning," she said, "and I told you that Mary Brennan-Taylor would return your call."

"She didn't return my call," I said. The woman said that Mary Brennan-Taylor was out to lunch and that she would return my call when she got back, and if she didn't, well, then I could call her back again. Mary Brennan-Taylor didn't call in the afternoon, nor did anyone else from Congressman LaFalce's office, so late in the day I called again. I identified myself to the woman who answered and said I was calling back. She then told me that she wasn't the woman

I'd talked to the other two times and she didn't know what I was talking about. I told her that *Artvoice* wanted to interview Congressman LaFalce about his position on the Peace Bridge.

She said I'd have to talk to Mary Brennan-Taylor about that and Mary wasn't there, she'd just left. I said, "Oh, really?" Maybe there was something in my voice: she said she'd try to catch her. She came back and said, "Mary got caught by another phone call but she put her hand over the mouthpiece and said she'd faxed the Washington office and would get back to you."

That was last Thursday afternoon. It's Tuesday now and I've heard nothing. Six days and I still haven't even been able to talk to the lady to whom you have to talk to make an appointment to talk to the congressman.

CANADA AND THE CRUISE MISSILE OPTION

There are two interesting hypotheses making the rounds, both having to do with why the Canadian government opposes a signature bridge between Buffalo and Fort Erie. I've made no more headway verifying or disproving them than I have connecting with Congressman LaFalce or finding out about his ambassadorial ambitions.

One has it that the reason the Canadian government insists on a bland bridge here is it doesn't want anything to distract or deflect tourists from the huge Canadian development around the Casino Niagara project. The casino is a cash cow (mostly American cash), it's expanding, and hotel space in Niagara Falls is expanding accordingly. Why put a gorgeous structure in the Buffalo area that might encourage visitors to spend part of their time away from the Niagara Falls tables and slots and sleeping rooms?

I asked Canadian Consul General Mark Romoff about this and he said there was nothing to it because the casino operation was inconsequential compared to the value of goods crossing the bridge every day. At the time that seemed reasonable enough, but later it occurred to me that the two issues have no connection with one another. Truck traffic is unaffected by bridge aesthetics; it is dependent only on bridge access. The hundred million dollars worth of goods trucked across the Peace Bridge every day have nothing to do with the millions of dollars lost by Americans at Casino Niagara. Canada can be equally interested in smooth truck flow and making sure as many people as possible leave their US dollars at the government-owned casino in Niagara Falls.

A savvy lawyer friend I call Deep Boat thinks things aren't so simple. "I'd think they'd want the signature bridge because everything that draws people to this area will increase the number of people who visit the casino. People will come to Buffalo to see the signature bridge and then they'll drive twenty minutes on the Canadian side into Niagara Falls. The Canadians have to have some other reason for wanting to kill the signature bridge. The question is, What?"

Then Deep Boat answered his own question: "Ottawa is fully aware that the Freschi-Lin bridge would be cheaper and faster, but I bet the politicians are more worried about performance lawsuits from firms with which the PBA has signed contracts or to which the PBA has made documented promises than they are about how long it will take to finish this thing or what it looks like." If the twin span is built as planned, Deep Boat said, it will be financed completely out of bonds paid back out of ordinary bridge revenues. If there is a shift to the signature bridge, there may be nonperformance lawsuits that will result in huge attorneys' fees and large default penalties, which the Canadian government will have to pay. "The Canadian government doesn't want to put a penny into a bridge in Fort Erie. They'll put money into Niagara Falls, which promises a fabulous return on investment, but as far as Ottawa is concerned Fort Erie is just a truck port, it is a place that should be producing money, not costing money." Prettiness or ugliness are equally irrelevant, he said, all that matters is that the trucks can come and go and that the bridge makes money.

The *New York Times* said last week that we're engaged in a border war. They're right. It's quiet, very few people know it's going on or what the terms or targets are, but we are having a border war. Given that condition, I suggest we do what we nearly always do when we want to end a war: ship a lot of money to the other side to prove we're really friends now. If we do that, maybe our good friends the Canadians won't be so fretful about losing money to plaintiffs in nonperformance lawsuits and maybe they'll forget whatever else is bothering them that we don't know about, and maybe they'll tell their five representatives on the Buffalo and Fort Erie Public Bridge Authority that it's now in the national interest to behave rationally and reasonably. It's worth a try. For not much more than a few cruise missiles ($2 million) and way way less than a B-2 bomber ($2 billion) we can settle the only border war in this entire region since the

Americans and British torched Fort Erie and Buffalo in the war the Peace Bridge was built and named to commemorate.

P.S.: Part of the problem in this entire affair is that so many things have been kept secret, ambiguous, sometimes even lied about. Too many questions, too much secrecy, so few answers. Next week I'll try answering one question central to all of this: Who owns the Peace Bridge?

9 THE TROJAN PEACE

Two proposals for resolving the Peace Bridge War emerged last week, one by Senator Charles Schumer, the other by John Maloney (Member of Parliament, Erie) and Representative John LaFalce (29th district, NY). Schumer outlined a plan that might, for the first time, give everyone a fair voice in the procedure. The Maloney-LaFalce proposal represents the first time any Canadian official has expressed any willingness to consider any alternative to the twin span; it is also the first time Congressman LaFalce has advocated a procedure in which signature span advocates could be heard.

These are important statements from three public officials critically involved in this entire process. Here are the two proposals in their entirety, followed by comments on both.

SENATOR CHARLES SCHUMER'S PLAN

(A letter dated May 4, 1999)

> Due to recent events, the plans to build a news Peace Bridge have reached an impasse. I am concerned that the process will become increasingly acrimonious and construction will be delayed indefinitely as the parties become tangled in lawsuits. While there is no consensus on what type of bridge to build, there is a consensus that nothing will be built unless all sides of the arguments- American and Canadian, signature span advocates and twin span advocates-can come to a consensus. The fact is, each side has the power to stop progress completely.
>
> To resolve this dispute, I propose the creation of a binational, independent commission to decide whether a signature bridge can be built at similar cost, in a similar time frame and with similar environmental impact to a companion span. The commission would be comprised of two Canadian members and two American members, with a supporter of a

companion bridge and a supporter of a signature bridge from each nation. The commission would have a fixed time frame, at most 120 days, in which to reach a conclusion. During that time, the commission would hold public hearings on both sides of the bridge and listen to area residents, environmental groups, public officials and concerned citizens. The commission would choose independent experts to answer technical questions about the financial, time line, and environmental impact of the twin and signature spans.

Before the commission began its work, but after the four commission members were chosen, all major parties, the relevant Canadian and American elected officials, the advocacy groups, the members of the Peace Bridge Authority and others would have to agree to abide by the decision of the commission, which would have to be unanimous In addition, all parties would agree at the outset not to file any lawsuits during the commission's 120-day debate period and thereafter, provided the commission reached a unanimous decision.

No part of this proposal is set in stone. I look forward to discussing this idea with you. We need the participation of everyone involved in this process – Canadians and Americans, citizens of Buffalo and Fort Erie, members of the Peace Bridge authority, and business and environmental groups. Buffalo and Fort Erie need a new bridge; let's work together to build one.

Whether you prefer a signature span as I do, or the twin span, I am sure you agree, no action would be the worst decision of all.

THE JOHN MALONEY-JOHN LAFALCE PLAN

(A joint undated statement titled "A Proposal", issued two days after Schumer's plan was made public.)

With a view to resolving the impasse that has developed respecting competing proposals of the bridge link between Buffalo, New York and Fort Erie, Ontario, we suggest that the Peace Bridge Authority agree to delay start of construction of the companion span to allow a 90-day period for an independent study to immediately begin detailed examination of the current Freschi-Lin plan for a single cable-stay bridge (including the demolition of the existing Peace Bridge) in order to determine whether the proposed single cable-stay bridge can be built equal in quality to the companion span, equal in both vehicular and pedestrian capacity to the companion span in combination with the existing Bridge within the same or similar time and cost (including the cost for demolition of the Peace Bridge) and with a likelihood of meeting all required environmental concerns and obtaining all necessary permits for both the construction of the Freschi-Lin bridge and demolition of the Peace Bridge.

We further suggest that if the report resulting from that study satisfied the listed conditions, the Authority seriously reconsider its plans, and in good faith review in detail and further develop the single cable-stay bridge plan, and put the existing companion span construction 'on hold' pending the results of such a detailed review.

We understand that during the period of the 90-day study, and any detailed review which follows, the Authority might well wish to protect its legal interests by either commencing required actions, or, preferably, entering into a statute of limitations tolling agreement. However, we would expect that the Authority would not prosecute any legal claims it believes it has during the time frame we have proposed In exchange for that commitment of the Authority, when the 90-day study period beings, the City of Buffalo and all other Governmental Agencies should place in escrow all required easements and consents, and delay or suspend any legal action seeking to prevent construction of the companion span, all of which would be released to the Peace Bridge Authority in the event the single cable-stay bridge cannot meet any of the specified conditions.

We also understand that other Governmental Agencies might wish to commence legal action to protect their interests, or, preferably, enter into a statute of limitations tolling agreement. However we would also propose that no government entity prosecute a claim against the Peace Bridge Authority during this study period.

We also suggest that the initial 90-day study period work we describe be overseen jointly by Transport Canada and NYSDOT; that two (2) independent and reputable and

> experienced bridge engineering firms with no previous or potential future affiliation with either the companion span or the single cable-stay bridge be retained to undertake this study-one from Canada and one from the USA; that any architects and/or experts in the necessary environmental and permitting processes should have similar credentials; and that the study group consult with public officials, supporters and authors of the single-cable-stay span proposal and with the Authority.
>
> The technical criteria and the scope of the study products should be established at the outset of the study.
>
> It is our hope that the foregoing process, or something similar, can produce an objective evaluation of designs while maintaining this border's long tradition of mutual respect for beliefs and cooperative work toward solutions.

PLAN A, PLAN B

Both proposals ask everyone to hold off suing while the process goes on, and both are versions of single combat: those agents take on the battle and we all agree to abide by the results. Both proposals assume good will on both sides and that the Public Bridge Authority cares about what anybody thinks. Neither proposal gives any consideration to the plaza design and construction, a critical factor in all of this. That's where the similarities end.

> Maloney-LaFalce would have the study done by two bridge engineering firms picked by Transport Canada and New York State Department of Transportation. The Canadian Minister of Transport just reappointed three adamant twin span supporters to the PBA. The head of NYSDOT recently removed from the PBA one of his senior aides, a person familiar with transportation issues in the state, and replaced him with steel-factory owner Brian Lipke. Lipke is also a twin span supporter. I'd have a difficult time assuming the two engineering firms were being picked for their ability to remain neutral.
>
> There would be no question where the members of Schumer's commission stood because he wants people with strong positions to be asking the questions. The

people getting the data and preparing the report and voting would be people who care, who are frank and up front about their positions. No sneaking around, no hidden agendas. (Democracy isn't having people with no opinions on anything making the decision; it's having all the vital opinions heard and represented. I like that.)

Schumer says he's willing to discuss the whole thing, that what he's putting forward is a proposal and nothing more. Maloney-LaFalce invites no comment, suggests no possibility of change, it's just out there: do this.

Maloney-LaFalce has the city putting all its powers in a box that the PBA gets to open should those two engineering firms find the signature span idea faulty in any way. Those firms will come to a decision and that's it. Schumer's plan requires two strong advocates to change position, to be convinced by the data. If no change occurs, we're back to where we are today with nothing lost; if one side is so convinced by the data that it crosses to the other side, we've got a bridge going up over the river.

Maloney-LaFalce gives less time to the study – 90 days versus Schumer's 120 days – but that's not surprising given that theirs is just an engineering study while Schumer's suggests including the human factors as well.

Maloney-LaFalce includes a study of the cost of demolishing the old bridge. Schumer doesn't mention that, but his proposal is so open-ended it easily could be. (The PBA did its own study of tearing down the old bridge in 1967. If you correct for inflation and their curious decision to exclude income from recovered steel, which is considerable, their projected cost comes to about $10 million in current dollars, only $1 million above the estimates provided by the consultants quoted by Bruno Freschi and T.Y. Lin.)

Schumer's plan is reasonable, it is based on consensus, on letting the daylight in. It is also dangerous for the

> same reasons: it threatens to put all the facts in the hands of a citizen's group, it threatens to involve the community, something the PBA has stoutly resisted all along. That four-person committee would be like a grand jury, able to call anyone who might provide useful information. Maloney-LaFalce would turn it all over to two consultants. It would be like submitting to binding arbitration.
>
> Schumer's plan is written in ordinary English; you can understand everything he's saying. Maloney-LaFalce is very legalistic, very technical. I bet it was written by an attorney rather than by either legislator.

CHANGING HEARTS

I have no idea why John Maloney decided to seem conciliatory at this time. He has consistently opposed the signature plan and a day after his statement with LaFalce came out he again said that he thought the twin span was the only way to go. Perhaps it was Stephen Handelman's excellent article on the Peace Bridge battle that appeared in the Canadian *Time*. Canada came out looking sluggish and obstructionist in that piece.

But I can think of three reasons why John LaFalce's name is on that document, why he might appear to have had a change of heart.

> *First, it must have been lonely to be the only politician of stature on this side of the Niagara River still fighting the signature bridge.*

For a while, it seemed possible to dismiss the signature span and its supporters. In a handwritten note to US Transportation Secretary Rodney Slater in mid-February, LaFalce referred to the questions raised by senators Moynihan and Schumer as "spurious allegations." A delegation from the New Millennium Group that met with him a few weeks ago to enlist his support got nowhere. A member of the group wrote me that LaFalce "basically said that he had been around the issue for a long time and claimed that the process had recently become politicized. He danced around the issue (full of sound and fury, signifying nothing) but ended up defending the PBA's efforts and their plan. He also encouraged us to drop the issue and find another of greater importance."

But the battle lines have shifted and even Governor George Pataki has come out for the signature bridge. It's no longer a few politicians and a group of citizens futilely trying to get the PBA to pay attention. Now it's Americans versus Canadians, with the PBA sitting silently in the middle. If, as many people think, LaFalce is hoping to be named ambassador to Canada, he's building trouble for himself by being on the Canadian rather than the American side of an issue about which so many people care so much.

I think that's what he meant when he told the NMG delegation that the "process had recently become politicized." It was *always* political, but now it's getting political in a way that might cause him trouble.

Second, Hillary, still poised about running for Pat Moynihan's senate seat, came to Buffalo.

The first thing she did last Friday morning was keynote a $1000 a ticket fund-raiser for John LaFalce at the Delaware Park casino. She praised him for being someone great at bringing people together. That would have rung pretty hollow had there been pickets across the street with signs naming LaFalce the only Democrat trying to block the popular will. Hillary is a great supporter of inspiring public symbols and community involvement and here's John LaFalce the only Democrat opposing both. I don't know if one of Hillary's campaign people called LaFalce and said, "Do you know how much damage you're going to do us up there? Everybody knows you and the First Lady are buddies, you brag about it all the time, you're always being photographed with her. You're going to cost us votes in November. We've got Rudy Giuliani breathing down our necks. Get your house in order." It probably wasn't that specific. LaFalce and Hillary are both sophisticated people. It probably took no more than a question: "So, John, what's with this Peace Bridge we keep hearing about?"

Third, the Maloney-LaFalce proposal isn't so much a solution to the bridge stalemate as it is a way of killing Chuck Schumer's proposal.

I don't know if the people who are heavily invested in the twin span project asked LaFalce to do something to neutralize what Schumer started, but that would be the effect if his plan were accepted. Maloney-LaFalce looks enormously simpler, neater. Instead of all these hearings and people trooping in and out speaking their minds, all these varied consultants, it's just two organizations hired by two government agencies that have already told us where *they* are on this issue, working on their own, two final reports.

A DIGRESSION: WHAT HAPPENED AT TROY

"LaFalce didn't deliver a peace plan," an attorney very much involved in all of this said. "He delivered the Trojan Horse."

That metaphor is thrown around so much I thought it might be useful to remind you what happened at Troy, how *that* battled ended. Homer's *Iliad* ends before the war does. You have to go to the second book of Virgil's *Aeneid* for this story. The wandering Aeneas is in Carthage and he tells the story to the city's queen, Dido:

One morning in the tenth year of the war that started after Paris, son of the Trojan king Priam, stole the wife of Menelaus, the Trojans awoke to find the entire Greek army gone. The ships, the tents, the troops: gone. Not far from the city gates they saw a huge horse made of wood. While they debated what to do, a Greek named Sinon appeared. He said he'd run away from his former friends because they were going to sacrifice him to ensure fair winds on their current trip to get new supplies and troops. The Trojans asked Sinon about the wooden horse. It was a gift to the gods, he said, and if it were ever brought inside the city of Troy the Greeks were doomed. Even though some of their wise men argued against it, the Trojans immediately set about doing what the Greek forces led by Agamemnon had been unable to do in a decade: they destroyed part of their wall and dragged the horse inside. There had been a prophecy that the Greeks would never triumph so long as Troy's walls remained intact, but what need to worry about that now that the Greeks were gone?

The Greeks weren't gone. They were hiding on the other side of an island not far from shore. And the horse wasn't empty. Late that night Sinon opened its belly and a group of Greek soldiers led by Ulysses (the Greek Odysseus) silently climbed out, opened the city gates from the inside, and admitted the Greek army. They killed or enslaved nearly everyone in the city. Only a few escaped.

The Trojan War happened about 1200 B.C., the *Iliad* was composed in late 8th or early 7th century B.C. Virgil's poem was unfinished at his death in 19 B.C. The end of the Trojan Horse story comes thirteen hundred years later, in cantos XXVI and XXX of Dante's *Inferno.*

Dante puts Ulysses in the eighth pit, or bolgia, of the eighth circle of Lower Hell. This is the bolgia of the deceivers. It is located between the thieves in bolgia seven and the sowers of discord in bolgia nine. Ulysses is on fire, and he will burn forever. That's because he thought up the idea of the Trojan Horse. Sinon is even lower, in bolgia ten, the place of the falsifiers, and he suffers more: he burns from a fire within so hot his body smokes, he cannot move, and he stinks.

For Dante Alighieri, inventing a deceptive plan is bad, but far worse is convincing people who think you're a friend into accepting it.

LET'S MAKE A DATE

So what happens now?

First, we wait to see if and the Public Bridge Authority responds to either of the proposals. The *Buffalo News* reported that some members of the PBA were looking favorably on the Maloney-LaFalce plan, which is hardly surprising. I can't imagine Maloney and LaFalce issuing that letter if they hadn't first run it by their friends on the PBA. And I'm sure that the Schumer plan would utterly terrify them.

We still need to know more than we have been allowed to know. As we reported last week, LaFalce's office has refused even to discuss setting up an interview to discuss these matters. Nothing has changed: we've still gotten no response to our request. We don't like doing all this speculating about the motives of a congressman who has served this area well for many years, but his inaccessibility leaves us no choice.

If John LaFalce wants to be Canadian ambassador when he's tired of being a member of Congress, he should be Canadian ambassador. He has many friends up there and he knows many of the border issues well. I'm sure he'll make a fine ambassador. If and when he's ambassador he'll have conversations with his Canadian counterparts and he'll report to and take instructions from the American Secretary of State. In the interim, he's still working for us and he really ought to talk to us about an issue we think is so important. Even if it's just to tell us why he thinks it's not so important – he should talk to us.

Maybe our messages haven't gotten through the stone wall of the staff surrounding him, so we'll reissue the invitation here:

Dear John:
How about sitting down for a conversation about the Peace Bridge issue and your position on it?

Bruce

We'll sweeten it: Let's have the conversation and *Artvoice* will print a transcript of the whole tape with no cuts, no editing. All the questions, comments, arguments, responses, requests for more coffee, everything. You'll get to have your say and we'll get to ask our questions and our readers will get to read what you want them to know about what you're thinking. Your office has my phone number – I left it every time I called.

10 THE PEACE BRIDGE PLAZA DOG & PONY SHOW

If you doubted that the efforts of signature bridge advocates were having an effect on the Public Bridge Authority, consider the dog & pony show they put on last Wednesday for the local press. So far as I've been able to learn, the only newspaper regularly covering the Peace Bridge story that did not get invited to the meeting was *Artvoice*. I was doing research in the Peace Bridge office three of the four mornings before the meeting, and they faxed us a press release immediately after the meeting, so it wasn't that they didn't know how to reach us.

Vincent Lamb, executive vice president of Parsons Transportation (the parent company of the pba's longtime engineering firm, DeLeuw Cather), the group responsible for the twin span design, presented his firm's analysis of the cost and timeline for a signature span. Parsons is a huge corporation. Lamb offered three major conclusions: the signature bridge would cost twice as much as we've been told, the plaza designed by Freschi would require acquisition of 140 buildings on the West Side, and the complete signature bridge and plaza project would take ten years. He also said there were unresolved environmental concerns, like what a signature span would do to bird migration patterns and the status of the bridge as an historical structure.

It must have been a real performance. I assume there were charts, perhaps slides. I don't know because, as I said, *Artvoice* wasn't invited to the dance. I do know that Lamb had no evidence for any of his major assertions, and that his firm elected not to look at a good deal of evidence readily available to them. According to Peace Bridge staff, he consulted only units within Parsons Transportation Group, people who had no direct knowledge of the signature bridge at all. Which suggests that the conclusions are made up, fiction, smoke and mirrors, jive.

Had *Artvoice* been admitted to the Parsons Transportation presentation, these are the questions I would have hoped to have heard Vincent Lamb answer:

1. Given that Parsons Transportation's estimate of construction costs for the bridge it designed after spending somewhere between three and six million dollars of pba money was off by nearly sixty percent (they estimated $57 million, the lowest bid came in at $89 million), why should we expect that your estimate of the cost of building the Freschi-Lin bridge is any more accurate?

2. When you were costing out your vision or version of the Freschi-Lin bridge, why didn't you ask to look at any of the financial documents prepared by Cannon, Freschi, Lin, and the several consultants who did detailed work for them? That is, why did you use hypothetical numbers from your own staff, which knew nothing about the Freschi-Lin design, when there were real numbers available for you to have worked with?

3. You claim that Freschi's plaza would require demolishing 140 privately-owned properties. Freschi says it's fewer than 30. Why should we have confidence in your far larger number when you never talked to Freschi or asked to see any of his plans? How can you predict with such confidence the footprint of a design you know virtually nothing about?

4. How much was Parsons Transportation paid to do this study?

5. The press release says you presented "the conclusion of an investigation." According to Bridge staff and Collins & Company (the pba's press agent) you spent three weeks to a month working on it. Is that truly long enough to have designed multiple alternatives to the Freschi-Lin bridge and cost them out? Does Parsons Transportation ordinarily present cost and impact conclusions based on three or four weeks of work, with no examination of available primary documents?

6. According to *Business First,* you said that the Freschi-Lin plan "lacked the proper design to mesh with the New York

State Thruway." Dean Freschi says their plan utilizes all existing Thruway connections. These seem to be contradictory statements. Why didn't anyone at Parsons Transportation try to find out which of you had it right?

7. No public or private agency or organization with responsibility for designating structures as having historical value has said the Peace Bridge was of any interest, and one – the Preservation Society – has said it ought to be torn down. No ornithological agency has ever expressed the slightest concern about what the signature bridge might do to bird traffic. The bird and historical issues are tired old things, and everyone around here knows that. Why did you resurrect them? Is there new evidence or, as seems most likely, are you just hauling stuff out of trash bins? If the former, what? If the latter, why would you do that?

8. Your own estimates put construction of the twin span, refurbishing the old bridge, design and building of a revamped plaza at ten years. You project Freschi-Lin at ten years. Even if you had examined data for Freschi-Lin and even if you're right, what point are you making, given that at your worst-time scenario for them is no longer than yours?

9. Even to a corporation as huge as Parsons Transportation, a $200-300-million project is attractive, desirable, worth fighting for. What steps, if any, did Parsons take to ensure that its evaluation of the competitive design was fair, objective, and free of self-interest?

10. How much more of Front Park will your new plaza consume? What other local properties will your new plaza and access roads consume? Which of the two plaza designs will serve Buffalo better and harm its citizens less?

Alas, we have no answers to any of those questions, nor others like them. No one in the pba board room saw fit to ask them. That doesn't mean the questions go away; it means only that the pba and its prime contractor put on a damage control dog & pony show in which they tried to make such questions go away. That only works when you're providing answers to the questions people really have, and when your answers are as honest as you can make them, neither of which seems to have been the case at Peace Bridge Plaza last Wednesday morning.

And what happens when an organization like the Public Bridge Authority and its press agents and its other hired guns fail to bully away the reasonable questions? You get what happened Monday: the City of Buffalo, the Olmsted Parks Conservancy, and the Episcopal Church Home of Western New York all filed suit in state supreme court to force the pba to do what it would not do willingly: pay attention to the people and play by the rules.

Lately, it seems I've been asking more questions here than I'm answering. Well, that's not necessarily so bad. One thing I've learned in a lifetime of scholarship is that questions that don't get answered may be as informative and helpful as the ones that do. Next time, I'm going to try to deal with a whole range of questions, some with answers, some with partial answers, and some with no answers at all. Questions like

> Who owns the Peace Bridge?
> How much money do they make and what do they do with it?
> Why are they backing the Maloney-LaFalce evaluation plan and saying nothing about Senator Schumer's far more comprehensive and democratic plan?
> What single lie about public money has the pba and its representatives told more times than any other?
> What is going to happen?

Stay tuned.

11 The Great Summer Peace Bridge Q&A

(Every question in this article was asked by someone who wrote a letter, sent an email to me at bjackson@buffalo.edu, called on the phone, talked to me in a store, on the street, at school. They are real questions; I didn't interview myself. I'd like to thank everyone who helped me prepare this article, as well as the ten that preceded it, especially Jeff Belt, Mark Mitskovski, Bill Banas, Jeremy Toth, Mary Catherine Malley, Laura Monte, Jim Kane, Earl Rowe, Steve Mayer, Jim Pitts, Sam Hoyt, Tom Schofield, Bruno Freschi, Sam Savarino, Pamela Earl, Tony Fryer, Clint Brown, Mike Desmond, Don Wharf, Ross Robinson, and Jamie Moses.)

Who Owns the Peace Bridge?

The Buffalo and Fort Erie Public Bridge Authority, a public benefit corporation chartered by the State of New York, the United States Congress, and the Canadian Parliament.

There were conversations about and attempts at building a bridge between Fort Erie and Buffalo going back at least to 1851, but nothing happened until 1919 when a group of twenty-five Canadians and Americans set up the Buffalo and Fort Erie Bridge Corporation. They wanted a bridge that would get let them move between the two communities faster and more flexibly than did the ferries then available. (Did you ever wonder why West Ferry Street has that name?) They put up $50,000 of their own money to get the corporation going, and then set out to raise $4,500,000 in bonds for the actual construction. The bonds were mostly sold locally, and the offering was oversubscribed before the first offering day was out. People on both sides of the river really wanted the bridge, they liked the idea of the bridge, they were willing to invest their own money to have the bridge.

The first automobile to make an official crossing was driven by Edward J. Lupfer, the bridge company's chief engineer, on March 13, 1927. The Bridge opened to the public on June 1 of that year. At the August 7, 1927, opening ceremony, which was attended by the Prince of Wales and the Vice President of the United States, John W. Van

Allen, one of the original incorporators of the Peace Bridge said:

> Hereafter, this bridge belongs to the public. Our sole remaining function is to collect the tolls and pass the money back to those who advanced it. The construction problems are over; [there] remains now only its dedication to service, and we wish to all, great joy and the convenience in the use of it.

Would that it had been so. The public has yet to get ownership of the bridge, the people who run the bridge have and are doing far more than collecting tolls and giving money back to investors, and the construction problems have never been worse.

By law, the PBA board has five members appointed by the Canadian Minister of Transport in Ottawa, two appointed by the Governor of New York, and three Americans serving ex officio, which means they're on the board because of their jobs. The defined ex officio members are the board chair of the Niagara Frontier Transportation Authority, the New York attorney general, and the director of the New York Department of Transportation. The NFTA board chair usually assigns the NFTA's executive director to attend meetings, but the current board chair, Luiz Kahl, has been attending himself. The attorney general usually assigns the job to the chief of his Buffalo office. Attorney General Elliot Spitzer did that when he appointed Barbra Kavanaugh to the Board. His predecessor, Dennis Vacco, who was far more political in his appointments, gave that job to steel mill owner and Republican Party contributor Brian J. Lipke. When Spitzer replaced Lipke, NYSDOT head Joseph Boardman pulled his transportation expert off the board and gave Lipke his seat.

So seven of the ten are political appointees. In practice, the PBA is an insular group, taking advice only from staff, long-time consultants and a few people who have special access. One of those is Andrew Rudnick, CEO and president of the Buffalo and Niagara Partnership, who has worked closely with the PBA and has on occasion acted as spokesman for it. Close connections between the boards of the two groups probably make Rudnick's participation in PBA affairs possible. Luiz Kahl is a member of both boards. PBA board member Gary Blum was, until last year, chief financial officer of Buffalo Crushed Stone, which is owned by Richard E. Garman, who is chairman of the Partnership's board.

PBA board member and Fort Erie resident Deanna DiMartile told *Time Magazine* correspondent Stephen Handelman, "It's only when outside influences step in that things break down." This, I think, is the heart of the current problem with the PBA. It regards nearly everyone outside its own boardroom as "outside influences." Senators Moynihan and Schumer, Governor Pataki, the Buffalo Common Council, the region's delegation to Albany, you and me – we're "outside influences."

How did a private corporation become a public authority?

Because of the Great Depression and the end of Prohibition, the numbers of vehicles making the crossing plunged and there was a real danger that the company would default on its bonds. For a time, Frank B. Baird, president and prime mover of the project, put his own money into the struggling company but not even his great private wealth could keep it going in the face of declining income.

In 1933, the Bridge Company sought governmental salvation. Over the next year, three separate pieces of legislation in Ottawa, Albany, and Washington, D.C., created the Buffalo and Fort Erie Public Bridge Authority, a public benefit corporation. Public benefit corporations exist in a land of deliberate legal ambiguity: they aren't government agencies and neither are they private corporations. Their profits in theory belong to the governments that created them, but they behave more like ordinary corporations than an arm of government. They can partake of some of the benefits of government status – their property and bonds are tax exempt, for example – but they have separation from primary agencies of government not enjoyed by organizations that really belong to the public, such as SUNY or the New York Thruway. A public benefit corporation controls its own resources.

The Authority at first had nine members, six from the US and three from Canada. It acquired all the assets and debts of the bridge company. With the debts restructured and almost no taxes to pay, the Authority was on firm financial footing.

In 1957, New York State created the Niagara Frontier Port Authority (later the Niagara Frontier Transportation Authority) and tried to tuck the now-profitable Peace Bridge into it. The Bridge Authority balked. It asked New York Attorney General Jacob Javits for a ruling: could the New York Legislature take over an organization created by the government of New York, the government of Canada and the US Congress? Javits said no, the power play wouldn't hold up.

So the ownership papers were redrawn another time. There were four key changes:

> Total board membership would increase to ten, with five members from each country.
> The bridge would be directly tied to no other agency so it could remain fully independent.
> The two countries would divide excess revenues equally.
> And the sunset, the date everything would be turned over to the two governments, was extended from whenever the outstanding bonds were paid off to 1992.

By 1970, it was clear that bridge capacity would have to be expanded, which meant the Authority would have to issue new construction bonds. The governments extended the life of the Authority to 2020 and raised the debt limit. The Canadian government and the State of New York, instead of splitting whatever cash was left over after payments were made, would each get a flat $200,000 each year, and the Public Bridge Authority would keep the excess for development.

What do the members of the Buffalo and Fort Erie Public Bridge Authority really do?

When it's not engaged in a war with the community about a construction project gone sour, most of what the PBA does is very ordinary. If you go through the minutes of the Buffalo and Erie Public Bridge Authority, you read of decisions regarding who shall paint the bridge and how much shall bridge employees who have worked 20 years get as a bonus and what shall we do for the widows of PBA employees who died mid-year and how shall we apportion the funds to the Canadian and American banks and what tolls shall we set? You read about negotiations with the City of Buffalo to take more of Front Park for truck and customs sheds. You read about issuing and retiring bonds.

The primary change I noticed in the sixty-five years of minutes has to do with secrecy. In the early years the Board seemed open; in more recent years, the minutes suggest an organization on the defensive, an organization with secrets, an organization with the kind of we-they mentality reflected in Deanna DiMartile's statement above. In recent years, the board began doing much of its decision-making in executive session, which means no observers were permitted and no minutes were kept. There are minutes of some meetings that record nothing other than attendance, a call to order, a motion to go into executive session, a motion to come out of executive session, and a motion to adjourn. Things happened at those meetings, but not things you and I will know about.

What's the meaning of the bridge's name and what's the real name of the organization that operates the Peace Bridge?

The Peace Bridge is named to commemorate the century of peace between Canada and the United States following the end of the War of 1812. It would have gone up a decade earlier had not Grand Duke Ferdinand been assassinated in Sarajevo and the world erupted into the Great War.

The corporation that owns the bridge is named The Buffalo and Fort Erie Public Bridge Authority, but you'd hardly know that if you had any dealings with them. An April 15 press release from the Authority's Secretary/Treasurer Earl Rowe has "Peace Bridge Authority" under his name. A statement in it about increasing frequency of crossings is attributed to "Stephen F. Mayer, P.P.E., Operations Manager, Peace Bridge Authority." If you call the Authority's number (716/884-6744) the recorded message begins, "Thank you for calling the Peace Bridge Authority." The letters glued to their office door at the bridge plaza say "Peace Bridge Authority."

I think it important that they deleted the word "Public" wherever possible. They could just as easily have the taped message say, "Public Bridge Authority"; they could just as easily use "Public Bridge Authority" on their correspondence and door. The difference is this: "Peace Bridge Authority" connotes an object that this group controls; "Public Bridge Authority" connotes something belonging to the public for which this group is responsible. "Peace Bridge Authority" is about power, "Public Bridge Authority" is about service.

I doubt anyone with a nefarious bent of mind consciously thought up that nuanced shift in nomenclature. But it is in nuance that the rest of us sometimes learn out what happened in those rooms to which we are never admitted or from which we are ejected when it's executive session time. Speech reflects attitudes of mind, and shifts in nomenclature usually reflect something we ought to know about.

Who uses the peace bridge and what do they pay?

In 1992: 7,192,199 cars and 919,991 commercial vehicles.
In 1997: 6,345,622 cars and 1,319,163 commercial vehicles
In 1998: 6,288,318 cars and 1,381,496 commercial vehicles

Fewer and fewer people, and more and more trucks. Automobiles pay $2 per vehicle. Trucks pay $0.41 per ton. There are far more commercial tons than automobiles, so the Authority collects far more in tolls from trucks than from passenger cars.

The toll schedule has simplified from days of yore. 1953 tolls were twenty-five cents plus five cents per person for autos, with an extra quarter for trailers; five cents for pedestrians or bike riders; fifteen cents for a motorcycle with or without a sidecar; twenty cents for a person on horseback, twenty-five cents for a one-horse vehicle and forty cents for a two-horse vehicle. Vehicles in which someone was horizontal were charged more: forty cents for an ambulance and half a buck for a hearse. Kids under five were free whatever mode of transportation they utilized.

They stopped charging people on foot when their accountants discovered it cost more to collect the nickel tolls than the bridge made on them. I can understand why they'd add a nickel for the second horse, all that extra sweeping and scooping, but why is a hearse a dime more than an ambulance? The PBA's equivalent of the pennies on a dead man's eyes?

Does the Peace Bridge make money?

Yes. Huge amounts of it.

In 1998, the PBA reported income of $25,040,401 ($5,996,828 passenger tolls, $13,952,982 commercial tolls, $4,866,212 rentals, and $224,379 other) and operating expenses of $14,009,836. It also reported interest and other income totaling $4,829,458.

That results in what their accountant terms an "excess of revenues over expenses" (what you and I would refer to as "profit") of $9,180,378.After paying all operating expenses and debt service, the Bridge made a 36.6% profit. If the Peace Bridge were a private corporation, this would be a staggering rate of return.

Equally important is the fund balance, which is what they've accumulated. When you want to know how rich someone is, you don't look at what came in or what went out, you look at the fund balance. Last year the Bridge had a fund balance of $64,668,291, up from $55,487,913 the year before. I don't know where that money is. The report gives no information about who's controlling the money or what is being done with it; it records very little interest income. Some institution or institutions are making a great deal of money handling the PBA's money.

If the Authority paid real estate taxes, the way you and I do and the way any profit-making corporation does, it would have paid about $6 million last year instead of the $783,000 it divided equally between Buffalo and Fort Erie in lieu of taxes and the $200,000 it paid to the NFTA. Even if the Authority had paid real estate taxes and taxes on profits, it would still have registered a very respectable 12.7% profit. Not spectacular, but still very respectable.

With what it is making on truck traffic, the Authority could dispense entirely with the income from automobiles and still turn a profit.

Does the Bridge make money for us or take money from us?

There's a pretty little packet of six prints on heavy 8 ½" x 11 stock the PBA sometimes gives away titled "Construction Paintings of the Peace Bridge by H. H. Green." Green was one of the original incorporators. A note on the inside cover of that packet says, "No public funds have ever been granted or used for construction, operation, maintenance or for capital expenditure. All financing has been done from private and institutional funds." John A. Lopinski's Chairman's Report in the 1997 Annual Report begins, "No public funds have ever been granted or used for construction, operation, maintenance or for capital expenditure. All financing has been done from private and institutional funds." They say that a lot. It is perhaps the lie they have told more than any other.

The PBA exists entirely on public funds. Entirely. If the public wants to cross the river from the Buffalo into Fort Erie, the public has to pay the Buffalo and Fort Erie Public Bridge Authority $2 for the right to do it – unless the public is driving a truck, in which case it pays considerably more. Last year the PBA received almost $5 million in rental income, the largest portion of it for space rented at above market rates for use by Customs and Immigration – your tax dollars pay that rent.

The Authority has been tax exempt since 1934. Tax exemption means the government has decided that what an organization does accrues to the public good, so it doesn't take from its profits (profits, not income) the share everyone else pays. Symphony orchestras, social service organizations, museums, churches, schools – all such organizations receive tax exemptions. Taxes are monies that belong to the public. If an organization is declared tax exempt, the government is subsidizing that organization to the extent that the organization is permitted to keep and use for its own purposes the funds that would otherwise have been shared by all of us. All organizations that engage in commerce and are tax exempt are recipients of public money to exactly the extent of the taxes they would have paid had they been treated like everyone else. That includes not only the taxes on profit, but also sales tax and real estate tax from which they are exempt. Bonds of tax exempt organizations are themselves tax exempt, which means those bonds enter the market with a competitive edge over bonds from profit-making organizations. That enables the sellers of nonprofit organizations' bonds to offer them for a lower interest rate, which means the nonprofit gets to rent money more cheaply than you and I do.

It's moot now. However often they made that claim in the past, they can't make it any longer. Last week the Peace Bridge received two grants from the US government for a total of $2.76 million to develop a high-speed frequent-traveler lane and an electronic document transmission system.

Why are the lawsuits necessary? Won't they just slow things down?

How much and whether the lawsuits slow things down depends on the PBA, not on the groups filing the lawsuits. The PBA saying "You shouldn't sue because that slows up our project" is like someone who hit you with a car saying you shouldn't sue because that's going to cost him money. Yeah, it will, but who damaged whom here? Blaming the victim is a lawyer's tactic, nothing more. If the PBA had done the right thing in the first place – considered all the alternatives rather than locking themselves into a single choice and then pretending they were considering alternatives, or if they'd just be willing to consider them now – the lawsuits would not be necessary. Furthermore, had the lawsuits not been filed now certain legal rights would have been lost for good because of the statute of limitations.

A lawsuit is the citizens' recourse to benign force and it is sometimes the only way to get government to behave decently. I came across this passage in a letter from Samuel Adams to Thomas Jefferson dated October 9, 1787: "I have long been settled in my own opinion, that neither Philosophy, nor Religion, nor Morality, nor Wisdom, nor Interest, will ever govern nations or Parties, against their Vanity, their Pride, their Resentment or Revenge, or their Avarice or Ambition. Nothing but Force and Power and Strength can restrain them." The lawsuits filed by the city of Buffalo, the Episcopal Church Home and the Olmsted Parks Conservancy were necessary because they was the only way to get the PBA to pay attention.

You keep saying the LaFalce-Maloney bridge evaluation plan, the one endorsed by the Public Bridge Authority, is a sham. Why?

LaFalce-Maloney would have two engineering firms, one American and one Canadian, estimate the cost and timeline for constructing a companion span and the Freschi-Lin signature bridge.

Its key defect is that it locks out consideration of two vital issues: construction of the plaza on the American side and maintenance costs for whichever bridge or bridges are up. The first goes to time, the second to money.

I'm assuming that LaFalce's and Maloney's goal isn't just to get a new bridge up; it's to have six fully functional lanes available as quickly and efficiently and economically as possible. The kicker is "fully functional." A bridge doesn't float in space; it lands somewhere. The speed with which it handles traffic, and its impact on the community, is determined by the entirety, the bridge and the plazas at either end. The plaza on the Canadian side is already there; the PBA says its new bridge will feed into it and Freschi says his design will land there without disruption. So the Canadian plaza is not a problem.

The American plaza is. The PBA hasn't told us what plaza plans it's been developing, but they have said they won't begin construction on it until after the new bridge is completed and the old bridge rehabilitated. By its own estimate, that project will take a full decade. A full decade before those six lanes and two plazas are capable of delivering traffic. Freschi says he and T.Y. Lin haven't designed just a bridge; they've designed a bridge system that articulates with the plaza on the Canadian side and contains a plaza on the American side more useful to the city of Buffalo than anything suggested by the PBA. He's also said that his entire system will be ready to use earlier and will cost less than the companion span and expanded plaza system advocated by the PBA. The PBA says Freschi is wrong, but it has never given his design serious consideration.

The only meaningful comparison is between the two entire bridge systems. Anything else is grounded in a syllogism, flawed from the start.

Once a bridge is up, it's got to be maintained. The concrete cable-stayed bridge is nearly maintenance free. Over their useful lifetimes, the maintenance costs for the two companion spans will be more than constructing a companion span and rehabilitating the old bridge. Whatever the initial cost of construction, the long-term cost, the real cost, will be far greater for the steel companion span than it would be for the concrete signature span.

That's why looking at the bridges alone and comparing only them is a sham issue.

If everything you say is true, why is the Authority so adamant about the companion span? They're not stupid.

No, they're not, but they're doing what a lot of people do when they find they've made a wrong decision that has wasted a lot of money: they close their eyes, dig in their heels and hope it will go away.

About five years ago, the PBA decided that the present bridge could not be torn down. All else follows from their initial error. They spent millions of dollars designing a companion span and now, instead of confronting that erroneous initial assumption they're defending its consequences. It's quite human for them to want to avoid admitting they screwed up. To make matters worse, they've gotten all their engineering advice from one company, Parsons/-DeLeuw, their engineering partner for decades and the company that, were all alternative plans to go away, will get to build the companion span. Going to Parsons for advice on alternate options at this point is like asking the fox to evaluate the henhouse security system.

Some board members may really believe the twin span is a better idea, but I'm convinced that it's now primarily a matter of saving face. They are in no financial danger, however things turn out. At its meeting of April 28, 1995, the board voted to revise the by-laws by indemnifying all its members – past, present and future – against the costs of any lawsuit and any legal judgements arising from their connections with the Peace Bridge, except for actions that were purely criminal. The minutes of that meeting also include the word "twinning." So far as I know, that's the first time that word entered the official historical record of the Buffalo and Fort Erie Public Bridge Authority.

What's more important, commerce or aesthetics?

Beware the false dichotomy. There is no good reason why we must set one above the other. It may be true that any bridge at all wide enough for the truck traffic will benefit the region commercially. That doesn't displace the fact that a beautiful bridge between Buffalo and Fort Erie will benefit the region commercially to exactly the same degree and it will benefit the region in other ways as well. If people tell us we must choose one or the other and they provide no compelling reason why such a choice must be made, we must conclude they are people who don't give a damn about the quality of life for people in this area or they have unstated economic interests. (None of the Canadian members of the board, for example, have said whether or how much the rigidity of their position has been influenced by the fact that a large Canadian construction firm has the contract to build the companion span.)

It's apparently easy to document the economic impact of a wider bridge. I write "apparently" because the businessmen and trade diplomats can tally tonnage of goods shipped back and forth and the number of jobs in the region involved in packaging and shipping and weighing and tolling and inspecting that might go elsewhere were a wide enough bridge not available, but they do not tally the costs to the region in pollution, environmental degradation, wear on the infrastructure, consumption of land that might be used for other purposes and so forth. But whatever economic benefits accrue to this region or the two nations will accrue whatever bridge is built. The kind of bridge built has nothing at all to do with the value of six fully functional lanes.

It is not so easy to document the aesthetic impact of a beautiful bridge. How can you tally in advance the wages of the aesthetic in ways the businessmen and trade diplomats can appreciate?

Maybe by looking elsewhere. A few months ago, radio station WNYC's "New York Kids" show sponsored a poetry contest in which kids in grades three through six were invited to enter poems on the subject of "New York Dreams." This poem was written by Kristin McMurrer, a ten-year-old fifth grader from Brooklyn:

My New York Dream

In my dream
I see the Verrazano Bridge
Bright silvery light forms a shimmering swing
I am drawn
Drawn by the light's welcome,
"Come Swing!"

I soar towards the shimmering lights
Through the cool starlit sky.
I hop on the long swing
Pump my legs faster and faster
The wind pushes between my legs
Billowing my skirt like a parachute.

Before me, the Statue of Liberty
Stands tall holding her flaming torch.
I am proud to be an American, a free American
Proud to be a New Yorker.

Above me, twinkling stars like little flashlights
Wink at me
The full moon casts a pathway shadow on the water below.
In the distance a tiny tug boat
Tugs a cruise liner under the bridge.

This is my freedom.
This is my city.
This is my bridge.
This is my dream.
This is my New York.

How much would it be worth if our kids in Buffalo and Fort Erie could have dreams like that inspired by the bridge we built across the waters that join our two cities? Wayne Redekop, John Lopinski, Andrew Rudnick, Mark Romoff: what's it worth to you? Brian Lipke, John LaFalce, John Maloney: what's it worth to you? Anything? Anything at all?

How will it end?

We'll get the signature bridge.

July 8, 1999

12 TROY/ROME/LONDON/SAIGON/BUFFALO

EXPLAINING THE PEACE BRIDGE WAR

I was trying to explain the Peace Bridge War to Warren Bennis, an old friend who now lives in Santa Monica. Warren was a provost at UB in the late '60s and he likes to hear about what's going on in town. He's a social psychologist and a keen analyst of management; most of his work has to do with making sense of institutional behavior. I thought he might help me understand what's been going on here.

I told Warren that we've got a public service agency – the Buffalo and Fort Erie Public Bridge Authority – in charge of bridge repairs and tolls that has taken on the business of designing major public works, though not one of its membership has any competence or expertise in these matters. I told him that the Authority and our congressman want to construct a bridge that is ugly, expensive, and anachronistic, one that will consume far more land surface on both sides of the river and will be far less flexible in use than its signature bridge alternative. I told him that Senator Charles Schumer has proposed a commission that would evaluate the two designs and take this all out of the area of personal pique, but the Authority has refused to cooperate with anyone or anything, save for one plan (LaFalce-Maloney) so narrow in design it will resolve nothing.

There was a sudden burst of static on the line. It stopped, started again, and then became a continuing crackle. For a moment we tried to decide whether it was in Buffalo or Santa Monica. I picked up a hard-wired phone, turned off the portable I'd been using, and the noise was gone. In the background I could hear Grace, Warren's wife, playing violin. I told Warren I couldn't quite make out what she was playing. He listened for a moment and told me what it was, then said she was really happy with the violin, which she'd just gotten after a two-year search. Then he told me to go on about the bridge. Warren is really good at focus.

I said the city of Buffalo had told the Authority it thought the Authority's design so hurtful to the city it would refuse to issue the permits the Authority needed to begin construction, and that the city had, in collaboration with two very respectable community organizations, filed suit to force the state agencies that had authorized the commencement of construction to rethink their decisions. And I said that not only had the PBA ignored all the requests from the community that it reconsider its decision – not *change* its decision, just *reconsider* it – but it had filed suit against the city to force it to issue the construction permits.

Basically, I told Warren, the Public Bridge Authority and its attorneys had examined all the requests and pleas from the community, had read all the legal briefs, had considered the Authority's role as a community agency, and had responded: "Fuck you."

I said I didn't think members of the PBA were like those tobacco executives and industrial polluters who are making so much money from their decisions they bury in ice all notions of social responsibility and all prodding of conscience. That kind of self-enriching position, I told Warren, I could understand, vile though it might be. But nothing I could think of made sense of the PBA position: they want to spend more money over more time to make something ugly that will give the public less service. As much as I've studied the Peace Bridge affair, I said, that decision finally doesn't make any sense to me.

"But why," Warren asked, "do you think it has to make sense?"

"Everything makes sense at some level," I said. "You just have to find it."

"Not so," Warren said. "Some things don't make any sense at all."

"So how do you explain them? How do you make sense of them?"

"You don't. You try to see them for what they are."

We talked more about the difficulty of dealing with behavior that doesn't make sense. Then Warren said, "There's a good book about this."

"Bad design and stonewalling public agencies?"

"Refusal by people in power to consider viable alternatives when what they're doing is clearly not in the best interests of the organization or groups they represent or in their own best interests either. Behavior that doesn't make sense. It's Barbara Tuchman's *The March of Folly.* It's about the Trojan War and the Renaissance popes and how the British lost their American colonies and how America got mired in Vietnam. And I think it may be about what's been going on in Buffalo."

ONE HORSE, SIX POPES, ONE CONTINENT, ONE WAR

The full title is *The March of Folly: From Troy to Vietnam*. It was published in 1984 but it's one of those classic inquiries that doesn't go out of date. Tuchman, who won two Pulitzer Prizes (for *The Guns of August* in 1962 and for *Stilwell and the American Experience in China* in 1971) addresses the assumption that people in power act rationally. When they do something we wouldn't do or of which we disapprove, we look for a logical reason: misinformation, corruption, an error in addition, madness, stupidity. We assume that the results, however good or evil or beneficial or detrimental, occur in service of some plan, are directed to some end. We try to figure out why we might make such choices if we were in that position. That is, even when we disagree with a position we tend to assume that reason plays a central process in achieving it.

Not necessarily so, Tuchman says in this superb description of and meditation on the way people in power make and adhere to decisions against their self-interest and the interest of the political organizations or communities they represent or control. Her concern isn't with decisions that are merely crooked or lousy, but rather with those that occur when there is a sufficiency of information available indicating they are stupid or dysfunctional, when competent people tell the decision makers that something is very wrong and the likely consequences will be costly and harmful, and when there are preferable alternatives clearly available.

She calls such behavior folly, and she identifies "block-headedness" as the primary cause. Block-headedness is the refusal or inability to listen.

She begins with an example from literature: the Trojans who took the Greek horse into their city without checking the contents, even though Laocoön and Cassandra warned against it and common sense should have told them the same thing. The horse was full of Greek warriors, who in the dark of night opened the city gates and admitted the Greek army, which proceeded to slaughter all the men, enslave the women and children, and destroy the city.

The main part of the book deals with the six Renaissance popes whose corruption, greed, sexual excess and general perversion led to the Protestant Reformation and the sack of Rome; Britain's loss of the American colonies; and the US involvement in Vietnam.

THE RENAISSANCE POPES

The six popes who governed the Church from 1470 to 1530 might have been scripted by Larry Flint, Ken Russell, and John Gotti. "Their governance," writes Tuchman, "dismayed the faithful, brought the Holy See into disrepute, left unanswered the cry for reform, ignored all protests, warnings and signs of rising revolt, and ended by breaking apart the unity of Christendom and losing half the papal constituency to the Protestant secession. Theirs was a folly of perversity, perhaps the most consequential in Western history, if measured by its result in centuries of ensuing hostility and fratricidal war."

These were popes who sold everything or who gave it away to their lovers, children, or other relatives. She tells of one man made a cardinal by Clement VII who visited his cathedral only once: the day of his funeral. In 1480, Cardinal Pietro Riario hosted a banquet at the papal court "featuring a whole roasted bear holding a staff in its jaws, stags reconstructed in their skins, herons and peacocks in their feathers, and orgiastic behavior by the guests appropriate to the Roman model."

As the popes sunk deeper and deeper into depravity, greed, militancy, and ambition, they became increasingly deaf to cries for reform from outside their own protected circles of power. And they became ever more violent: when Leo X (1513-21) wanted to bring Perugia into the Papal States he invited its dynastic ruler, Gianpaolo Baglioni, to Rome, promising him safety in transit and while there; as soon as Baglioni arrived, Leo had him arrested, tortured, and beheaded.

On 6 May 1527 a Spanish-German army invaded Rome and carried on like Slobodan Milosevic's troops in Kosovo:

> Massacre, plunder, fire and rape raged out of control.... The soldiers looted house by house killing anyone who offered resistance. Women were violated regardless of age. Screams and groans filled every quarter; the Tiber floated with dead bodies.... Ransoms were fixed on the wealthy and atrocious tortures devised to make them pay; if they could not they were killed. Priests, monks and other clergy were victimized with extra brutality; nuns dragged to brothels or sold to soldiers in the streets. Palaces were plundered and left in flames; churches and monasteries sacked for their treasures, relics trampled after being stripped of jeweled covers, tombs broken open in the search for more treasure, the Vatican

> used as a stable. Archives and libraries were burned, their contents scattered or used as bedding for horses....
>
> ...The first wave of carnage lasted eight days. For weeks Rome smoked and stank of unburied corpses gnawed by dogs. The occupation lasted nine months, inflicting irreparable damage. Two thousand bodies were estimated to have been thrown into the Tiber, 9800 buried, loot and ransoms estimated at between three and four million ducats. Only when plague appeared and food vanished, leaving famine, did the drunken satiated hordes recede from the "stinking slaughterhouse" they had made of Rome.

The sack of Rome ended but the other result of the papal folly did not: it occasioned the rise of Protestantism, from which the Church never recovered.

LOSING AMERICA

Eighteenth century Britain was governed by a small inbred ruling class generally taking counsel from no one but themselves. They imposed ever more onerous taxes and trade regulations on their American colonies. For example, they required every colonial legal document to have a stamp, the revenue from which went to the British government. It quickly became clear that the Stamp Act would have a stifling effect on trade. Trade generated more than two million pounds per year, the stamps at most sixty thousand. Nonetheless, the ruling class for years refused to revoke or even modify the Stamp Act.

When they finally did repeal it, Charles Townshend, Chancellor of the Exchequer, proposed and Parliament voted import duties for the colonies on glass, paint, lead, paper and tea. Townshend's duties were likely to generate even less income than the Stamp Act, and likely to engender greater rage in the colonists, but Parliament gave it to him anyway. You know what famous party the tea tax produced. After Townshend's death, his successors continued imposing the tax. Thomas Pownall, who over seven years had been an administrator in four of the colonies argued that the Act would encourage American manufacture rather than purchase of British goods and once that happened the status would not be reversible. Pownall's experience of the colonies was not considered of any import whatsoever and his advice was ignored.

After the Boston Tea Party, the Cabinet prepared a bill closing the port of Boston until the East India Company was compensated for its loss. They expected that shipping to ports in other colonies would make up for the loss of the closed port and that the other colonies would simply let Boston take the punishment. It didn't happen: the other colonies supported Massachusetts and told England to bugger off. "Wooden-headedness," Tuchman writes, "enjoyed no finer hour."

What we call the Revolutionary War dragged on for years. Even after inner members of the British government began to understand that they could not win a war so far away in so hostile a battleground, that pursuing it was counter-productive, England could not cut its losses and sue for peace. Some perverse sense of national honor was at stake and the war continued at terrible cost. This selfsame error, Tuchman points out, the United States would itself repeat two hundred years later in Vietnam.

THE VIETNAM TARBABY

American involvement in Vietnam began at the end of World War II, when the United States helped France regain control of its former colony, what was then called Indochina. Roosevelt had considered French rule in Indochina colonialism at its worst and he would not have permitted any involvement in reestablishment of that rule. But FDR died of a stroke in 1944 and his successor, Harry S Truman, had no knowledge of, directions about, or interest in the matter. Truman's generals were preoccupied with the threat of international Communism. French president Charles de Gaulle, ever the working hustler, threatened that France might go Communist if the US didn't help it reestablish its base in southeast Asia.

The US eventually underwrote eighty percent of the French effort there but the French never did reestablish their colonial power. Marshall Jacques Leclerq, whom de Gaulle put in charge of his military operation in Indochina, said in 1946, "It would take 500,000 men to do it and even then it could not be done." Leclerq was a visionary. "In one sentence," Tuchman says, Leclerq "laid out the future, and his estimate would still be valid when 500,000 American soldiers were actually in the field two decades later."

For nearly a decade the French fought Vietnamese nationalism, they brought all the power of their great industrialized country to bear on a peasant economy – and they were beaten. After they were driven out, the US went in and made exactly the same mistakes, lost exactly the same war, only it took longer and cost more in money and lives. "The question raised," Tuchman writes, "is why did the policy-makers close their minds to the evidence and its implications? This is the classic symptom of folly: refusal to draw conclusions from the evidence, addiction to the counter-productive."

America was driven by the Domino Theory, first articulated by Eisenhower's Secretary of State John Foster Dulles: if Vietnam went communist, so would all of southeast Asia, so would the South Pacific, so would everybody. There was no evidence to support the Domino Theory and as time went by there was more and more evidence against it, but the United States, through the Kennedy, Johnson and Nixon administrations, never abandoned it. In the most recent volume of his memoirs Henry Kissinger admits he knew Vietnam was a lost cause in 1965, yet until the very end of the war a decade later he argued for us to continue fighting there, for us to increase the severity of our bombing there. Why? "War," Tuchman writes, "is a procedure from which there can be no turning back without acknowledging defeat. This was the self-laid trap into which America had walked. Only with the greatest difficulty and rarest success, as belligerents mired in futility have often discovered, can combat be terminated in favor of compromise. Because it is a final resort to destruction and death, war has traditionally been accompanied by the solemn statement of justification...."

Nixon came into office in 1969 with a promise of a plan to end the war. It was a lie. He didn't get out until five years later, after his reelection. Like Johnson before him, domestic politics and fear of a backlash locked him into pursuit of a military policy he and his Dr. Strangelove – Henry Kissinger – long knew was futile.

The peace treaty Kissinger and Nixon negotiated was no different from the settlement agreed to in Geneva almost twenty years earlier. They could have had it the first day of their administration. Lyndon Johnson could have had it. John Kennedy could have had it. Reason doesn't explain why they all avoided it so assiduously and at such abominable human and economic cost. "If pursuing disadvantage after the disadvantage has become obvious is irrational," Tuchman writes, "then rejection of reason is the prime characteristic of folly.... Although the structure of human thought is based on logical procedure from premise to conclusion, it is not proof against the frailties and the passions."

"THE STARK IMPOSSIBILITY OF THINKING *THAT*"

When I finished *The March of Folly* I recalled the passage in the Preface to *The Order of Things* where Michel Foucault tells of the laughter that seized him when he came across a passage in an essay by Borges:

> This passage quotes a 'certain Chinese encyclopedia' in which it is written that 'animals are divided into: (a) belonging to the Emperor, (b) embalmed, (c) tame, (d) sucking pigs, (e) sirens, (f) fabulous, (g) stray dogs, (h) included in the present classification, (i) frenzied, (j) innumerable, (k) drawn with a very fine camelhair brush, (l) *et cetera*, (m) having just broken the water pitcher, (n) that from a long way off look like flies.' In the wonderment of this taxonomy, the thing we apprehend in one great leap, the thing that by means of the fable, is demonstrated as the exotic charm of another system of thought, is the limitation of our own, the stark impossibility of thinking *that*.

"The stark impossibility of thinking *that*." What a splendid description of the problem we always have with people who make choices that make no sense: "the stark impossibility of thinking *that*."

But it's not only a matter of people having an entirely different model of thought. Sometimes it's a matter of them not engaging in thought at all. Tuchman quotes a wonderful caveat from Ralph Waldo Emerson's *Journals*:

> "In analyzing history do not be too profound, for often the causes are quite superficial." This is a factor usually overlooked by political scientists who, in discussing the nature of power, always treat it, even when negatively, with immense respect. They fail to see it as sometimes a matter of ordinary men walking into water over their heads....

Finally, there is the sad factor of what both physicians and attorneys call "curing": all of these grim events could have been cured, fixed, avoided almost anywhere along the way. The Trojans could have looked inside the horse before dragging it inside their city or, after they made that blunder, before they went to bed that fatal night. Any one of the six decadent Renaissance popes could have discovered God and brought the Church along with him. Any one of the American presidents could have gotten out of Vietnam at any time on exactly the terms Richard Nixon settled for in his fifth year in office. People worked very had to ensure those disasters. Each could have been avoided had the people in power been capable of listening to the voices around them and of reconsidering choices already made. Each of these stories could have the same epitaph: It didn't have to be that way.

The March of Folly ought to be read by anyone in power, whether business, government, private foundation, public agency. It ought to be read by administrators of public institutions and members of public commissions and authorities. It ought to be read by anyone in a position of civic responsibility who thinks responding to community requests for a fair hearing with "Fuck you" is all the community deserves or needs. It ought to be read by the attorneys who serve or service such individuals and the organizations they comprise. It ought to be read by anyone trying to understand why otherwise competent and decent people conduct themselves in ways that at best seem to make no sense at all and at worst seem stupid or venal or block-headed.

13 The New Woodrow Wilson Memorial Bridge: Taking the High Road to Public Works

An Agency with Enemies

It's no secret that the Buffalo and Fort Erie Public Bridge Authority considers community groups on this side of the river an inconvenience to be avoided whenever possible and fended off any way that seems legal. They have consistently refused to cooperate with citizen's groups, and the only public hearings they've permitted on the new bridge have been shams with a few invited innocents who thought the game was for real, suckers who didn't know that the key decisions had already been made in secret.

The PBA has stonewalled every attempt to involve them in any look at the issue other than their own staged affairs or affairs managed by hired hands and collaborators, like executives from the engineering firm of Parsons and Andrew Rudnick of the Buffalo Niagara Partnership.* They carry on as if the Peace Bridge were a private business operation, the American public their enemy, and the Canadian public a bunch of fools who can be forever conned into accepting trash and bad planning.

A Better Way

That animosity and xenophobia exist only because the PBA is so hermetic and self-serving it can't or won't see any other options. But there are other options, options that get good work done, serve the public well, and leave everyone feeling good about what went on. It takes more work than the PBA has been willing to put forth and more trust in the public than it can muster, but some public officials feel it is reasonable for them to do a little more work and have a little more trust if the result is a bridge that works well and a public that feels its public officials have served it well and fairly.

A case in point is the project to modernize the Woodrow Wilson Memorial Bridge, which crosses the Potomac from Maryland into Virginia just south of Alexandria as part of the southern segment of the Capitol Beltway, I-495/I-95.

The six-lane Wilson bridge was built in 1959. The traffic load then was 19,000 vehicles per day. The designers built it to handle what seemed to them more than would be needed in the lifetime of the bridge – up to 75,000 vehicles per day – but their estimates weren't even close. By 1980 more than 100,000 vehicles crossed the bridge every day and now the rate is 200,000 vehicles per day. The problem was compounded by the nature of the traffic: a large portion of the increase was heavy trucks, so the bridge has deteriorated at a far more rapid rate than anyone expected.

Sound familiar? Sure, the same thing happened to the Peace Bridge, with slightly different numbers. But there is one significant difference: the public agency empowered to modernize the Wilson bridge decided to involve the community, rather than patronize and deceive it, and they've come up with a design people like after engaging in a process all involve parties respect.

EVERYBODY'S ENVIRONMENTAL IMPACT STUDIES

The Woodrow Wilson Memorial Bridge is owned by the Federal Highway Administration. In 1991, the FHA conducted a draft environmental impact study. The Buffalo and Fort Erie Public Bridge Authority has consistently refused to do any kind of EIS, and has instead engaged in vigorous and snaky legal maneuvering to avoid its responsibility to do one. One key conclusion of the Wilson Bridge preliminary EIS was that more community organizations had to be involved – exactly the reason the PBA has avoided doing one here. Unlike the PBA, The Federal Highway Authority took the position that its responsibility went beyond accommodating truckers; it had a responsibility to the communities on either side of the river through which those trucks moved.

They created a Coordination Committee composed of state and local elected officials and senior government executives from the region. The committee was charged with finding a solution that would ameliorate the traffic problem and address community and environmental concerns. They opened a site office and a Study and Design Center, and engaged in extensive public outreach. After many public and interagency meetings, workshops, and symposia, they issued a supplemental draft EIS in January 1996. They sought reviews and comments from all concerned. On August 12 of the same year, they issued a second supplemental draft EIS, and submitted that document to review and comment from everyone concerned.

As here, it wasn't just a problem of fixing a superannuated bridge: there was also the problem of what the bridge connected to. Here we've got the two plazas, there they've got the highway interchange systems. The Coordination Committee recognized the total project would have major effects on the surrounding communities; the PBA has segmented the bridge from the plaza construction in an effort to pretend that they can do one separately from the other. They're not that dumb, they know that's absurd. They separated the projects as part of their legal maneuvering to avoid the EIS, which would have forced them to consider community concerns.

Bear in mind that the increase in Peace Bridge traffic underlying this whole affair is trucks. Auto traffic is down. We're a casualty of NAFTA. There is federal help out there, but the PBA has refused to ask for it because that would have forced them to engage in the environmental impact study, which would have allowed for serious consideration of alternative designs, which neither they nor their design and maintenance consultants want.

Instead of hiding from the public, the Federal Highway Administration and the Coordination Committee sought it out. Instead of seeing the public as their enemy, they saw the public as their client. Instead of satisfying their own needs and desires, they asked the public what were its needs and desires. Instead of seeing their job as exercising power, they saw their job as engaging in public service.

They had ten Town Hall meetings in various locations. They had numerous open houses, citizen work groups (design, environmental issues, interchange issues, housing and neighborhood improvement, regional perspectives, traffic projections, trucking issues, transit issues). They did more than 50 displays to civil and professional associations.

The Coordination Committee considered some 350 concepts and came up with seven possibilities, out of which they selected one. The first item in their list of reasons for their choice among the seven alternatives was "highest level of public support." Another was that it had the "lowest overall impact to parks and historic sites." The Buffalo and Fort Erie Public Bridge Authority has ignored public opinion and has chosen the option that will destroy the most parkland.

After the Coordination Committee made their choice they again invited public suggestions and comments. It wasn't just window-dressing, like the *Buffalo News* charettes here, now referred to as "the charades."

But the Coordination Committee wasn't finished. It prepared a final environmental impact study, which was released for public and agency review on September 2, 1997. That document addressed all the concerns raised at the various stages leading to the preliminary environmental impact studies.

All along, they maintained a community center where anyone could come in, look at documents, make suggestions, complain, and they also maintained a web site that kept the public informed of where everything was all along the way. That web site is still there; it helped me prepare this article. You can look at it too: ***www.wilsonbridge.com.*** It's an interesting, intelligent, and friendly web site. If you do visit it, I suggest you follow with a visit to the web site maintained by the Buffalo and Fort Erie Public Bridge Authority: ***www.peacebridge.com***. You'll find the differences informative.

WHAT PARSONS DIDN'T TELL

Once the Coordination Committee decided what they wanted and what would be acceptable, they had a design competition: they invited serious submissions and underwrote part of the cost involved in preparing the most interesting of them. In contrast, the PBA and Parsons decided what ***they*** wanted and then they had sham public hearings. They never considered any alternative to the companion bride plan they had before the public hearings began. They never considered community concerns because of their Big Lie: that the bridge and plaza projects were separate. (The PBA may have stepped on its own foot two weeks ago in its lawsuit trying to force the city of Buffalo to issue construction permits. One of their arguments was that they had spent a good deal of money on what would be part of the new plaza. It is, to my knowledge, the first time that the PBA has admitted in public that it was deceiving us all along about the separability of the projects.)

The design for the new Wilson bridge is actually for two matched bridges, real twins, but unlike the companion bridge the PBA would impose on us here, these fit the landscape, they aren't ugly, they reduce the number of piers in the water by two-thirds, and they harmonize the other nearby bridges on the Potomac in the D.C. area. Which is to say, the design is appropriate to the environment rather than an insult to it.

Once the Wilson bridge design was chosen, there were more Stakeholder Participation Panels. The bridge authorities defined their relationship with the community as continuing, not occasional. Not only were there government agencies and officials, but also the Sierra Club, AAA, bicycler user groups, and others. Their comments will be used to help design the interchanges and do final tuneups on the bridge design itself.

Here's the shocker: the successful design was submitted by Parsons. Yes: the same outfit that gave us the profoundly ugly twin span adopted by the Buffalo and Fort Erie Public Bridge Authority.

One division of Parsons has been the PBA's technical consultant for more than 30 years. They've also had the contract for bridge maintenance. All projections indicate that the cost of maintenance of the twin spans will far exceed the cost of building the second span and rehabilitating the old bridge. If the concrete signature bridge design were to replace the Parsons companion span design, Parsons would lose millions in long term profits from maintaining two anachronistic steel bridges, one of them already decrepit. We'll never know if Parsons told the PBA all it learned about community involvement while working on the Wilson Bridge project or if it decided to keep very very quiet, protecting those maintenance millions.

LEARNING FROM EXPERIENCE

I don't know if the members of the PBA appreciate or understand how considerately the Woodrow Wilson Memorial Bridge project was handled. If they don't, I recommend a visit to the Wilson web site and that they read Henry Petroski's description of the competition process, "Drawing Bridges," *American Scientist,* vol 87 July-August 1999, 302-306. I suspect they neither know nor care, that they will continue their stonewalling and keeping their eyes wide shut.

But the various groups mobilizing to challenge the PBA position or evaluate alternatives can take heart from what happened in Washington: it *is* possible to involve all groups having a legitimate interest in a major public works project and wind up with a result that benefits everybody. All it takes is a willingness to treat the public as if it were as important as a firm with a lucrative maintenance contract or a steel mill hoping to make a lot of money or a distant manufacturer shipping heavy things from one place to another.

*The former Chamber of Commerce. They have a curious web page: all kinds of information about their activities but no information at all about who their directors and officers are. If you look up "directors" on their search page you get a list of undertakers. Check it yourself: *www.gbpartnership.org/about.html.* We wrote and requested a list of current directors and officers, but got no reply.

July 31, 1999

14 TRUCKS, TALK, AND SUNSHINE

CANADA'S EXPANDED TRUCK DUMP

The Ontario provincial government Transportation Ministry has been floating trial balloons about a new hundred-kilometer eight-lane highway they'd like to build between Hamilton and Fort Erie. The QEW can't handle the present traffic, let alone the huge increase expected when the new Peace Bridge is finally operational, and it can't be widened to eight lanes because the farmland and commercial areas through which it passes are too valuable. They've got to build something new.

What does that mean for our side of the river? If there are eight lanes of new Canadian highway spilling into six lanes of the new Peace Bridge and that in turn empties into four lanes of our old Thruway – ouch! It doesn't take a physicist to know that the flow through any system is determined by the narrowest stricture. Anyone who's ever had a clogged sink has learned that lesson. When the new bridge is built, we'll either have a new mess along the 190, or that road will have to be expanded to help those trucks on their way out of town. All the griping you're hearing now about the inadequacy of the current Peace Bridge will be doubled when the new one backs up because there's nowhere for that increased truck traffic to go.

PLANNING FOR WHOM?

To avoid having to participate in an Environmental Impact Study, the Buffalo and Fort Erie Public Bridge Authority pretended that the new bridge project and the new American plaza project were totally separate from one another. I don't know if there's equivalent deception going on here, or if it's just that no one involved is doing any long-term thinking, but it's foolish of the planners and politicians to deal with this bridge expansion as if it's just a matter of widening an old bridge to fit current needs. And it's naive of us to think that once we decide whether to go with the ugly and expensive companion span or the beautiful and economical signature span we'll be free of problems with the Peace Bridge. It's a mess now and, however the design issue is resolved, it's going to be a mess for decades to come.

Those questions are of no interest to David Collenette, the Canadian Minister of Transport, or to his five hand-picked representatives on the PBA. The Canadian Minister of Transport is concerned about facilitating the exchange of goods made possible by NAFTA, he is concerned about those trucks; Buffalo isn't even a blip on his screen.

He may be aware of Fort Erie, but Fort Erie has everything to gain and nothing to lose by expanded truck traffic. Their highway design is such that the trucks are out of town in minutes, and they go near the beautiful water only at the truck plaza itself. Increased truck traffic will only mean more customs and immigration jobs for Fort Erie area residents. Remember a few months ago when Fort Erie Mayor Wayne Redekop said that it might be a good idea not to fix the bridge at all, just increase the numbers of customs inspectors to improve traffic flow and leave other things exactly as they were, and that if the truckers didn't like that they could just cross the border somewhere else? He didn't mean that for a minute: it was a hissy-fit, it was political posturing, nothing more.

There's no way in the world Fort Erie is going to let that truck traffic go somewhere else. Without that truck traffic Fort Erie's economy would implode. If all those customs and immigration jobs went, if all those brokerage houses moved elsewhere, if that huge 24-hour duty-free shop was doing half the business, Fort Erie's economy would rest on the gooch joints, bingo parlors, Chinese restaurants, and rich Americans picking up groceries on their way to their summer houses. Look at Fort Erie's gorgeous new civic center. What other town the size of Fort Erie on either side of the border has a civic center like that? You know where they got the money for it? The Buffalo and Fort Erie Public Bridge Authority paid for it. Of course Fort Erie loves those trucks and the Buffalo and Fort Erie Public Bridge Authority.

But should we? Are there substantial benefits to this city and this area from a huge increase in truck traffic? If the truck traffic hurts us, we ought to know; if it helps us, we ought to know that too. The PBA and the Greater Buffalo Partnership have long asserted that the trucks passing through Buffalo are good for us. Are they?

THE PUBLIC HEARINGS

Maybe we'll get some answers to those and similar questions during the first round of hearings conducted by the Public Consensus Review Panel that begin next Wednesday and Thursday, August 4 and 5. The Panel is chaired by Sister Denise Roche (president of D'Youville College) and Randolph Marks (Buffalo businessman). During this round, proponents of various Peace Bridge and American plaza designs will have a chance to tell us all what is on their minds. The hearings, which are open to the public, will take place at WNED-TV, 17 Lower Terrace, Buffalo; they will be aired lived by WNED (970 on the AM dial). The Wednesday sessions will run from two to ten p.m., the Thursday session from five to nine p.m.

The panel is funded by the City of Buffalo, County of Erie, Margaret L. Wendt Foundation, and the Community Foundation for Greater Buffalo. The project manager is David Carter, of Toronto. I'm not sure why they felt the need to hire someone from 90 miles away to run the project, unless it was to make sure that there was at least one Canadian involved in the enterprise: there are no representatives from Fort Erie or any other Canadians government agency on the large panel that will be listening to the presentations, and neither are there any members of the Buffalo and Fort Erie Public Bridge Authority. They were invited to be on the Panel but they all refused to participate. The Canadians didn't want to participate because they're not interested; the Public Bridge Authority didn't want to participate unless they were certain that at the end of the process the Panel would come out for their companion span.

The two-day schedule allows for six invited presenters, each getting forty minutes to talk and twenty minutes to field questions, and dozens of five-minute slots for anyone else who has an opinion or an idea. The invited presenters are the Buffalo and Fort Erie Public Bridge Authority, signature span advocates Bruno Freschi and T.Y. Lin, SuperSpan proposers Jack Cullen and Clint Brown, the Buffalo Niagara Partnership, UB civil engineering professor John Mander (whose students two years ago came up with three companion span designs and three signature span designs, all of them, according to Mander, better than anything the PBA and its consultants have shown us), and Bob Biniszkiewicz, who has some alternative plaza ideas.

When we went to press, the PBA still hadn't accepted the invitation to present its position at the public hearing, presumably for the same reason it didn't appoint a representative to the Panel: if they can't be sure of the outcome, they have no interest in participating. The space on the program is still saved for them, should they want to venture forth from their bunker down at Peace Bridge Plaza.

The Panel has also given the PBA's chief spokesman, Andrew Rudnick of the Buffalo Niagara Partnership, a full hour for a presentation, even though the Partnership has no bridge or plaza plan of its own, as do the other five invited presenters. That really means the Public Bridge Authority is getting a free ride: Rudnick comes in and presents their side of the case, but they don't have to face the public and defend it. I asked David Carter why the Partnership got to have a full hour and a the New Millennium Group (which has gathered impressive and reliable data about the impact on the community of the two bridge designs) got nothing. Carter answered that the New Millennium Group had a representative represented on the Panel. That was disingenuous: the Partnership has a representative on the Panel and it's getting the full hour as well. Why should Andrew Rudnick get two bites of the apple? It would make more sense to combine his presentation with the PBA's since there's not a bit of daylight between their positions. And that way we'll know in advance that the hour won't be wasted.

Someone told me that the schedule was set up this way so the Public Bridge Authority wouldn't feel the deck was stacked against them. They've got one-third of the prime presentation time. One third of the time for a single point of view. That seems more than fair. From here, it seems as if the Panel bent over backwards to make it possible for the PBA to join the conversation. If the PBA wanted to join the conversation. Which it doesn't.

GO ASK ALICE

The Public Consensus Review Panel is conducting itself exactly as responsible community members should before the community undertakes a major project with long-term consequences: it's listening to proponents, consulting experts, inviting community input and response, all those good things. But there's an Alice-in-Wonderland quality to the whole enterprise: none of what they do has any no meaning unless the Public Bridge Authority decides to listen. If the PBA keeps the cotton in its ears, then the Panel is just talking to itself. The Panel has no legal authority of any kind, its power is grounded entirely in the concept of goodwill.

Some Panel members say that they hope the PBA will listen to what goes on and, in the name of goodwill, take account of the concerns of the people. Nothing the PBA has done thus far has indicated any such willingness to listen, or any interest in goodwill. And nothing the Canadian government has done thus far has shown any interest in our area, save as a conduit for Canadian trucks.

As long as the PBA stonewalls and there's no participation from Minister of Transport David Collenette or any other Canadian officials, this is going to be fought out in the courts. We may learn a good deal from the Panel hearings, and what we learn may help us next time something like this comes along, but for now, the courts are our only hope for getting the PBA to behave responsibly. The courts are a lousy place to work out public policy, but the Canadians and the PBA have left us no other choice.

WHAT TO DO

It's time for serious reevaluation of the composition and function of the PBA. They dance only to private industrial and Canadian government drums and that isn't right. The PBA is an international agency, but other than Barbra Kavanaugh I can't think of one board member who represents public interests on this side of the river.

Peace Bridge construction is almost certainly on hold for a year: the lawsuits aren't going to be resolved in time for construction to start before winter. If the PBA doesn't start construction before December 31, its permit from the International Joint Commission expires and they may have to go through that part of the process all over again. It's noisy out there, but in all likelihood we're going to get a breather.

We just can't continue letting political appointees and businessmen make lousy shortsighted plans for us, plans that leave us playing desperate catch-up, as the Public Consensus Review Panel is doing now. We've got to get in front of it. Perhaps one way is a real regional planning agency involving governments and citizen's groups on both sides of the river. If the Canadians won't join us in looking at these long-term problems then we should start the examination ourselves and they can join us later. We can't force them to sit at any communal table but at some point they'll have no choice – like when they want to talk to us about widening route 190 so those trucks will move sweetly on their way to distant markets.

SUNSHINE KILLS GERMS

You can attend meetings of the Buffalo and Fort Erie Public Bridge Authority. Call their office (884-6744) and someone will give you the time and date of the next meeting. And you can read the minutes of any meeting you didn't attend: go to their office at Peace Bridge Plaza and tell the secretary what minutes you'd like to see. There's only one problem: whenever they talk about the issues we're concerned with here, they go into executive session. They didn't do that in the old days, but that do that now. If you're an observer and they go into executive session they'll all sit there quietly until you're out of there and the door has closed behind you. That's when they do their serious business. If you look at the minutes of that meeting you'll find nothing but an entry telling you they went into and came out of executive session.

The PBA was created in the Depression to bail out investors in the Peace Bridge. Those investors were saved decades ago; that problem has been taken care of. No reason we should have to be stuck with this relic from another political era. Community advocates and younger politicians, such as Kevin Gaughan and Sam Hoyt, have called for a reconfiguration or total abolition of the PBA and a new way of distributing of its huge profits. They're right: we should not have to suffer decisions about public works made by owners of steel mills who get to make such decisions solely because they are major contributors to the Republican party.

We can't do anything about what happens on the other side of the river – if the Canadian Minister of Transport wants to ignore Buffalo and this region and focus only trucks, that's his decision – but we should at least have people on our side who represent our interests, and we should have a structure that lets us have a voice in what goes on. It would be good for all of us if the Canadians were willing to join us in serious conversation about such matters, but if they refuse, that's their choice and we needn't be immobilized by it.

Decisions of this importance shouldn't be discussed and made behind closed doors and shuttered windows. The whole process should be public. They concern the public and the public shouldn't be banished from the discussions. The City of Buffalo, the County of Erie, the Community Foundation and the Wendt Foundation did a fine thing in setting up the Public Consensus Review Panel, but it's a scandal and a shame that they were forced to do it. The Buffalo and Fort Erie Public Bridge Authority should have done it. They could have done it if they wanted us to know what they were doing and why. They didn't. It's time for a change. It's time to let the sunshine in.

PS: Fairness to All Department

I got a call last week from a friend of Andrew Rudnick, President and CEO of the Buffalo Niagara Partnership. The caller said that Andrew Rudnick is personally hurt that we think there is anything wrong with his advocacy of the twin span and he'd like us to stop saying there is. The caller said that Andrew Rudnick had no personal position at all on the issue, that he was only representing the wishes of the membership of the Buffalo Niagara Partnership.

The caller was someone I've known for many years, so I told him that what he said seemed to me totally absurd. Rudnick has been an effective advocate of the companion span for more than two years and, more important, *Artvoice*'s survey of the entire Buffalo Niagara Partnership membership directory indicates the membership overwhelmingly favors the Freschi-Lin signature span. A miserly 22% of the membership favored the twin span. My caller said that Rudnick felt he wasn't being represented fairly, whatever the facts were.

Artvoice wants to make things right and so do I, and we certainly want to give all quarters on this key community issue the opportunity to tell our readers what they think and what they know and why they've taken the positions they have. If Andrew Rudnick would like to write an article or note for *Artvoice* about why he wants the twin span and why he believes the majority of the membership of the Buffalo Niagara Partnership want the companion span, we will be happy to print it.

15 TRUTH AND LIES: PEACE BRIDGE REVIEW PANEL HEARING #1

The Peace Bridge Public Consensus Review Panel – created and funded by the City of Buffalo, Erie County, the Wendt Foundation, and the Community Foundation to bring light to the Peace Bridge expansion project – held the first of three scheduled public hearings in WNED's Studio One last Wednesday and Thursday. Fifty-five individuals and groups made statements or presentations over thirteen hours, all of it broadcast live over WNED-AM.

The Review Panel scheduled six major 40-minute presentations, each followed by 20 minutes for questions and answers. The Panel members listened in silence, save for a few introductory remarks, housekeeping work by the co-chairs Sister Denise Roche and Randoph Marks, and the response of Joseph Ryan (Buffalo Commissioner of Community Development) to a member of the audience who lamented the absence of Canadians. Ryan said the organizers had, at every stage of the process, invited participation by representatives from Fort Erie, the Ontario government, the Canadian National government, but the Canadians had rejected all the invitations; Ryan said those same Canadian officials had received copies of all the Panel's planning and programming documents, and those had been ignored.

THE NO-SHOW

The Panel hoped to open the hearings with a presentation by the Buffalo and Fort Erie Public Bridge Authority, but the PBA refused to participate. That was no surprise: they had also refused to engage in any preliminary discussions with the Panel co-chairs or staff, didn't respond to any letters or telephone calls, and refused to appoint representatives to the Panel itself or either of its two technical groups. Their only written response was a letter from PBA Chairman

John Lopinski to the Panel's project manager, David Carter that was delivered late Tuesday afternoon.

"At this time," Lopinski wrote, "we must regretfully decline your invitation to participate because we feel strongly that this process fails to bring bi-national consideration to the table and duplicates last years [sic] report issued by the Buffalo Niagara Partnership. We are taking this action with the understanding that the Peace Bridge Authority will be accused of not willing [sic] to participate in a public process. Nothing could be farther from the truth."

Well, it is *exactly* the truth. Of course they are unwilling to participate in the public process. The real hypocrisy, as state assemblyman Sam Hoyt pointed out, is that there is a bi-national body empowered by our two governments to represent us in these matters: the Buffalo and Fort Erie Public Bridge Authority. That is the sole job of the members of the Authority, representing their *two* countries in Peace Bridge matters. How can Lopinski honestly fault the Panel because it is not bi-national when the appropriate bi-national body is the BFEPBA, which he chairs, and which he refused to let participate?

(Not that Lopinski has ever been trustworthy on Bridge issues. Back on March 3, 1998, he told the Niagara Regional Council that "The essential government agencies on both sides of the border have signed off on the [twin span] project." That wasn't close to true, and he knew it. The US Coast Guard didn't give its approval until April 28, 1999; the International Joint Commission didn't give its approval until April 30, 1999. The BFEPBA still does not have the required easements from the City of Buffalo. In a "My View" piece in the *Buffalo News* on March 26, 1999, Lopinski wrote, "The Peace Bridge is a historic structure and any attempts to raze it would be met with stiff opposition by governmental agencies and citizens in both countries." Hogwash. When Lopinski wrote that, he knew that no government agency in either country had expressed any interest in the bridge as an historical object, and that the executive director of the Landmark Society of the Niagara Frontier had said it ought to be torn down.)

Five times in his three-page letter to David Carter, Lopinski refers to the "Peace Bridge Authority," which is not his agency's real name; he never refers to the "Public Bridge Authority," which is. As I've pointed out before, they don't like to use the word "Public" because it implies they have some accountability or responsibility to the area's citizens and communities.

Lopinski also claimed that the PBA decision is the correct one because, since the Buffalo Niagara Partnership conducted meetings on the bridge in 1998, "no one has provided any evidence to the contrary." Wrong again: there is a huge amount of evidence to the contrary, and the present hearings are a clear demonstration of that, but both the Partnership and the BFEPBA have refused to consider it.

Refusing to consider information or pretending that it doesn't exist does not mean the information doesn't exist or that it is not valid. I remember many years ago when one of my daughters, then two or three years old, developed a technique of covering her ears with her hands when something she didn't like was being said. Happily, she outgrew it. John Lopinski didn't.

THE MAJOR PRESENTATIONS

Four of the five major presentations were informative and straightforward. Bruno Freschi's team discussed the signature design that has energized much of the present interest in reconsidering the BFEPBA's plans. Bob Biniszkiewicz presented his plan for shifting the American plaza north, thereby freeing up all the park space that has been consumed by the PBA over the years. Most people there thought it a wonderful idea, one that could work with any of the bridge designs.

Thursday's session began with Andrew Rudnick, presenting the views of the Buffalo Niagara Partnership. I'll come back to him.

Clint Brown, Ross Robinson, Gary Burroughs, and John Cullen discussed their SuperSpan Upper Niagara project. SuperSpan isn't a bridge design; rather it's an idea about a bridge and plaza that are part of the city rather than an insult to it. Brown said they had contacted Eugene Figg, perhaps America's most important bridge designer. Figg has an impressive history of successful and beautiful bridges; he knows the technology, the problems, the politics, the economics. Figg drafted five possible signature designs for this site, all of them coming in at about the same cost as the Freschi/Lin design, and all of them taking only half the time of the plan developed by the Buffalo Niagara Partnership and the Public Bridge Authority.

The final major presentation was UB civil engineering professor John Mander, who described six bridge designs his graduate students developed two years ago – three companion spans and three signature spans. The students' companion spans appear to cost less and look better than the companion bridge the BFEPBA thought up, and their six-lane signature spans were all more economical than any companion to the geriatric Peace Bridge.

Mander spoke about the difference between *cost estimating* (what it will probably cost to build something) and *project economics* (what it will costs to build it and use it). Two cars may cost the same to buy, but if one gets 8 miles a gallon and the other 24 and you drive a lot, the project economics are very different. Mander's data indicate that whatever small differences there are in initial construction cost (and a difference of $10 million or so on a $150 million project with a life expectancy of 150 years is not a big deal), the differences in project economics are huge. Bottom line: the two steel bridges advocated by the PBA will cost many times the signature span.

What impressed me about all the presentations, save Rudnick's, is that none of them excludes the others. They're all moving in the same direction. The Freschi/Lin plan has a bridge most people seem to like and Bob Biniszkiewicz has an imaginative plaza idea that struck people as well worth exploring. No reason the panel can't consider both: Freschi's bridge can land in any kind of plaza and if someone comes up with a better plaza than he developed, better for all of us. The SuperSpan idea has the bridge land further south than anyone else has suggested, which would require a longer bridge, but what's really important isn't so much where their bridge lands but their insistence that we can have a beautiful bridge and a plaza that are part of the city rather than an assault upon it. The SuperSpan Upper Niagara group were telling people there's a better way to do this long before Bruno Freschi came up with a design that showed us what that better way might be. If it hadn't been for them taking action two years ago, there might be construction trucks backed up on Niagara street this summer.

VOX POPULI: THE SHORTER STATEMENTS

The Panel allocated about seven hours for brief statements by individuals and representatives of groups. With only a very few exceptions, the speakers were intelligent, informed, focused, and interesting. The prime evidence of that was the behavior of the Panel itself: none of the 15 or so members nodded off, doodled, seemed distracted, exchanged notes with one another, or shuffled papers. One, the representative of the Buffalo Niagara Partnership, chewed gum much of the time, sometimes slowly and evenly, sometimes vigorously, but I don't think that was in any way connected with the value of the presentations.

People talked about the importance of aesthetics, the possibility of resurrecting a neighborhood and a city, about the relation of all of this to life in our city, about heritage, about the future. Some had ideas about bridge or plaza design or about bridge operations. A few spoke favorably of the companion span idea. With a single exception, the audience applauded the remarks of every speaker.

"Fighting over a peace bridge," said attorney Kevin P. Gaughan "is no way to enter a new millennium." He said he had written the parties to the various bridge lawsuits currently before Judge Fahey suggesting they put the litigation on hold and give talking to one another one more try. Someone said to him afterwards, "The PBA isn't going to talk." "I know," Gaughan said, "but I thought someone should say to them, 'It doesn't have to end this way.'"

Common Council President James Pitts several times said what has become one of his favorite lines in these discussions: "Don't you built no ugly bridge." He also said that, "What we have to do is somehow address the Peace Bridge Authority and many of its members, and also yes, our neighbors across the stream, we have to get them to elevate their view of this issue. The signature bridge is not just bricks and mortar, steel and cable. It's a question of the future of this city and the hope that many of our young people have. I have never seen the kind of consensus and the kind of united action that has taken place around this issue. You would think that alone would allow the Peace Bridge Authority to reconsider their decision. There is not one issue that I know of that has ever galvanized us like this."

There were so many more: Catherine Schweitzer of the Baird Foundation quoting visionary words about Buffalo by Frederick Law Olmsted, Tony Fryer of the Landmark Society of the Niagara Frontier telling us that there is every reason to demolish the present bridge and plaza and replace them with a beautiful modern bridge and a new plaza in a new location, Fr. Jud Weiksnar of the Franciscan Center for Social Concern at St. Bonaventure saying why aesthetics really do matter, Pamela Earl eloquently speaking on behalf of a restored Front Park, business people, teachers, ordinary folks. It was eloquent, informative, considered, and very moving.

ANDREW RUDNICK: DENIAL AIN'T JUST A RIVER

Andrew Rudnick, president and CEO of the Buffalo Niagara Partnership, spent most of the direct portion of his time rehearsing the history of the Partnership's collaboration with the Buffalo and Fort Erie Public Bridge Authority, and telling us why the companion span was the only viable choice. He presented it as reasonable, inevitable, and perfect. Basically, he said they knew what was best for us and that should be good enough for us.

He several times referred to the importance of "trade, tourism and transportation related economic opportunity." He said that one of their major objections to the Freschi/Lin bridge is their concerns about"significant environmental issues associated with any cable stayed bridge," and that there were problems with a single pylon in the river. He said,"Minimizing the time and money spent…is key…." He said that the only site for the plaza is the current plaza site because consideration of any other site would need further study and further study takes time and money and we shouldn't spend any more time or money on this.

This is Rudnick's closing statement. I italicized words he seemed to be stressing:

> The Authority should *increase* bridge capacity as soon as possible while continuing our efforts to design the right US plaza gateway in its *current* location. Regarding the bridge there is no question that there is a real need to increase the current vehicular capacity at our border crossing and it is our belief that given the time and cost issues we have cited above there *still* seems to be only one viable *conceptual* alternative that enables this region to achieve economic development potential. That is a *second span in the same alignment* as the current Peace Bridge and landing on the existing US and Canadian plazas. However we continue to believe as we originally stated directly to the Authority in July 1998 that there *may* well be a better version of the second span option with a prestressed concrete second span, and replacing the Parker truss with an arch that matches the proposed arch design of the second span in the twin span concept.

When Rudnick was finished, only one person applauded – David Carter, project manager for the Panel. I assume Carter was applauding mechanically, the way a master of ceremonies does, but when it was clear no one was going to join in he quickly stopped.

The silence was impressive because every other speaker in the two days of hearings received applause from the room, and some received applause in the middle of their statements or responses to questions as well.

RUDNICK'S Q&A

The really important part of Rudnick's presentation occurred during the Q&A, because that was when he had to talk about the audience's concerns rather than his agenda. Rudnick got more questions from the audience than any other speaker or group of speakers. The questions were all sent up on cards to the co-chairs by the audience and members of the panel, so there was no opportunity for followup questions. That was not a problem with the other speakers, all of whom, so far as I could tell, were happy to engage the questions read by Randy Marks and Sister Denise, but many of Rudnick's answers were evasive, deceptive, and at least one was untrue. Here's a sampler:

> *Q: Your workshop never included the community. As a matter of fact, no one from the Partnership asked for our input in spite of our engagement.*
>
> **Rudnick:** That's an incorrect statement. Both community leaders and most importantly the district council member from the area immediately surrounding the plaza and the landing were invited to the meetings.

The two days of panel hearings last week were full of members of community groups who said they had not been permitted to take part in the Partnership's or the PBA's process. Many had described that exclusion in detail in their statements before Rudnick spoke. How does Rudnick handle their complaint? By saying it isn't so.

> *Q: Can you expand on or discuss the environmental issues associated with a cable-stayed bridge?*
>
> **Rudnick:** There are a number of them we can provide you in writing that were provided to us in our hearing. A good bit of them have to do with migratory flyway area, seagulls, ducks, geese, what have you. They're mostly DEC related.

Seagulls? Ducks? Geese? What have you? Nobody, *nobody*, has ever introduced a single serious concern about birdlife except the Bridge Authority and the Buffalo Niagara Partnership. Not one of the environmental agencies on either side of the border has said this would be a problem. It has not been a problem with any cable-stayed bridge anywhere in the world. And it's not like our birds haven't dealt with cables before: as architect Clint Brown said later in the evening, "Those cables carrying the 25-cycle power across the river next to the bridge have been there for about 100 years and no one's seen a bird skeleton in them yet, or had any reports of birds hitting them and falling in....It's silly."

> *Q: What are the specific concerns regarding the adverse impact of a single river pier design?*
>
> **Rudnick:** Well, I'd have to.... Again, it has to do with a bunch of DEC- and Coast Guard-related reviews that have already taken place. Water level, boating, the construction, actually, what's entailed in actually creating the pylons in the river. And, again, from our workshops, the material that we provided to you all at the beginning at the outset, at least conceiving project dealt with that specifically.

Do you find an answer in that reply? I don't. What about water level, boating, and the construction? Is he saying that the single pier in the Freschi/Lin design causes more problems than the ten piers in the BFEPBA design? If so, why not say it? If not, what is he saying? What's his data? This is just sidestepping by innuendo.

> *Q: If the twin span is going to take ten plus years to achieve increased capacity why not consider some other plan?*
>
> **Rudnick:** Again, I believe any plan that can be built within the same time frame and within the same budget is worth serious consideration. I do not believe the twin span itself will take ten years to complete. The project in its totality, which includes plaza and connecting improvements, may take ten years, but not the span itself.

No one, not even Bruno Freschi, has said the companion span would take ten years to complete. The BFEPBA itself, however, has said that its plan to build the companion span, refit the old bridge, and redo the plaza will take ten years to complete. That's their own estimate. That question was about the whole project and Rudnick knew it perfectly well. Why didn't he answer the question he knew was really being asked? Because he didn't have an answer he liked. Eugene Figg says a six-lane cable-stayed bridge and plaza can be designed, go through the environmental tests, and be built in a little more than half that time. Freschi/Lin/Cannon say the same thing. Every expert consulted, except the ones who work for the BFEPBA and the Partnership, have said the same thing.

> *Q: Would you please explain the willingness of the Partnership and PBA to segment the plaza project from the bridge project?*
>
> **Rudnick:** I understand the arguments that have been, the legal positions that have been taken on both sides of that issue. I'm proud to say in front of a panel that includes lawyers that I am not a lawyer, so I can't comment on any of those technical details.

This is a key question and a non-answer. So far as I can tell, Rudnick's response is, he understands the arguments and the positions but he won't tell us what he knows. If he does understand the arguments, as he claims, then the fact that he's not a lawyer is irrelevant.

> *Q: Does the Buffalo Niagara Partnership see any insurmountable obstacles to the signature bridge as a show stopper, so to speak?*
>
> **Rudnick:** That question implies that there is a common definition of what a signature bridge is and a specificity with regard to what the project elements of such a definition are. I don't believe that's the case.

This response is an illustration of the adjective "disingenuous" in action (my desk dictionary "disingenuous" as "not straightforward; not candid or frank; insincere"). Do you think for a moment that Andrew Rudnick didn't know exactly what bridge that questioner and every person in that room had in mind? Is there any other specific bridge that has been widely discussed that anyone has called a "signature bridge"? It gets worse: Rudnick was at that moment standing directly in front of eight huge full-color renderings and one three-dimensional model of the Freschi/Lin signature bridge and plaza. So how do you deal with a question you don't want to answer? Slip, slide, deny.

> *Q: If an environmental impact study on the plaza is not completed before the bridge is built, what situation does that leave the neighborhood in if more space is needed in 2006?*
>
> **Rudnick:** If an EIS is not complete is before the bridge is built?
>
> *Q: Yes.*
>
> **Rudnick:** Say the second half.
>
> *Q: If an environmental impact study on the plaza is not completed before the bridge is built, what situation does that leave the neighborhood in if more space is needed in 2006?*
>
> **Rudnick:** I don't believe there will be more space needed in 2006.

He sidestepped the previous question by pretending he had no idea what signature bridge might be at question. He sidesteps this one by dismissing it outright.

> *Q: You keep referring to cost as a major factor in your decision, when the cost assumption at the time was $65 million, not $90 million. Please explain.*
>
> **Rudnick:** I don't believe the difference between, I'm not sure what the final, if and when there is a twin span contract let by the Peace Bridge Authority, I don't know what that final number will be. I know the way in which the Canadian bid process works is that they really use that number as the starting point for negotiations, not the end point for negotiations. So our understanding of the technical process that the Authority follows is one for which that is a number that will, may be measurably reduced. I don't know that as a fact, but it may well be the case. In addition, the differential in cost

> figures that we are talking about with regard to any of those alternatives we believe makes that difference between 65 million and whatever the final number to be minute. That it's a dramatically different cost for any major conceptual alternative. And we continue to believe that it will be to the benefit of the project overall and long-term transportation-based economics in this community, to keep the costs low so you can keep the tolls low for as long a period as possible.

I think that's an exact transcription of what he said. It's a little difficult to follow, but it's worth the work because it goes to the heart of Rudnick's presentation. Nothing matters, no external data is worth considering. If Public Bridge Authority's cost projections were off by nearly 50%, as they were, they're still better than the numbers of everyone else in the bridge business. The Authority's projections don't matter anyway, Rudnick suggests, because the final bids may be less than the bids they actually received. Having trouble following that? You should, because it's double-talk. Then he goes on to say that the real cost of the plan he backed at the beginning of this process will still be vastly cheaper than any alternative, even though he hasn't considered the cost of any alternative.

That was Rudnick's last response to the final question. I haven't transcribed all of the questions and answers here. Everything he said wasn't evasive or snarky, but most of it was. There is one question I've saved for last, and not just because it was the question I sent up:

WHEN ALL ELSE FAILS, LIE

> *Q: Artvoice polled a large segment of the Partnership's membership. The results were: 49% wanted the signature span, 31% were undecided, didn't care, or wanted a tunnel, and 20% wanted the twin span. How do you justify your continued opposition, in the Partnership's name, to the signature span, when only a fifth of the membership supports your view?*

> **Rudnick:** There are two answers to that. The first is that the premise is false: such a survey was not taken. And the second is that our board of directors, elected by our membership, has periodically reiterated the position that I have stated today.

I'm always surprised when public figures lie in public situations. Do they figure we're so dumb we won't listen to the answer or so lazy we won't check the tape to be sure the guy actually said that? Rudnick could have said, "I haven't heard of any such survey and if there is one I'd have to know how it was done and what question was asked before I could respond to that. But, knowing my membership, I find it hard to believe." That would have been a reasonable way to field the question. But he didn't just field it; he lied flat out: "Such as survey was not taken." Well, folks, such a survey *was* taken, and the result are exactly as they were specified in the question.

What's important there isn't so much that he lied in answer to that question. Rather that the way he responded to it is illustrative of his way of dealing with all of us: he has no interest in new ideas, he attacks the credibility of people offering new information, and he lies about the fact of fresh information he finds it distasteful.

RUDNICK'S DRUMMERS

If it is true that Rudnick's board of directors at the Buffalo Niagara Partnership have endorsed or ordered his position, maybe it's time for us to ask them what they are up to. The question Rudnick avoided can be asked of them as well: "Given that your rank and file membership doesn't want this ugly and uneconomic bridge, why are you trying to get us to accept it?"

To whom should the question be addressed? We asked the Partnership for a list of its current directors, but they wouldn't provide it. We got some help from a recent article in the *Buffalo News* listing this year's Partnership officers and executive committee members. The officers are:

Robert T. Brady, chairman, president and CEO of Moog Inc., (chairman of the board of directors)
Andrew J. Rudnick, Partnership president and CEO
Randall L. Clark; chairman, Dunn Tire Corp. (vice chair)
Mark E. Hamister; chairman and CEO, National Health Care Affiliates (vice chair)
Howard Zemsky; president, Russer Foods (vice chair)
Robert M. Greene; CEO, Phillips Lytle Hitchock Blaine & Huber (secretary)
Brian E. Keating; regional president, HSBC Bank USA. (treasurer)

The other thirteen directors on the executive committee are:

Thomas E. Baker, executive director, John R. Oshei Foundation
William R. Greiner, president, University at Buffalo
Marsha S. Henderson, district president, Key Bank N.A.
Peter F. Hunt, president and CEO, Hunt Real Estate Corp.
Luiz F. Kahl, chairman, Niagara Frontier Transportation Authority
Stanford Lipsey, publisher and president, The *Buffalo News*
Dennis P. Murphy, president, InnVest Lodging Services
Stephen A. Odland, president and CEO, Tops Markets Inc.
Bill Ransom, president and general manager, WKBW-TV
Victor A. Rice, chairman of Buffalo Niagara Enterprise
Robert L. Stevenson, president and CEO, Eastman Worldwide
William E. Swan, president and CEO, Lockport Savings Bank
Robert G. Wilmers, chairman and CEO, M&T Bank

Luiz Kahl is also a member of the Public Bridge Authority, so it's no surprise that he backs the twin span plan. But why would the heads of all those banks want us to have that ugly bridge? Why would UB president William Greiner and *Buffalo News* publisher Stanford Lipsey and WKBW-TV president Bill Ransom want us to have that ugly bridge? Why would Tops Markets and Hunt Real Estate want us to have that ugly bridge? Why would Mark Hamister, who served with distinction on the Erie County Cultural Resources Board, and Victor Rice, who collects art, want us to have that ugly bridge?

It makes no sense at all. This is a smart and involved group of men and women, people of sharp business acumen and civic pride. Why would they adopt a policy that says "We know what's good for you, we made a decision two years ago, we won't listen to any new information." None of them would run their own businesses that way; you can't stay in business if you don't respond to new information and public concerns. So why do they let Andrew Rudnick run the Partnership on that basis?

Maybe there's some secret here they're not sharing, or some

subtle point we're not getting. We'll keep trying to find out. We're going to call every one of those executives and ask why they want us to have that ugly bridge and why they are closed to new information. We'll let you know what they say.

DEMOCRACY

The two days of hearings were democracy at work, and it was impressive and moving. Public hearings with open mikes usually draw a fair number of kooks and people passionately in love with their own voices. There was very little kookery or narcissism among the nearly three-score individuals and groups that gave testimony over the two days. Little wonder the Public Bridge Authority was afraid to participate: in the company of all these honestly concerned, well-informed and deeply committed people, how could they have maintained their pretense of being fair, objective, or concerned for the community?

The heart of the democratic process is an informed citizenry that can speak its mind without fear of government retaliation. That's why the framers of the Constitution made this the first item in the Bill of Rights:

> Congress shall make no law respecting an establishment of religion, or prohibiting the free exercise thereof; or abridging the freedom of speech, or of the press; or the right of the people peaceably to assemble, and to petition the government for a redress of grievances.

A long time ago the great *New Yorker* essayist A. J. Liebling wrote, "Freedom of the press belongs to those who own one." This event couldn't have happened without WNED. It wasn't just that they provided the space, though that was important, but also that they canceled their ordinary programming and aired all thirteen hours of testimonies, questions and statements. I don't know how many people listened to parts or all of it; what matters is that some people who weren't there got to hear as much of it as they wished. I'm sure the PBA and the Partnership would have preferred that those voices of the people and those questions to and answers by Andrew Rudnick never escaped that room. But escape they did. WNED is doing more: they're making sound recordings of the whole event available on their web site (www.wned.org), and they even made a videotape of the proceedings so there would be a full historical record. WNED was one of the fine Buffalo citizens in that room last Wednesday and Thursday. Hooray for them.

So many wonderful things were said in those two days. I could pick a dozen statements that would capture the spirit of what went on, but two in particular come to mind. One was by Colden resident Eli Mundy, whose Her brief remarks were interrupted several times by applause:

> Several years ago there was a popular television commercial that asked, "Where's the beef?" I think that citizens on both sides of the Niagara Frontier should ask ourselves, "Where is the Public Bridge Authority?" I think that it's important for all of us to remember that the original enabling legislation that set up that Authority said that this is the *Public* Bridge Authority, *not* the Peace Bridge Authority. Webster's dictionary defines "public" as "acting officially for the people." Ladies and gentlemen, by any reasonable definition, "public" means just that: acting officially for the people. Therefore, I believe it also means that we the people have the right and the obligation to respectfully request that the Public Bridge Authority attend the next meeting of this group, so that in an open and public forum the Public Bridge Authority can answer all the questions that have been raised here during the last two days of these hearings. And if the Public Bridge Authority continues to ignore these questions and these hearings, I believe that we the people should also request that Governor Pataki should remove his four appointees to the Public Bridge Authority so that the governor can reappoint new members and that reconstructed new membership of the Public Bridge Authority will represent the people.

The other statement was in the corridor during a break in the proceedings. Anna Kay France, an old friend who teaches theater at UB, was at the preliminary public information meeting of the Panel a few weeks ago and both sessions last week. She didn't speak at any of them; she just sat there, obviously attentive. I asked her why she was there. "I was planning on being on vacation this week," Anna Kay said, "but I went to the opening session and changed my mind.

This is too important. The passion and the eloquence of those statements, both from the people representing organizations and from people just talking about life in their neighborhood: I've never seen anything like this in Buffalo. This really matters to us."

October 21, 1999

16 Endgame at the Peace Bridge

There were three major interlinked developments in the Peace Bridge War last week:

> The Buffalo and Fort Erie Public Bridge Authority decided to join the study of bridge and plaza alternatives now underway and to abide by the decision of the Peace Bridge Public Consensus Review Panel, which was set up last summer by the City of Buffalo, Erie County, the Community Foundation and the Margaret L. Wendt Foundation.
>
> The Public Consensus Review Panel had two sessions in which the public responded to the Panel's October 6 interim report, in the course of which the major defect in the process became clear.
>
> Judge Eugene M. Fahey released his long-awaited opinions in the gaggle of lawsuits filed by the City of Buffalo, Episcopal Church Home of Western New York, the Olmsted Parks Conservancy, and Buffalo and Fort Erie Public Bridge Authority.

MEET THE PRESS

Shortly after it set to work last summer, the Public Consensus Review Panel announced that its engineers would report on their preliminary examination of bridge expansion and plaza location alternatives in a public meeting at WNED-TV on October 6 and that there would be two evenings the following week at which the public would be invited to respond to and comment upon that preliminary report.

On October 1, the Public Bridge Authority, which had refused to participate in the Public Review, announced that it would conduct its own public meeting in Fort Erie the following day to report on its companion span progress and to let its own engineers talk about how they thought things should go. PBA officials have for some time been unwilling to answer questions from the public about their expansion plans. I suspect this unexpected meeting was called because of the success the New Millennium Group was having with its recent informational meetings and open discussions in Fort Erie. Because of those meetings, many Canadians were for the first time learning the rationale behind the American opposition to the companion span.

On Monday, October 4, the PBA announced without explanation that the October 7 meeting would instead take place on Thursday, October 14, the same time as the second of the two nights scheduled for public response to the Review Panel's October 6 preliminary report. No explanation was given for the change in schedule.

It didn't seem to make any sense. It couldn't be to steal some of the audience because hardly anyone from Canada has been present at any of the Review Panel sessions anyway. It couldn't be to steal the press coverage because press coverage of response to a bland week-old engineer's report wasn't important enough to steal. It's possible that this was just careless planning: sometimes things that appear sinister or snaky are just klutzy.

On Tuesday October 12 Mayor Masiello's staff announced that there would be a press conference on the Peace Bridge in his office at four p.m. the next day. The word was that a face-saving deal had been reached with the Public Bridge Authority. The next morning, the PBA cancelled the October 14 Fort Erie meeting, saying it would be rescheduled soon.

The press conference wasn't actually in the mayor's office. It was in the mayor's new Press Room, around the corner and about eight doors down the hall. About a dozen reporters were there, as were some people from the New Millennium Group, Canadian Consul-General Mark Romoff, Public Bridge Authority secretary/treasurer Earl Rowe, a few members of the Authority, political staff members, and members of some community groups. When I saw Romoff I knew this was serious business: Romoff has been involved in this from the beginning, but he's never been at any of the public events since the issue got contentious. He had a big smile on his face and seemed happy to see everybody.

Buffalo Common Council President James Pitts arrived. I asked him what part Judge Fahey's opinion on the lawsuits, scheduled for two days hence, might have played in what the mayor was about to announce. He shrugged and said, in the voice he uses when he's quoting someone else, "Death destroys a man but fear of death saves him." Jim Pitts is really good at putting things into practical perspective.

A few minutes after four p.m. Mayor Masiello – flanked on the left by Erie County Commissioner of Environment & Planning Rich Tobe and Jim Pitts, and on the right by Buffalo and Fort Erie Public Bridge Commission chairman John Lopinski and Gail Johnstone of the Community Foundation for Greater Buffalo – read a statement announcing that the PBA had decided to join the process of examining and evaluating bridge and plaza alternatives begun last summer by the Peace Bridge Public Review.

"Up until today," the mayor said, "our process had lacked direct participation and input from our Canadian friends and counterparts on the other side of the bridge. Members of the steering committee of the Public Consensus Review Panel and members of the Peace Bridge Authority are here today to announce that we have come to an agreement regarding full Canadian participation in a study of short-listed alternatives and acting as a bi-national team they will act to produce a consensus recommendation by January 24 of 2000."

The mayor was speaking as if some distant Canadians had stonewalled us thus far and were now joining us in a burst of international cooperation. In fact, it was the five Canadians and the four George Pataki appointees to the Public Bridge Administration who had long ignored all pleas from the community for consultation and involvement. What was accomplished by this agreement wasn't just Canadian participation, but PBA participation, which is more important because it's the PBA that decides what's to be done with the Peace Bridge.

When the mayor was done, John Lopinski, Rich Tobe and Gail Johnstone each spoke briefly (Tobe spoke on behalf of his boss, Erie County Executive Dennis Gorski, who was elsewhere). Jim Pitts did not speak, but he stood there listening to the others, his fingers interlaced, a slight smile on his face. He is, as everyone in that room knew perfectly well, the Buffalo political figure who argued longer and more passionately than any other that the twin span was a lousy idea and the region deserved better. The Peace Bridge Public Review is an outgrowth of the Peace Bridge Task Force he set up more than a year ago. It was appropriate that he was there for this announcement.

The mayor thanked nearly everyone who made this happy event possible. He even thanked Congressman John LaFalce, which I thought odd: LaFalce was the only major New York elected official who endorsed the twin span and tried to squelch the introduction of any other idea. In collaboration with Canadian Member of Parliament John Maloney, he came up with a plan that would have limited discussion to a cost-based comparison of a second span and a signature span, cutting out all questions of plazas, systems, aesthetics and human factors, which is to say, he wanted an inquiry that would have been a lock for the PBA and its twin span concept. But perhaps LaFalce was involved in the behind-the-scenes negotiations that resulted in this agreement. If so, I wish they'd tell us so he wouldn't have to bear the burden of having been on the wrong side of this important issue.

In his remarks from the lectern, the mayor did not thank Senator Daniel Patrick Moynihan, who, like Mr. Pitts, endorsed the signature bridge early on. After the press conference was over someone must have told him of his lapse because he yelled from the far side of the room, as the last reporters were about to leave, that he also wanted to thank Senator Pat Moynihan.

(An ironic note: while the mayor's press conference was going on, an Allied moving van was loading the belongings of Bruno Freschi, whose signature design fired the imagination of most people who've come to oppose the PBA's companion span idea. The next morning, Freschi left Buffalo and moved to Washington, D.C. Presumably he'll be available for hire should his signature span be the choice of the Review Panel. Coincidentally, or maybe not, the famous bridge designer Gene Figg is due in town for a two-day visit later this week.)

WHY NOW?

Part was surely Judge Fahey's pending decision. Part has been an almost constant stream of negative publicity: even in the Canadian press, PBA has been widely portrayed as closed, tasteless, and bullying. Part has been a real deterioration in what had been a very good relation between Canadians and Americans on this border. Part is Canada's very aggressive interest in getting Americans to cross the border to dump money in its gambling joints. Part is a great deal of behind-the-scenes work by several people over the past month. And part is recognition by Canadian Minister of Transport David Collenette that without some kind of agreement the court cases were likely to tie up this construction project for years, which would make life difficult for his beloved NAFTA truckers.

I don't think any one or two of those factors would have done it, but the increasing pressure of all them pushed things to a point where a radical shift in policy was unavoidable. Most people close to the affair were certain a break would have to come, but few if any knew when or how it would come about. It got me thinking about an anecdote I heard in my metallurgy class in engineering school and in a sociology class when I moved over to the liberal arts. It has to do with the metal bar that is bent, made straight, bent, and made straight again. For a long time nothing much seems to be happening, then the place where the metal is being bent gets warm, and then, as the bending and flexing and bending and flexing continue, the bend gets hot. And then, at a moment no one but a metallurgist (my sociology professor said) might predict and no one at all (my metallurgy professor said) can predict, the metal bar separates, it simply breaks apart. Was it the last bend that did it? The fifteenth? The hundredth? Who knows? All we really know is that the bend gets warm and then hot and then things break open.

THE AGREEMENT

The document handed out at the press conference was titled "Agreement Proposed by the Peace Bridge Authority." It says that Transport Canada will name a Canadian engineering firm to work with the American firms hired by the Review Panel. The two teams of engineers will evaluate the various bridge and plaza proposals, will discuss this with the Review Panel, and, after an "interactive process" with the Panel, will eventually recommend a specific system. The PBA will accept whatever bridge and plaza system is ratified by the Review Panel steering committee. Once that happens, the mayor will "move the necessary easements forward."

Perhaps the most important thing about this agreement is the fact that it exists at all: this is the first time the PBA has agreed to join any process having to do with the bridge not totally in its control or in the control of its longtime dance partner, the Buffalo Niagara Partnership guided by Andrew Rudnick. (Rudnick was the only major player absent from the mayor's press conference, I assume because he didn't want to answer questions about his own role in trying to keep this collaboration from coming about.)

The agreement lists six issues that are "the critical criteria in the evaluation and comparison of alternative bridge design and plaza locations." They are: costs and funding, implementation schedule, environmental impacts, economic impacts, functionality of the system, and aesthetics.

This list introduces key subjects fiercely excluded by the PBA in the past. The PBA has long insisted that the bridge could be designed and built independently of the plaza, that the Olmsted park destroyed by the bridge and its roads and plaza was not its concern, and that what the bridge looked like was unimportant. It has spoken of the need to service the NAFTA trucks and ignored the economic consequences of plaza design to the city of Buffalo. All of those are now admitted to the conversation.

The agreement several times refers to the extensive work done by the PBA's consultants (Parsons) and insists that the data accumulated in the course of that fifteen-million dollar inquiry be admitted. The danger several people see here is that the Review Panel's engineering consultants will be overwhelmed by that huge mass of data and will not be able to consider other options in light of it.

It is indeed a danger, but not necessarily fatal. The two American consultants – Ammann & Whitney and the Louis Berger Group – are both major engineering firms. Both of them, along with Parsons Design and T.Y. Lin International are in the *Engineering News-Record's* list of the top 25 bridge design firms for 1999. I think it unlikely that they'll roll over and play dead simply because Parsons has spent a huge amount of the PBA's money thus far and will come to the meetings with a ton of printouts. They know that Parsons never considered a six lane bridge or any other plaza location, which means most of the Parsons data is, in this expanded inquiry, irrelevant. I wouldn't be surprised to learn that Parsons itself might be looking for a way to get out from under what more and more is becoming an untenable design position. This affair has gotten a good deal of international publicity and to many observers Parsons is looking like paid whores in a grungy political process. They probably don't like that.

I have far less confidence in the ability of the engineering firms to deal with the aesthetic and social aspects in that list of six criteria. More on this below.

Things are scheduled to move fairly quickly now. The Canadian Government is supposed to announce the name of the Canadian engineering firm this week. That firm has three weeks to familiarize itself with the project, then it joins the American engineering consultants and engages in the collaborative process with the Review Panel, delivering a final report and recommendation to the Panel by January 10, 2000. The Steering Committee of the Public Consensus Review Panel will inform the PBA of its preferred bridge design and plaza location by January 24, 2000.

What if the American and Canadian engineers disagree? What if the two engineering teams have different responses to the Parsons data? What if the Canadian engineers are in the PBA's pocket and come to the table with closed minds? What if they won't play fair? The document, which seems hurried and is sometimes a little fuzzy, doesn't say.

Some lawyers say a contract is as good as the lawyers who drew it up; others say it's as good as the parties signing it. I don't know who drafted the agreement for the PBA but I know who the eight signers are. There are four signers for the PBA: John A. Lopinski, Brian J. Lipke, Luiz F. Kahl, and Roderick H. McDowell. And there are four signers for the American public: Mayor Masiello, County Executive Gorski, Gail Johnstone and Robert Kresse.

Two of those eight signatures give me confidence. Gail Johnstone is acutely aware of the complex issues here, as is Robert Kresse, who is a very astute lawyer. Both are members of the Panel steering committee and both committed major resources from their respective foundations to make this process possible. If there's fuzziness in the text it's because the neither Johnstone nor Kresse believed it was of an order that would allow the PBA to renege without great cost to itself. I trust their judgment and I think this is a good agreement.

THE PUBLIC RESPONSE TO THE ENGINEERS' REPORT

The consulting engineers' report of October 6 was basically a consolidation and clarification of what everyone else has suggested thus far. The public document that accompanied that report also included a summary of the statements made in the Panel's August sessions. The first of the two public response sessions began only a few minutes after the mayor's press conference ended.

Only a few of the statements those two evenings were directed to the report itself. That's because there was little in it one could say much about: it had all been said before. Most of the statements were either about what was needed or wanted, partly a rehash of what had been said in the August hearings, and some were reminders to the Panel and the engineers that important factors were still being ignored or left out.

There was some microphone ego-tripping, but not very much. The most egregious was a thirtyish attorney who went on for 11 minutes and 31 seconds, half again the allotted time. Three times co-chair Randy Marks politely tried to get him to conclude, but he kept on talking. It wouldn't have been so bad had there been substance to what he was saying but there wasn't any substance. He was patronizing, preachy and irrelevant. He told the panel (which includes several superb lawyers) what it takes to be a great lawyer. He talked about the need for "hard work and innovation"and about what a swell job his wife did packing the trunk when they went on vacation.

What impressed me was that all the time he was preening at the lectern the entire panel listened to him attentively. They had invited people to talk, they said they wanted to listen, and by all appearances that's exactly what they did. Most of the time what they listened to was far more interesting and sometimes it was innovative.

One speaker asked if it mightn't be possible to put the whole fume-belching truck operation indoors, maybe even below grade, with the exhausts filtered before they were released into the atmosphere so there wouldn't be so much poison spewed into the West Side air. UB engineering professor John Mander offered an interesting efficient bridge design different from any of those presently under consideration. Bob Biniszkiewicz outlined the current version of his plaza design. It's an impressive design grounded in a profound vision of how the plaza might work for the city's benefit. I've heard his presentation before, but this time there were key differences in central design aspects. Bob has done what the PBA has refused to do: listened to what other people were saying and adapted and improved his ideas.

Buffalo architect Clinton Brown pointed out that the report sees the "bridge system" as bridge, plazas and connecting roadways, but omits parks, neighborhoods, jobs – all the human factors. Brown questioned the inclusion of parks and communities in the report's list of "key environmental constraints." He said "When is a park a constraint? In all due respect to my colleagues, it's when you're an engineer that wants to speed traffic through on a ribbon of concrete the park gets in your way. ...When do we call communities constraints? Why do we call the people who've bought houses on Busti a constraint?" Good questions.

The responses were all given by Seth Grady, an engineer with Louis Berger International and co-chair of the Technical Review Committee. His responses to questions about plans already on the table seemed astute and lucid. His responses to fresh ideas seemed politely resistant. When he was making those responses I thought of what panel member James Kane, Senator Moynihan's representative, referred to as "not thinking out of the box." Engineers don't usually think out of the box: they make boxes, but they don't involve themselves in why they're needed or what alternatives to them might exist. Engineers rarely think socially or aesthetically.

The technical consultants haven't, so far as I can tell, taken a fresh look at anything, nor do they intend to. Rather, they've consolidated what everyone else has said and have given all those suggestions a look to see if any might be so improbable or difficult they could be dismissed now. Into that category, for example, fell the two-level plaza developed by Bruno Freschi and Cannon.

In the course of their work so far, they didn't seem to challenge anyone's assumptions or constraints: if the PBA or Customs said this was a requirement, then it was accepted as such. Instead of looking at the whole issue afresh, they picked it up at the ragged edge. That may be all that is possible given the time constraints, but it places an increased burden on the members of the Review Panel to interrogate the binational teams of engineers carefully to be sure that important options aren't being ignored just because they're not simple enough to fit easily into the high velocity study.

Grady several times said that his group had to compare "apples and apples." It seemed almost a mantra for how they would behave in their research. He seemed to mean that whole systems should be compared to similar whole systems and segments should be compared similar segments. It seemed so neat and orderly.

Neat and orderly, but probably not a very good idea. "Apples to apples" means you're locked into what's already in the cart, which is what the PBA would like. It's starting at the wrong end, falling into the trap that already consumed so much of the PBA's time and money. Instead of deciding among options already on the table, we should be thinking about what would serve us all best and how do we get it. What's worrisome here is, the PBA carefully excluded certain systems and ideas from its planning work for years and its design consultants were bound by those initial conditions and assumptions. There are, for example, no calculations in the PBA's charts for the social and economic consequences of the various designs on the citizens of Buffalo. Will the alternative ideas be weakened in the consultants' eyes because they don't match the PBA's applecart? If so, then the entire exercise is pointless. I think "apples and apples" is a crippling metaphor and the sooner the consultants abandon it the better for everybody.

The consulting engineers have produced an engineering report.

Nothing inherently wrong or surprising in that: you don't go to engineers for goals, only for implementation. It's good to have this preliminary engineering report and it will be good to have the later reports from the binational engineering teams, but they will not in themselves be sufficient to answer the questions being asked by the Review Panel.

As Bill Banas, a leader of the New Millennium group said, "What we're missing is a bit of vision. I'm an engineer and I know how engineers work. Engineers generally only do what they're told to do. It's very easy to determine the constraints.... It's more difficult to determine what the public wants. That's the entire point of these hearings."

WHAT'S MISSING FROM THE REVIEW PANEL'S PLAN

In response to questions raised by Catharine Schweitzer, Seth Grady said, "The emphasis was on function rather than on its impact ultimately on the community, on parks and all the other things that are of interest." He said that in the next stage they would look at how each bridge and plaza design would have effects on aesthetics and such.

Every time someone asked how the questions of aesthetics were going to be addressed Grady said that was something they'd have to address further on down the line. One time he said, "That's something that we have not yet fully discussed with our technical subcommittee and with the review panel, exactly the way it will be applied." This is too vague to be acceptable. It's good that aesthetics and social issues are now in the list; it's not good that there is a procedure for handling questions about concrete and steel and no procedure for handling human questions.

Grady seemed to be saying "Trust us." I trust his engineering skills and his honesty, but not his ability to develop skills he admits he doesn't have to deal with issues he admits he doesn't understand. But how will such competence be introduced to the process and by whom? Will the Review Panel provide it? If so, how will that influence the reports and decisions of the American and Canadian engineers? Should aesthetics be included in their report at all, or should it be presented separately to the Review Panel? If the panel doesn't quickly find a way to make sure aesthetics and social factors are included at the same high level of sophistication as the engineering factors, they'll leak away.

Perhaps the Review Panel should set up another advisory group, this one dealing with the human issues of this project, so the advice it receives isn't only technical. Many members of the Review Panel are educated in and concerned with these issues, but that data shouldn't come into the discussion only in response to technical reports, and neither should it come into the conversation only in comments from the floor at public sessions. These are central issues, as important as steel and concrete, and they should be treated with the same seriousness.

JUDGE FAHEY KEEPS THEM HONEST

Any thoughts the PBA might have held about still having some wiggle room if things didn't go the way it wished at the end of the binational process probably vanished in Judge Fahey's court Friday morning.

The PBA had asked the judge to toss out the suits by the city, the Episcopal Church Home and the Olmsted Parks Conservancy that would force the state DEC to rescind the permits it had issued to the PBA earlier in the year. The PBA had also asked the judge to force the city of Buffalo to give it easements so it could begin construction now.

The judge rejected all the PBA requests, and along the way pointed out that nearly all the cases its lawyers cited were totally inappropriate or significantly misinterpreted. The PBA tried to kick the city out of the suit on a technicality, but the judge ruled that it was the PBA which had bungled the technicality by not raising it in time. He pointed out that the Common Council authorized the mayor to enter into an agreement, but it did not order him to enter into one, and since he never signed off on it, the agreement never existed. "By signing the resolution, the Mayor accepted the grant of authority to act, but he did not obligate himself to act in a specific manner. He did not give up the inherent discretion lodged in his office to later decide not to execute the grant."

He reserved judgment on the Episcopal Church Home, the Olmsted Parks Conservancy, and the city of Buffalo lawsuits against the DEC. He said that if any of those organizations were not satisfied with the results of the binational process now going on, they were to come back to his court and their cases would go forward.

There's a deal, but Judge Fahey is keeping the players honest. He could have done nothing, he could have said, "They seem to be agreeing, I'll wait until they're done." He didn't. He dismissed every one of the PBA's defensive maneuvers, he said their lawyers had mostly raised irrelevant caselaw, and he kept very much alive the lawsuits that would keep the PBA from putting a shovel in the ground. If at the end of all of this the engineering consultants are befuddled by Parsons' fifteen-million-dollar mound of data and the Review Panel can't think of a way to get around a bad report from the engineering consultants, neither of which I think is very likely, then the city, the Episcopal Church home and the Olmsted Conservancy can veto the whole deal. Talk about checks and balances! After all of this, it comes down to the veto power resting in the people, in an old folks home and in the custodians of Frederick Law Olmsted's vision. It doesn't get any better than that.

LOOKING AHEAD

Will the Buffalo and Fort Erie Public Bridge Authority deal with the process honorably or will they slip and slide? Are they in the process honestly or are they just playing games in the face of a difficult American legal situation?

I have no idea. Who can predict what anyone will do? If you cross the river, you get fumes of bad faith. According to *Fort Erie Times* reporter Mike Robinson,

> PBA chairman John Lopinski told The *Times* that, should the signature plan come out on top, the PBA is committed, by the agreement, to "pursuing" the recommended bridge plan, not "building" it. And, after five years of study that has resulted in verifiable facts, figures and permits, he doubts that anyone can come up with a comparable, viable alternative in three months.

This reflects either terrific confidence in what they've done thus far or astonishing cynicism about where they are right now. I suspect both. It will be a long time before we know how this apparent deal will really play out.

But that's not where it ends.

Much was learned in the Peace Bridge War and the question now is, will any of it apply to the next similar event? Many of the same players are involved in the proposed convention center – our elected politicians, the *Buffalo News*, the Buffalo Niagara Partnership, a variety of companies hoping to make a fat buck on whatever deal goes down. Will they again try to buffalo through a deal made behind closed doors? Will they give the people a voice at the beginning rather than the end? Thus far, it looks like they've learned nothing from the Peace Bridge experience: the pols and developers tell us they know what's best for us and the *Buffalo News* tells us to shut up and sit down.

It's slower if you listen to people, but it's slower still if you ignore them and they have to take you to court to get you to pay attention. That's one key lesson of the Peace Bridge War. I think people don't mind losing nearly so much as they mind being ignored. If the convention center developers and the *Buffalo News* and the Buffalo Niagara Partnership maintain that arrogant posture – 'we've giving this our full attention and we've made a decision and we know what's good for you' – we're going to be going through this whole dreadful dance again in the not very distant future. But there will be one key difference: next time they won't be able to pretend it's just the Canadians who've been screwing us.

October 28, 1999

17 THE MAN WHO LOVES BRIDGES: A CONVERSATION WITH EUGENE C. FIGG, JR.

Eugene Figg, Jr., loves bridges. His company, the Figg Engineering Group, of Tallahassee, Florida, is the only national engineering firm in America that does nothing but bridges. He loves to talk about the bridges he's built, how they're faring, how the people who own them feel about them now.

He's proud of the ones that came in early and under budget (like the Natchez Trace Parkway Arches, budgeted for $15 million, brought in for $11 million). He's equally proud of the ones that won major design awards. The National Endowment for the Arts began giving Presidential Design Awards in 1984. A total of 41 awards have been given, only five for bridges, and Figg got three of those: Lin Cove Viaduct in North Carolina (1984), Sunshine Sky Bridge in Florida (1988) and the Natchez Trace Parkway Arches in Tennessee (1995). Figg's pride in his bridges doesn't come off like vanity; it's more like a parent talking to anyone who'll listen about a child who is doing well in the world.

He was in Buffalo last week for a conference of the Association for Bridge Construction and Design, where he spoke about bridge permitting and community involvement issues, and about the community design charettes for which he has become famous. He also managed to talk with a good number of people involved in the Peace Bridge expansion: Buffalo Development Commissioner Joseph Ryan, Common Council President James Pitts, the *Buffalo News* editorial board, the Public Bridge Review Panel's Technical Review Subcommittee, and about 75 people at D'Youville College, a meeting incorporated into one of the New Millennium Group's informational sessions.

We talked for about an hour on Friday morning. His enthusiasm, energy, experience and creative approach to the world of bridges reminded me that building a bridge doesn't have to be about institutional ego, ratification of sunk costs, political favors, money-grubbing, international rivalry, local second-ratism, and blind service to distant masters. He's convinced that Fort Erie and Buffalo can have, in a reasonable time, a bridge system that works, that serves us, that makes us feel good.

Here are parts of our conversation:

You've been to Buffalo several times now. What's of particular interest about this project?

This is definitely an opportunity for Buffalo to make a statement about the city. It's an opportunity really to do something nice. You're building a bridge for a hundred years, you're going to see this bridge for a hundred years, so it needs to be the best you can get. It needs to be a piece of art.

And the construction will employ a lot of people. If it ends up being a concrete bridge there will be even more people employed right here in Buffalo and in Canada because the contractor will make all of this concrete right here. The contractors that build a bridge like this will come in with a small group of their own people and the rest will be hired locally. The work will be right at the site, or close to the site, it will all be right here. It will employ local people.

How did you get interested in the Peace Bridge?

I got a call from Jack Cullen with SuperSpan Niagara, maybe two years ago. He said there was a bridge that they needed some help with. He found out about what I do – exclusively design bridges – and he wanted me to come and talk to him about it. I asked him to send me some information. I'm always interested in bridges. At the same time a man named Stewart Watson called me to talk about the bridge. Stewart Watson was the owner of Watson-Bowman, which made bridge bearings and expansion joints. I've known and have been friends with Stewart for 25 or 30 years.

When I had both these people calling I knew there was something happening in Buffalo. So I came and found out about the project. They set up a number of meetings for me to talk to different people. And it worked. The chemistry worked and it's something I want to do.

You've visited the site. What do you think about it?

It looks to me like the bridge should be moved to a better location so you can build a plaza that's going to last for a hundred years. The one they have now, the area's too small and it's going to be a big problem, it looks like to me, for the city to be able to do all the things needed to do in that spot. So that really hurts people.

What would you do with the plaza?

I would move it. I would move it so that you can get it into a location where you can really design something. The plaza should be an award-winning design also. The plaza should be something that really belongs to the city. Whatever happens there should be something the people would really enjoy having rather than something that's just functional for the bridge itself.

The bridge authority people say, 'We've spent $15 million developing our design. How can somebody come in and quickly come up with a better design, more quickly?' What do you say to that?

I haven't any idea what they did. I just know that you could build a brand-new bridge of six lanes for about $90 million and it would take about five years from the time you started with the design and environmental documents to when it was completed. I have no idea what they did for $15 million.

Let's leave them out. Can you talk a little bit about the timeline for the bridge and the timeline for the plaza?

They both would be on the same timeline. The bridge and the plaza, the whole thing would have to be finished in about five years. The five years would include the fact that you've got some environmental documents you've got to get approved, but you would work on that simultaneously with the design phase of the work. And then the construction phase would be included. But the plaza needs to be done exactly the same time as the bridge. It has to be one unit.

Can you come into a place where you haven't been and get up to speed quickly enough to get it done in five years?

Sure we can. We work all over the country. We've worked in 30 states. We have work in New York City right now, two major projects for the Port Authority. One is 8.7 miles of elevated structures for mass transit for JFK airport that's under construction. It's a $100-million project. The other is a bridge we designed to Staten Island, a cable-stayed bridge. We're about 30% finished with it. They stopped the project temporarily but they'll come back and that bridge will be built. That project is about $150 million total. We have $185 million under construction in Boston right now. Another $46 million bridge being built in Maine.

So we worked all over the country and so there's no problem with going quick. We're bridge people. We understand bridges. We understand what it takes to design one, what it takes to get the environmental documents done.

We have about 90 people. We're doing about $1.4 billion either under construction or in design. As far as I know, we're the only firm in America that designs just bridges. So we can be highly specialized and this is why we do varying types of things. We even do the financing of bridges. Just like the Ambassador group: we could come in and do the same thing they do. We have several bridges – we don't own the bridges but we help finance them and develop them for authorities, small authorities. We just finished one in Pensacola, Florida, $96 million. The bonds were sold by Paine-Webber. It an 18,425-foot-long bridge. It was finished in 29 months, one month ahead of schedule. At one point we did seven spans in seven days. 980 feet a week.

How do you do that?

It's a system. It's our design system at work. Right now at JFK Airport where we're doing that 8.7 miles of elevated structure, we have to build in the existing median and the existing shoulder, that's all we have to work with to build an elevated bridge to take care of the mass transit. We have four trusses working while the project's going on, and the contractor is building 800 or 880 feet a week on those bridges. The project has 5195 segments on it and all the segments are being made down in the southern part of Virginia, Chesapeake Bay, by Bayshore Concrete. They're barged up and then they're put on trucks and brought from New Jersey to JFK Airport. It's a process. It's going on right now, as we speak, constantly bringing these up, and they're probably building today three or four spans. It's the system that works. The system becomes kind of monolithic, it becomes almost seamless. It all goes together. And you don't have to interrupt traffic either. We would never interrupt boat traffic or vehicle traffic.

Could you talk about the way you work, this process of how you involve the community and what that does to the timeline?

We call them design charettes. The bottom line is, it saves time because it comes to closure quicker on what the bridge should look like. And it also unifies the community so they are focused on a conclusion and then support the design the rest of the way. So it saves time. It takes approximately six months from the time we have notice to proceed to be able to conclude the design charette.

The process is to have the customer – whoever that may be – and ourselves work together on discerning how many people we think ought to be involved. The most I've done is a hundred. It's been anywhere from 25 to 100. I like it in the range of 75, because you want a good cross-section. I ask them to invite – let's just say we're going to have 100 – I want to invite 300 people they think are the best cross section. And the first 100 that respond and say that "I can spend two days with Gene Figg and his staff initially and one day about eight weeks later" then become voting members of the design charette. I want that to be a cross section. I would like to have the press there. I prefer two newspapers because if you have two newspapers, then you get the right story. I like to have all the other media involved if they'll do it.

We spend approximately five to six hours up front educating the people as to what they're getting ready to do, helping them learn as many things as they can about bridges. If it's a bridge over water, we want to take a boat ride and get them really focused on it. Even though they've seen this area a lot, they need to focus for this day or two days on what would be the best bridge for that site. Then we vote on a number of elements. We vote on span lengths, we vote on the shapes of the piers, on lighting, landscaping, everything. We vote on maybe 75 or 80 items, but not all at one time. Usually the first vote happens about one or two o'clock in the afternoon on the first day.

The voting process is on a one-to-ten scale. It's a consensus vote. If you like something a lot, you vote a ten or eight or nine. If you don't like it at all you vote below five. If they don't have any opinion they give it a five; if they don't vote we give it a five. Five is a neutral vote. It's not a yes or no, it's not a beauty contest, it's really about the things that can be achieved. Within a few minutes after a vote is finished, we electronically give them the average. Some item might be a 7.6 or a 3.2. in the vote of all 75 people. We work with everything that's above five.

We have two hand-held mikes. Only one person can talk at a time. We have lots of conversation. And lots of displays in the room of all the things they're going to vote on. There'll be lighting displays with different kinds of lighting options. This gives them a chance to learn before they vote.

The best part about the voting process with the community is, when we vote even if we don't win, we accept what happens. Even if people are on the losing side of a vote, they become advocates like everybody else after a few votes.

About eight weeks after we've had this two-day session we come back with a design that fits the things that they want. Then we can take the whole design to a public meeting of maybe three or four hundred people, and now we've got fifty to seventy-five advocates who can answer most of the questions that are asked by the rest of the public that comes, instead of us as engineers answering it. They'll give the reason why a certain choice was made because they were part of the decision.

The process has worked every time we've done it.

How many times have you done it?

I have no idea. We started this in '88, '89. We use the process for more than just designing signature bridges. We do it for location studies, a lot of different things. But for the bridges, most of we do is for the bridge designs themselves at difficult locations where people have strong opinions one way or another what a bridge should look like. You get it done quicker because you come to a conclusion, instead of it dragging out for years with people taking shots at a design all the time. You bring it to conclusion because they've actually voted. And no one else in America is doing this, we're the only ones doing this.

Do you come in with a range of designs and options? Is that the starting point? Or do you talk first and then come in with the options?

We only offer to the people what they can have in the way of the budget. We always start out with the budget and we always start out with those things that are the parts of the project that can't be changed, such as a vertical clearance of a channel, horizontal width to a channel, any number of things that can't be changed. We work diligently to give the people those options that will fit that particular site. Some sites, they're too small, you can't put in a cable-stayed bridge or something like that, or an arch, or you can't afford it. In some cases, maybe it's a different type of long span that needs to be used. So all these things are studied. But only what can be afforded.

We don't come in with any preconceived design. We know what will work at that site. We spend that time up front before we have the first meetings. We bring those things that can work at the site. We're not coming in with a predetermined idea what the bridge should look like. But we have an idea of a number of things that can be achieved at that site. It's the details and the way you put it together that makes an award-winning bridge. That's why we won 124 design awards and three Presidential awards.

But I've heard that you have sketches for what might be done here.

I wanted to confirm what the costs would be, so we made about six different conceptual ideas to try to confirm the costs. I wanted to be sure we were talking about something that was feasible. So we did about six different ways you could build or design the bridge. There are other ways. When we did a cost estimate on each one of them they all came in this range of about $85 to $100 million.

I do this everywhere we go. I don't want to come in and say we can do a bridge and it doesn't fit the budget or doesn't fit the conditions. It's just sketches so far because it's not time to design it yet. You first have to do the sketches and make sure we're in the ballpark and then find out what the customer wants before we start designing it.

The primary situation here is that you have a chance to build or design a really outstanding structure. This structure is something that will be in Buffalo for a hundred years. It's a chance not just to get a bridge but an award-winning bridge, a bridge that probably will be revered by lots of people for a long period of time. If you just do that and not look for the ordinary, this could be an important part of the skyline.

Some people have worried about a huge pylon competing with the city's skyline.

I don't think it's an issue because that's just one of the ways you can design this bridge. With a cable-stayed bridge, the reason for the height of the pylon is the span length. It's a direct ratio: the longer the span length, the higher the pylon. We have some unique pylons that are really very small, very slender, even with a long span.

One of the design elements in this bridge that needs to be considered is the fact that you have the water moving fast. You want to have it unrestricted as much as possible. It costs more to build in the water. A new bridge could certainly have fewer piers than they have in the present bridge. Probably you could even design it with two piers in the water, that's one of the concepts we worked on. But that may be too severe, it may be three in the water.

You've met the Bridge Authority. Do you see any particular problems with this approach here?

I have not met the Bridge Authority. I made a presentation to some of them, but it wasn't to the whole Authority. It was to staff and some Authority members two or three years ago. It was just an information meeting. That's as much as I've done there. It'll work anywhere. The process will work anywhere. Buffalo or St. Paul, Minnesota, or Fort Lauderdale, Florida. It works anywhere.

Have you done bridges in the north?

Sure. I have bridges in Maine. I have one that was built in '83 in Maine. It's in great shape today. We have one under construction in Boston, We have St. Paul, Minnesota, which has almost as severe a climate as you have in Buffalo.

Worse.

Or worse.

It's right in the center of the city and it crosses the Mississippi. That bridge is excellent. It's an excellent bridge, an excellent location. We worked all through the winter with no problems. With our designs, you can work all year around. That makes them interesting. That bridge is right *in* the downtown. We have major sidewalks on it because you walk across it. There are high banks and they're trying to rejuvenate that part of St. Paul, downtown, and they're doing a good job of it. But they do have flooding that occurs there, so it's designed to deal with that. It was just a super job.

[He shows me photos of projects in a 1999 company calendar and says, as he turns the pages:]

This is as close as we've gotten to a brochure. We should do a brochure. There's never been enough time.

...This is a bridge in Dauphin Island, Alabama, completed in 1982. A great project. I visited it about a year ago. It's in terrific condition. It's a 400-foot span. It was to replace the bridge that hurricane Frederick knocked out in 1979. They hired us immediately to redesign another bridge for them. We did the whole project in thirty months. It was a great project.

...Here's one place we built right over the traffic. It's the Hanging Lake Viaduct in Glenwood Canyon. There's the traffic right underneath the bridge. That's the only place you could put the traffic. So we built it above. Once we got it finished, the traffic went on top. The most we ever stopped the traffic was about thirty minutes during each of those times we jacked up the pier. You have to be versatile. You have to be very versatile.

...This is a 400-foot span in Charleston, South Carolina. Hurricane Hugo ran right on top. We've had our bridges hit by eight hurricanes and no problems with any of them. And two tornadoes.

...This is a cable-stayed bridge in Columbia, South America that this year in January was hit by a Richter scale measurement of six, acceleration factor of point two-nine. Everything around it was almost totaled. But there was no damage to this bridge. So as far as I know we have the only cable-stayed bridge that's ever been through an earthquake.

... This is the Wiscasset Bridge, completed in '83, between the cities of Wiscasset and Edgecomb, Maine. It's right on the coast on the Sheepscot river. Very severe environment. More severe than Buffalo, much more. It's a terrific bridge.

It's beautiful.

It's not a big bridge, but it works fine and the people love it. They love that bridge. It's no maintenance. It's a very interesting project. We sent people up there and they talked to the people who maintain the bridge and they're like the Maytag guy.

We're almost passionate when it comes to talking about bridges.

It seems like you're having a great deal of fun.

If you're not having fun you shouldn't be in the business. You should try another business.

My colleague Robert Creeley has a poem that begins, "If it isn't fun, don't do it. You'll have to do enough that isn't."

We love bridges. What we really want is getting them built. We like our bridges built. That's the key. We're not interested in just designing things and have them never be built. We want them built. That's been our focus, really: to build bridges that win awards and are economical all at the same time. Anybody can build something that costs a lot of money. But you have to build it so it's within a budget. And that's what we do. In fact, most of our bridges are built at least costs because we're competing with another design. The Skyway was in competition with another design. The Chesapeake-Delaware Canal Bridge, we were competing on that project with a cable-stayed bridge against a steel design. Our design was ten percent less. It saved $6 million on the design alone. And then it won a lot of awards too.

And we want it to live a hundred years, so we do owner's manuals. Like you buy a car you get an owners' manual, right? You get our bridge, you get an owners' manual, and you find out how to take care of it. We design it so you can inspect all of it easily. You can inspect every piece of our bridge. It's easy to inspect it, it's easy to maintain it. That's another important part of what we're doing. I want it to be there a long time. Much longer than me.

18 Unanswered Questions

Peace Bridge Secrets

Every week I receive letters and email messages that begin with lines like, "You've written all these articles about the Bridge, how come you've never written about . . . " or "I don't want to sound paranoid, but . . . " or "This may be a naive question, but I'd like to know . . . " The questions are just about never naive or paranoid, and the only reason I don't answer them is because I haven't been able to learn the answers.

The Buffalo and Fort Erie Public Bridge Authority and the Peace Bridge Public Consensus Review Panel agreed last month on a process that may bring the acrimonious Peace Bridge expansion project to a decent resolution. But many important questions remain unanswered – some because nobody knows the answers and some because the people who do know aren't willing to tell the rest of us.

Perhaps some readers who know more than we do can help. My email address is bjackson@buffalo.edu and I welcome informed advice. In any case, it's important to keep asking questions, about things that matter, even if the answers are not immediately forthcoming. Knowing what we don't know is itself an important thing to know. Here are a few of the current questions:

1. Is Transport Canada trying to sabotage the Peace Bridge review process agreed to by the Public Bridge Authority last month?

After months of public stonewalling and behind-the-scenes negotiations, the Buffalo and Fort Erie Public Bridge Authority agreed last month to take part in and abide by the Peace Bridge Public Consensus Review Panel's evaluation of alternatives to the unpopular twin span idea. A few days after the agreement was signed, the Fort Erie newspaper quoted PBA chairman John Lopinski as saying they'd promised to "pursue" but not to "build" anything other than his pet twin span. It seemed like a cynical waffle, but everyone I've talked to about it says that was just Lopinski showing off to a reporter, that at the end neither he nor any of the other Canadian PBA members would want to be caught having acted so cynically and hypocritically.

The deal's lynchpin was the involvement in the design evaluation process of a major Canadian architectural firm, which was to have been named by Transport Canada Minister David Collenette five days after the October 13 press conference in Buffalo city hall at which Mayor Masiello announced the agreement. It was to be all cooperation and modern science now. The Canadian engineers would have a specified time to get up to speed on the project, then would work jointly with the American engineers. The Review Panel pushed the project date for its final report into early next year to accommodate the Canadian team. The name of the likely firm – a highly-respected Vancouver group – was unofficially bruited about within hours of the city hall press conference.

The problem is, Transport Canada still hasn't made it official: David Collenette never named the Canadian engineers who would work with the American engineers. That puts everything on hold but the clock. Collenette doesn't return phone calls on the question.

This may be what Texas convicts call a "slow-buck." A slow-buck is screwing something up by working so slowly nothing really gets done. You don't ever say "I'm not gonna do that." You move around as if you really were doing something useful, but you're not. It's silent sabotage. Is that what David Collenette is doing?

Why would a high Canadian government official sabotage by deliberate inaction an agreement drafted primarily by his own five appointees? It might be because those appointees never cleared the deal with him or with any official of the Canadian government or Fort Erie town government. Collenette has never been ambiguous about his position on the US in general or the Peace Bridge in particular: he couldn't care less about either of them, and he probably doesn't care very much about Fort Erie, Ontario, either. Trucks and their contents are his concern, and perhaps the large profits to the Canadian steel mill that was promised the lucrative contract if an old-fashioned steel suspension bridge was built here. Will he back up his appointees to the PBA or will he leave them dangling in the wind, looking foolish and precipitous?

2. *Where's the Buffalo Niagara Partnership now that a resolution seems to*

be a possibility?

For well over a year the Buffalo Niagara Partnership and its CEO Andrew Rudnick did everything possible to squelch consideration of any design other than the twin span. The Partnership even contributed a very large sum of money in an attempt to defeat Common Council President James Pitts in last month's Democratic primary. Pitts is the single local politician who insisted the twin span was a lousy idea all along and he has consistently refused to buckle when the Partnership pressured politicians to shut up and go along. Andrew Rudnick was the only leader of any community group to advocate the twin span and oppose consideration of any alternatives at the public hearings of the Review Panel last August.

Now there is the possibility of a happy resolution to a process that was starting to have nasty international implications. How does that Buffalo Niagara Partnership feel about this Canadian-American partnership? Are they happy that resolution is possible? Are they working behind the scenes to scuttle it? They were so vocal for so long saying they had only the community's interests in mind. Now that the community has found a way to solve the problem, why are they silent?

3. What happened to Cannon?

Opposition to the twin span idea took form when Jack Cullen and Clint Brown came up with their SuperSpan idea, but Cullen and Brown never had a bridge design, all they had was the good sense to see that something much more useful and imaginative could be done in that construction project. Public opposition took fire only when Bruno Freschi and T.Y. Lin came up with their gorgeous curving concrete cable-stayed design. Freschi did his initial design working with students at U.B.'s School of Architecture, but once the project started getting complicated he moved it to Cannon, a large architectural firm based on Grand Island. Cannon provided artwork, cost projections, and engineering analyses. The combination of Cannon here and T.Y. Lin's firm in San Francisco took the air out of the claims by Andrew Rudnick and other twin span advocates that Freschi's plans were too sketchy to deserve serious consideration.

At the first public session of the Review Panel last August, Cannon provided an articulate and enthusiastic engineer to answer the complex technical questions that Freschi, who is an architect and not an engineer, would have been hard-pressed to answer on his own. It was a solid, lucid and impressive presentation.

Since that evening, Cannon has been absent from the entire process, at least from any public aspect of it. There were, so far as I could tell, no Cannon representatives at the Review Panel hearings in October. There have been no public statements about the bridge question from anyone at Cannon.

But Cannon president Mark Mendell was appointed to the board of directors of the Buffalo Niagara Partnership, and Bruno Freschi moved to Washington, where he now lives in a condo only a few blocks from Cannon's small office, in which he works on some of Cannon's international projects.

Did Cannon back away from the Peace Bridge project because other new projects – the proposed convention center, for instance – were more attractive and a contract was more likely if Cannon made nice-nice with the People-in-Power? Was Mark Mendell's appointment to a board that, according to Rudnick, fiercely opposed Cannon's most interesting project in this area in years mere coincidence, an indication that the Partnership was giving up its opposition to a signature span, or an indication that Cannon was abandoning its support of a signature span?

4. Why the silence from the American members of the PBA?

On several occasions last year, Governor George Pataki said he favored the signature span and was going to instruct his four appointees to the Public Bridge Authority to advocate it. The fifth American member is appointed by New York's attorney general, Elliott Spitzer, an outspoken advocate of the signature span. Why hasn't one of those five Americans spoken out about what's been going on in the PBA for the past year? It's not a secret society, it's not a privately-held corporation, it's not the Buffalo Niagara Partnership or the Saturn Club. So what's with the secrecy? Even Barbra Kavanaugh, Spitzer's appointee and a signature span advocate herself, has never said anything publicly about what goes on in those closed meeting rooms down at Peace Bridge Plaza. Why not?

Why are they silent about location of the plaza, restoration of Front Park, the two bridge plans, and the Canadian insistence on a bridge-plaza plan that will cost more, take longer to build, and look lousy? Why have they been silent about a plan that would run massive truck traffic through Buffalo's West Side residential areas for most of a decade? They're appointed by the governor and the attorney general, but not as personal representatives of the dukes. They're supposed to represent the public interest. What do they think that interest is? Why do they keep all those secrets? Why won't they ever come out and talk to us?

5. Why does the web site of the Public Consensus Review Panel contain so little useful information?

Beats me.

They keep promising to include all kinds of links and data, but so far what's gone up is trivia and anachronism: the consulting engineers' preliminary report, the text of the agreement proposed by the Public Bridge Authority, directions to the studios of WNED-TV for the August and October public sessions, the names of the panel members and consultants, and some PR stuff about the consulting engineers. That's it. The Review Panel's executive director, David Carter, promised much for this web site at the informational sessions last summer, but thus far they've delivered virtually nothing. The opening screen is ugly and uninviting, with telephone and fax numbers in the teensiest possible type. In a decade when any high school kid knows how to put up a sexy web site and half of them have, this underutilization of a potentially powerful informational resource makes no sense at all. Check it out yourself: www.peacebridgereview.org.

6. Will Joel Giambra's defeat of Dennis Gorski for Erie County Executive unravel the Peace Bridge peace process?

Gorski has long been a strong advocate for a signature bridge and his staff – Rich Tobe in particular – was instrumental in getting the Public Bridge Administration to act as if it had community responsibilities. Joel Giambra assiduously avoided taking a position on the bridge issue in his recent campaign. Was it because he had no position, because he didn't want to offend anyone who might vote for him or give him money, because he was in already someone's pocket and didn't want to let the other side know, or because he has real ideas about the bridge issue but didn't want to announce them before he was in a position to do anything about them? I wish we knew. His silence on this issue is really scary.

7. Will the process undertaken by the Review Panel accomplish anything useful?

This one I can answer.

If the various participants engage the process honestly and decently, then whatever the result we'll all feel that we got to it fairly and reasonably. If the various participants engage the process deceptively and dishonorably, then we'll know what kind of people they really are, and so will Judge Fahey when he revisits the three pending lawsuits next January, and so will whatever other judges are involved in this affair further down the line. It's not just New York appellate judges who may get to view this mess: all of these same questions about inadequate environmental review can be raised anew in the Federal courts.

I remain convinced that if any impartial body examines the issues and the facts, and if the data and process are public, we'll get a gorgeous bridge out of it. You only get dogs like that companion span when a small group of people talks only to itself and with a small gaggle of political and corporate cronies in rooms from which serious questions and high-level professional competence are excluded.

We'll get something good out of this, so long as the process isn't driven by hidden agendas, sabotage, and secret deals. And if there are hidden agendas, sabotage, and secret deals, we'll learn about them, and then we'll decide what to do about the people who so cynically betrayed their public interest. My hope, however, is that decency will prevail.

That is always the hope of nearly all of us. Hopes like that aren't fulfilled at random. That's why we've got to keep asking hard questions – right up to the end of this.

19 FRESHWATER SHARKS

Three interesting Peace Bridge events in the past week:

> Representatives of the Ambassador Bridge organization in Detroit told the Public Bridge Review Panel why New York and Ontario should turn over to them operation, control, and ownership of all border operations in this area. The Panel overwhelmingly rejected the offer a week later.
>
> Fort Erie Mayor Wayne Redekop told the Public Consensus Review Panel why he opposed anything other than the Buffalo and Fort Erie Public Bridge Authority's twin span plan idea.
>
> The Canadian government and the Public Bridge Authority, after a month-long delay, announced that Buckland & Taylor, Ltd., of Vancouver, would work with American engineers in evaluating bridge and plaza options.

TRUST US

In a public session at the studios of WNED, on Wednesday, November 10, two executives from the firm that operates the Ambassador Bridge linking Detroit, Michigan, and Windsor, Ontario, and two engineers from a Florida consulting firm they brought along for support, made an elaborate pitch to the Public Consensus Review Panel. As they talked, they showed one PowerPoint slide after another, and answered and deflected questions. I kept thinking of four freshwater sharks who had swum over from Detroit, drawn by the scent of money in our waters, energized by the possibility of feeding long and well at our expense.

The two executives were Dan Stamper, president, and Remo Mancini, corporate vice-president, of the Detroit International Bridge Company, operators of the Ambassador Bridge. The two consultants were Scott Korpi and Brian Mirson, both of American Consulting Engineers, based in Land O'Lakes, Florida.

Mancini began the presentation with a tacky eight-minute video in which their friends and some truckers told us what a swell bridge the Ambassador Bridge was and what a swell duty free shop it had and how the bridge really got you from one side of the river to the other side of the river. The music track was execrable. It reminded me of those commercials for plays in which they have "ordinary people" in the lobby telling us with oppressive enthusiasm what a great and inspiring family experience and theatrical experience it was.

Then they posed several good questions, none of them new. It was like they'd looked at all the articles in *Artvoice* and the *Buffalo News* and the videos of the Public Consensus Review Panel hearings and extracted the key issues raised again and again by the Olmsted Parks Conservancy, the Episcopal Church Home, and all the other community groups that have been so passionate and articulate about the various aspects of this issue. (Their only fresh idea – which seems screwy to me – is to relocate everything so it runs along the International railroad bridge, but in the question and answer period they didn't seem the least bit attached to that, so I assume it was just something tossed out to get our attention.)

The really interesting thing about their presentation was having this one group articulate the concerns of so many community organizations all at once. It was as if many of the most important parts of last August's Review Panel's public hearings had been compressed into one brief utterance. That was refreshing: it reminded everybody what this was all about, what got us here, what was being asked, what problems stood in the way of rational answers. They asked such questions as:

> What's the rush to build a new bridge? Can't some of the present problems be solved by better traffic management and more staff at the toll, customs and immigration stations?
>
> Why don't we do an environmental impact study to find out what the real needs and limitations are (exactly the basis of the lawsuits in state supreme court)?
>
> Why are we dealing with the bridge as if it were a standalone issue, without implications on parkland, the environment, the trade corridor?

They didn't seem to have any specific ideas about what they would do when they took over, other than that they'd build a signature bridge somewhere around here and would apply their management skills so well we wouldn't have to worry about anything ever again. They said that a smoothly-operating trade corridor was necessary for the region's economic and mental health. To nearly every question about design, location, cost, and capacity they said we'd have to wait on the environmental impact study.

All they seem to ask of us is that we grant them complete control of the whole operation. They'll run it as a profit-making enterprise and they promise us various unspecified benefits when they develop a lucrative new bridge and get money from the two governments to give trucks and cars access to it. Their basic attitude, finally, seemed paternal: you kids in Buffalo and Fort Erie can't get your acts together so Big Daddy from Detroit will take care of it. All you kids have to do is agree to give Big Daddy all the money. And control.

WHO IS DADDY?

The Ambassador Bridge is wholly owned by CenTra Inc, a holding company owned by reclusive Grosse Pointe businessman Manuel Moroun. CenTra was started by Moroun's father, Tufick. Moroun owns a lot of property in and around Detroit and several trucking companies. He seems to be very good at leveraging his large contributions to the Republican Party into public funding for access roads to his properties. He went to the same high school as novelist Elmore Leonard – University of Detroit Jesuit High School and Academy.

He recently settled, after ten weeks of nasty trial, a lawsuit filed by his two sisters seven years ago. It was about control of CenTra. His sisters were seeking $53 million; the settlement was not disclosed. The sisters had accused their brother of "shareholder oppression and keeping them out of the business."

At the Public Consensus Review Panel here last week, Remo Mancini wouldn't answer with any specificity a question about the company's profits earned and taxes paid, but he did say that "Our property taxes in both Windsor and Detroit are quite healthy. Our corporate taxes are quite healthy.... We're proud taxpayers and proud to be good corporate citizens." He didn't mention Moroun's recent lawsuit attempting to force the state to give him refunds for more than $4 million in state taxes on gas sold at his duty free Ammex store, which is part of the Ambassador Bridge operation. A lower court rejected Moroun's original claim and a three-judge appellate panel upheld that decision last month. Moroun's lawyers said they would appeal.

THE SOUNDS OF SILENCE

That's not the only issue the Detroit guys avoided. "Heard melodies are sweet, but those unheard are sweeter," wrote John Keats, musing about painted figures frozen forever on an old bit of pottery. Translate that to what went on at the Review Panel last Wednesday and it reads: "Those guys talked loudly, but what they didn't say shouts."

They discussed increased auto traffic across the Ambassador Bridge in the past seven years and ascribed it all to brilliant management. Not one of them mentioned the presence and impact of Casino Windsor, a huge gambling establishment just beyond the Canadian terminus of their bridge. When Peter Cammarata pointed out that the opening of slot machines in Fort Erie in October increased Peace Bridge traffic 7% the first month, they had no reply. (According to Public Bridge Authority officials, that 7% rose to 13% in November.) Sure, some of their traffic flow is attributable to good management, but surely some is attributable to that immortal Kevin Costner line, "If you build it they will come." In this case, "it" was a casino.

But what happens if lots and lots of them come? Ambassador Bridge president Dan Stamper said, "In the late 80s we had congestion at the Ambassador Bridge and there was outcry that there was a need for another bridge. And there were a lot of people who were quoted in the paper and on tv saying 'We need another bridge, we need another bridge now.' We doubled our traffic since then without adding one additional lane over the water.... We did that in a matter of months, not in a matter of years, by managing the traffic and working with the Customs and Immigrations agencies to improve their process."

That's all no doubt true, but what he didn't mention, and neither did any of his three associates, is that his firm is now busy trying to get a twin span constructed, that the Canadian town across the river is saying there are obstacles because of road capacity and environmental concerns, and that there is another guy in town saying the bridge should be located further up the river. They've got many of the same arguments, issues and problems there that we've got here, but they never said that to the Panel.

They weren't much more direct in their responses to questions from the audience. When someone asked, "Where else have you built and operated a bridge?" Stamper said they operated the "Ambassador Bridge between Detroit and Windsor," but he wouldn't admit that they hadn't ever built any bridge anywhere.

The first time Stamper spoke at last Wednesday night's session he said, "We're here today because we were asked to be here." Two times questioners from the audience asked, "Who asked you to be here?" Both times they avoided giving a straight answer. The second time they were asked the question, Stamper looked left and right, looked down, made a gesture with his head, after which Mancini got up and said, "We've been invited by this committee. I have a letter signed by Mr. David Carter that I received about a week ago with some followup phone calls. And I want to sincerely thank this committee for having us. It's been an honor to appear before you."

What patronizing horseshit.

Stamper and Mancini both knew that question wasn't about why they were there that night but why they were in Buffalo at all. I'd asked Mancini the same question when he was in Buffalo the previous week and got evasion. When *Buffalo News* Washington reporter Doug Turner asked them the same question in October they "declined to identify the person who aroused the company's interest in the issue."

Mancini's answer Wednesday night reminded me of the scene in Christopher Marlowe's *Jew of Malta* when Barabas, on his way home from setting fire to a nunnery, is challenged on the road by two friars. One of the friars says "Thou hast committed – " Barabas interrupts him and says, "Fornication. But that was in another Country: and besides, the Wench is dead." That's the Barabas Gambit: when they're asking you questions about something you don't want to cop to, answer a different question and hope that will keep them so busy they'll forget about what they really wanted to know.

People hardly ever forget what it was they really wanted to know. I don't know what possesses salesmen to think they can sell evasiveness when they're in front of a panel of lawyers, business people, politicians, and community activists. Even if they the Ambassador Bridge people had good ideas about what to do here, that kind of evasion would turn people against them. Why would you turn over a cash cow like the Peace Bridge to someone whose major line is "Trust me" and they won't tell you the truth about things you can find out anyway?

THE POLS

The Detroit guys were sponsored by our US senators. The Review Panel gave them that airtime because the staffs of senators Moynihan and Schumer asked for it. The week before the Panel event, the Detroit guys were in town meeting nearly everybody having anything to do with the bridge issue: politicians, community groups, reporters. The meeting I went to was organized by Senator Moynihan's office. After the Detroit guys left town, Senator Schumer issued a statement saying that their ideas deserved our consideration. But the local politicians had a different reaction: Buffalo Mayor Anthony Masiello said they were "a big waste of time" and Fort Erie Mayor Wayne Redekop said considering their ideas was delaying everything even more.

Redekop is wrong: considering their ideas isn't delaying anything because the collaborative engineering project is continuing on schedule. The only delay there was occasioned by the month-long delay of the Public Bridge Authority and Transport Canada in naming the Canadian engineers. The fact that the Panel heard these Detroit developers in a public session has nothing to do with that process.

Someone in the audience said after it was over, "I bet Moynihan invited them in last summer before the panel got going and was stuck with giving them a venue. Otherwise this doesn't make any sense. They're just not up to his good sense on this." That's a possibility. Someone else suggested that the two senators wanted an opportunity to consolidate the key questions raised by various community groups and critical journalists in town, which the Detroit guys did, and to embarrass the Public Bridge Authority for its short-sighted management of the bridge, which the Detroit guys also did.

But why would we want to have the Detroit guys come in and take it all over, apply corporate efficiency, and then take all the money home to Manuel Moroun? If the two senators are backing that horse they're wrong; if they were trying to make a point, well, they made it and it's time to move on. Our problems aren't going to be solved by turning everything over to a secretive rich guy in Grosse Pointe, Michigan.

We've got to get off the dime on the bridge and we've got to see that the bridge is part of a much larger regional development issue, and we've got to understand that we're the only people who can deal with all of these issues in a way that serves both sides of the river well. The Review Panel knows that, which is why on Wednesday, November 16, all but two members voted to reject the proposal. The two exceptions were Jim Kane and Jack O'Donnell, the representatives of Senators Moynihan and Schumer.

The only pol right on the money this time was Tony Masiello.

20 A Matter of State: Tom Toles, Wayne Redekop, Madeleine Albright, & the Niagara Border War

Perhaps the most lucid recent statement about what's going on with the Canadian-American imbroglio over the Peace Bridge was Tom Toles's November 14 cartoon in the *Buffalo News*. Toles drew a cockamamie bridge that combined all the past and present design ideas: cable-stayed on the right, steel-arched on the left, and topped dead center by a perfectly irrelevant and utterly useless Parker truss. Running the entire length of this goofy bridge is a thin frame supporting script letters that say *"We don't know what we're doing."* A large caption across the lower part of the drawing identifies it as, *"Our signature bridge."*

Everyone I know who saw that cartoon said Toles was once again right on the money. No one said, "Tom's wrong this time." Whatever side of the bridge question the speakers were on, they thought Tom got it right. Had almost anything else been written across the goofy bridge that consensus would have been great. With the script saying "We don't know what we're doing," well, it's not great. It's screwy.

The Junior Senator from New York

The screwiest bridge moment of the past two weeks was maybe last Friday afternoon when Jack O'Donnell, Senator Charles Schumer's Buffalo office chief, read a statement from the Senator to the Buffalo Common Council's SuperSpan Signature Bridge Task Force. The senator said he regretted the refusal of the Public Bridge Review Panel to embrace the offer of Detroit's Ambassador Bridge sales team that visited Buffalo earlier in the month. The Ambassador team had said that if we turned the bridge operation over to them, they'd provide a better and more efficient operation than we have now. All they asked in return was total control of the bridge, cooperation of all governments concerned in building roads wherever they wanted to locate whatever bridges they elected to put up, and no limit on what tolls they might collect and keep. Their president, Dan Stamper, said designing and building bridges weren't difficult, all you had to do was write checks and hire people who knew how to do such things. (In a page one story last Sunday, the *Buffalo News* said the Detroit group had built the Black Rock Canal, but that's not true. They've never built any bridge anywhere.)

Senator Schumer wrote, "Whether you support a signature span or a twin span, I think we can all agree that the discussions on the future of the Peace Bridge have reached somewhat of an impasse." We can't agree on anything of the kind. The Public Bridge Review Panel only last month got the Public Bridge Authority, the agency that maintains the Peace Bridge, to abide by the design recommendation endorsed by its executive committee. That selection process is now going on. It got off to a bumpy start because the PBA and Transport Canada squabbled over who would pay for the Canadian consulting engineers, but they worked that out and the engineers are off doing what consulting engineers do.

Why would Schumer want to sabotage the Public Bridge Review Panel? "I believe this plan merits our consideration as we work together to build a bridge for the next one hundred years," he wrote. That's the mantra of the Detroit pitchmen: "the next one hundred years." Every time they make a presentation they say it over and over. Schumer's been visited by the Detroit pitchmen and he's picked up their lines.

The statement Jack O'Donnell read seemed sadly out-of-touch. One of the things that's been really exciting about Schumer since he moved up from the House has been his presence in upstate affairs: he's been a key player, for example, in the cruel airfare wars and his pressure has really helped us. But this time, it is as if his local information sources dried up so he was getting direction from the Detroit developers. Keep in mind that he didn't join and he didn't appoint a representative to the Review Panel in its key stages because he was pursuing a plan of his own. It was a good and thoughtful plan, but his advocacy of it kept him out of the conversation for months. It's only recently that Jack O'Donnell has occupied the seat the Review Panel has kept open for Schumer since it was organized last summer.

"To turn this proposal down summarily makes no sense," Schumer wrote. In fact, it makes a great deal of sense. There never was a viable proposal from Detroit, other than turn everything over to us and trust us. There was no plan for a bridge, no explanation of what would be done about the money, nothing other than a claim that they'd tuned up a badly managed bridge operation in Detroit and now wanted to take this one over as well, giving a single private family firm total control of the two key bridges at both ends of Lake Erie.

Chuck Schumer needs to start visiting Buffalo again. He shouldn't be dancing to a drumbeat from Detroit.

MOYNIHAN'S MAN

Moynihan's Buffalo office chief, Jim Kane, has been involved in the Bridge issue from the beginning, which is hardly surprising given that Moynihan and Buffalo Common Council President James Pitts were the two officials who longest and loudest said we could do better than the twin span advocated by the PBA, the Buffalo Niagara Partnership, and the *Buffalo News*.

When O'Donnell finished reading Schumer's statement, Kane addressed meatier matters. He's fearful that the agreement signed by the Public Bridge Authority and the Review Panel last month will help the PBA continue to avoid an environmental impact study, which is the reason we're in the mess we're in now. Had the PBA done an EIS and involved the community in its plans, there never would have been need for the lawsuits filed against it by the city of Buffalo, the Episcopal Church Home and the Olmsted Parks Conservancy. "What kind of model is this where you let them sidestep the EIS?" Kane asked. "You do an EIS. It's the law of the land. You don't give them a choice. Not doing it is against the law." We're all locked, Kane said, into a process started the other side of the river. Whatever the Canadian concerns are, he said, they don't have the right to screw up an American city for their convenience. "They started a project on the Canadian side and they're telling the American public you're stuck. That's not right. They can't do that."

Jim Kane is right, and that's another reason this whole Detroit thing is a diversion. They've introduced nothing to our conversation except maybe an expression of outside greed and, as I wrote here last week, a neat summary of the key issues that had been raised by all the community groups and individuals involved in this difficult affair since the PBA went public with its twin span design a few years back. People will use the occasion of the Detroit offer to make speeches and get press, but the Detroit salesmen will soon be gone and we'll be back to the real issues and the real players. Half of whom are across the Niagara River, a side that has been very very quiet through all of this.

WHAT FORT ERIE'S MAYOR WAYNE REDEKOP SAID

That is why Fort Erie Mayor Wayne Redekop's presentation to the Review Panel two weeks ago is so important. It is, so far as I know, the first time any Canadian official has been willing to make an extended position statement on the Peace Bridge expansion question, let alone answer questions about that statement. Because of that, we'll quote from his remarks at length.

Redekop began by talking about Alonzo Mather who in 1893

> put together a visionary concept of a bridge to cross the Niagara River between Fort Erie and Buffalo, and that bridge provided for not only motorized vehicle traffic but also train traffic and pedestrian traffic and trolley traffic and that bridge was, many people would agree, years ahead of its time. In fact, it was so far ahead of its time that ten years after trying to get the approvals necessary to construct that bridge he gave up. It was a senator from New York State who had interests in the hydroelectricity generating in Niagara Falls who was able to stymie construction of that bridge primarily. And it wasn't until 1923 that a group of people on both sides of the river came together because by that time it was absolutely imperative that Buffalo in particular and Fort Erie as well to have a fixed link across that river. As we all know, in 1927 construction was completed with respect to that bridge. I fear that we're in the middle of a similar process now at the end of the 20th century and I'm concerned that what happened in 1893 to 1903 is going to happen now to 2003.

What analogy was he making? That Americans were again screwing up a grand Canadian idea that's ahead of its time, and they're doing it for venal and personal reasons? Or was he saying, You Americans screwed us at the end of the 19th century and we're doing it to you at the end of the 20th?

> I grew up in Fort Erie and for me there are three major

> landmarks that Fort Erie has. Those are symbols that identify our community to people from other parts of our country and perhaps from other parts of your country. The racetrack, which has been on its site since 1898. The old British fort . It's called 'Historic Fort Erie,' but if you grew up in Fort Erie you called it 'the old fort.' The one that's there was constructed during the Depression but it's based on a fort that was constructed in 1812 and quite coincidentally destroyed by the Americans in 1814. And then of course we have the Peace Bridge, which is a symbol of peace and the nature of the relationship between our two countries and it's been in existence for over 70 years.... Those symbols help to define my community. And for people like me, every day that I'm in Fort Erie, literally almost every day that I'm in Fort Erie, I come in contact with that bridge. I see the bridge, I pass the bridge, sometimes I cross over the bridge. I'm not untypical of the people that live in the eastern part of the town of Fort Erie. So for us, it's a real symbol, it's part of our past, part of our story about who we are and how we got to where we are.

More important than symbols, Redekop said, is how fast trucks and cars can get through Buffalo and Fort Erie:

> People don't care if Fort Erie is a nice little town. People don't care if Buffalo was a magnificent center fifty or sixty or a hundred years ago. It doesn't make any difference whether you've got some magnificent rate architecture here, a first rate museum, it doesn't matter. What matters is how quickly can we get from point A to point B. And if you're manufacturing goods in Toronto and you want to get them into the United States, all you're concerned about is 'What is the fastest and the cheapest route I can take.' And I fear that we're going to lose the opportunity to retain and expand the trade that's coming out of the Golden Horseshoe both ways in crossing the river at Buffalo.
>
> It's the same thing with tourists. Tourists don't want to sit in line on a hot summer day, even if they've got air conditioning, particularly if they've got children in the care bouncing off the roof. They want to get from where they've come to where they're going and they want to get there as quickly

as possible.

So our concern in Fort Erie is every day we don't have expanded capacity on the bridge, whatever the bridge looks like, we lose opportunity. The opportunity that we lose is revenue, investment, jobs. We're losing them, you're losing them. In other words, our future prosperity is at risk. The prosperity of Fort Erie and Buffalo, the Niagara region, western New York, Ontario, New York State. It's that simple. And that's a major concern for us.

The Peace Bridge Authority has a plan and the plan is ready to go.

For a moment he seemed to abandon local aesthetics entirely:

When I go from Fort Erie to Buffalo, it really doesn't matter what the bridge looks like. I could care less, to be perfectly honest with you.

But then he reversed himself and went back to where he'd begun:

It has an attractiveness to me.... It is a symbol. One of the concerns we have in Fort Erie is that we have this nagging suspicion that whatever bridge is constructed, if it's a signature bridge or a superspan or some other magnificent structure, that equates to the demolition of the Peace Bridge, and that is a major concern. Because we consider that to be not only to be a symbol and a landmark, but it's a heritage symbol and a heritage landmark. And the notion that you're going to keep that peace bridge for a bike path or a recreation trail, we just don't believe that. We think that what you're talking about is putting up a monument to something other than what we cherish....

We have a Peace Bridge plan that, aesthetics aside, meets the requirements of the residents of Fort Erie. We can get across the twin span and the existing Peace Bridge safely, efficiently and affordable, that and that's important to commercial traffic: affordability....

It is a plan, it's a known plan, it's a plan that's ready to go, and all the ducks primarily are in order. That bridge could have started construction earlier this year, but it hasn't. And it's been delayed by a process in 1998. I participated

> then in a process that the Buffalo Partnership conducted. And it's being delayed by this process. That's fair enough. I understand that people's needs and concerns have to be addressed, I don't begrudge them that. I certainly don't begrudge the people in Buffalo wanting to be satisfied what's being done in their community because I'm concerned with what's done in my community. But you can't just pick up the plans and say okay we'll move the plaza here, there or whatever....
>
> Time is money and every day that goes by that we don't expand the Peace Bridge capacity we lose opportunity. And once those opportunities are gone, once the manufacturers in Toronto decide that they're going to ship through Windsor, they're not going to go back through Fort Erie and Buffalo once we get our act organized whether it's five years from now or ten years from now or twenty years from now.

THE MAYOR AND THE AUDIENCE

When the mayor finished his prepared statement someone in the audience asked, "If the plan that comes out of this panel's work is faster and cheaper than the PBA's, will you support it?"

He said he hadn't seen a plan with hard numbers so he couldn't respond. "I'm open-minded, I like to think that I'm open-minded. I like to think I'm a realist. On the other hand I like to think I'm a bit of a poet and I'd like to see something significant done. But it doesn't matter to me. You can run the risk of losing economic opportunity and future prosperity or you can have a nice fancy bridge somewhere, someday. Well, My choice is economic growth prosperity now and the rest of the stuff will take care of itself. You know, a bridge across a river does not define a community. It doesn't say that Buffalo will be a better place because you have a signature bridge. That's just about the goofiest notion I've ever heard. Just like the notion that people will travel long distances to see a bridge. That doesn't make any sense to me either."

Someone else asked if he or his delegates would join the Review Panel. He said this was an American problem, he had no interest in it, and he was too busy with local affairs on his side of the river to get involved with it:

> The Peace Bridge Authority has a plan. We've gone through

> a process. I participated in the process last year that the Buffalo Partnership operated. Thought that this thing was finished. I can't do this indefinitely.... You're the people that have to sort through this. The bridge does land in two countries. We don't have a problem. We're at peace in Fort Erie and Niagara. We're at peace. You're not. I'd like to help you to sort that out and I think I've done that by giving you at least some understanding of how we feel about this and how we feel that our prosperity is tied to yours and how the delays are going to affect both of our communities.

So it came down to an icon that couldn't be changed from their side, commercial traffic to distant places that had to be accelerated, and a plan that couldn't be improved upon. Redekop's bottom line was, there was no need to do anything other than what the Public Bridge Authority said should be done.

When he was done answering questions from the panel and audience, I followed him into the corridor, hoping to ask a few questions about what he'd just said. But a mid-thirtyish beery guy had him cornered. The guy was hammering at him, shouting that he has his rights to something or other. Redekop said "You can write whatever you like." The guy said, "Goddamned right I do. So what makes you think your rights should come before mine? I have my rights!" It was just awful. The guy was either drunk or performing for two of his friends, who were standing nearby and seeming to nod encouragingly. I wanted to apologize to Redekop for the abandonment of civility, but he moved out of the building fairly quickly after that. He went back home to his own country.

A MATTER OF STATE

What can we do when our partners the other side of the border won't listen to, don't understand, or don't care about our concerns?

How can a citizens' panel on this side of the border, whatever its decency and good intentions and competence, have any influence on an international construction project if the other side of the border won't come out and talk, won't express any good will whatsoever, doesn't care?

It can't.

How can the mayor of Buffalo influence the mayor of Fort Erie if the mayor of Fort Erie won't pay attention to any of the questions that matter this side of the border?

He can't.

The Peace Bridge, as Mayor Redekop so eloquently pointed out, is one object but it has very different meanings on the two sides of the Niagara River. Failure to understand those differences means more waste – not just of construction time and money as now, but of human resources, which are more valuable than either. Just because people use the same words doesn't mean they're always speaking the same language.

I no longer think the real failure here is entirely local. The Public Bridge Authority screwed us all, but the responsibility doesn't stop there. This is an international affair, a major disagreement in border policy between two countries. The officials on the Canadian side, from the mayor of a small town up to a cabinet minister, are turning deaf ears to officials and the public on the American side, from the people of Buffalo up to our representatives in the United States Senate.

It's a stalemate: the Canadians won't consider anything that might alter the operation of their truck stops, duty free shop, brokerage houses and other jobs for half of Fort Erie's employed adults, and we'll keep going to court to keep them from forcing on us a bridge that makes life here worse than it already is.

The sharks from Detroit maybe had it right: maybe we do need outsiders to help us resolve this issue. If the PBA isn't going to honor its agreement with the Review Panel (as its chairman John Lopinski recently suggested) and if the town of Fort Erie is going to oppose anything that changes anything, then we do need outside help. But the outside help shouldn't come from secretive millionaires from Grosse Pointe who never give interviews. It should come from our government officials specifically charged with handling international affairs. Mayors can't do that, congressmen can't do that, senators can't do that, and Review Panels can't do that. This is what we have national governments for.

It's time for Madeleine Albright to look north and face the fact that all troublesome borders aren't halfway around the world. The Canadians are making economic war on us here in Buffalo. You can bet that Canada's foreign ministry has been very much involved in this affair from the beginning; indeed, all their representatives to the Public Bridge Authority are appointees of the Canadian national government, not the Province of Ontario. How about our State Department telling them that instead of making war it's time to play ball?

THE RAGGED EDGES OF DEMOCRACY

Redekop and other Canadian officials have complained about the interference by the city of Buffalo, various citizens' groups, and the Public Consensus Review Panel in the smooth process planned by the PBA. They're right: those groups *have* slowed things down, and a good thing it is.

Autocracy and dictatorships are fast and efficient: in those systems the boys in the closed rooms decide what's going to be done and then they go and do it. If they're benevolent and wise the public isn't hurt too much; if they're not benevolent or wise the public is hurt a great deal. Mussolini got the trains to run on time – and he nearly destroyed Italy.

Democracy is inefficient. It's ragged at the edges. It's not simple. So what? Why are velocity and smooth edges and a single voice any better? That's what Wayne Redekop is saying, it's what Dan Stamper and his Detroit sales team would have us believe and it seems to be what Senator Schumer is now thinking. I'd say to them all: relax. The Public Consensus Review Panel is moving forward; the Common Council's Task Force is watchdogging the Panel. Will there be time to do it all? I don't know, and neither does anyone else. But before we jump ship, let's see if the ship stays afloat.

One thing is clear: after the Review Panel announces its choice of bridge design, we'll have to take a long and hard look at the operation and constitution of the Public Bridge Authority. We'd have none of this mess if it had been willing to act as a public agency from the beginning, if it hadn't engaged in phoney segmentation to avoid the EIS required by law. We shouldn't have to go to court every time we want that public agency to act lawfully. If that agency is structured so it can get away with this callous disregard for the people and the law, then that agency has to be restructured. On both sides of this contentious border.

21 The PBA Comes to the Table

The Border Problem

Truck traffic across the Peace Bridge has increased enormously since NAFTA and bridge officials say in few years there will be so much traffic on the bridge that traffic will clog even if there are no barriers of any kind on the American side. Maybe. In any case, we're not there yet and we're not close to it.

Most days a line of trucks stretches halfway across the Peace Bridge; some days the line goes all the way to the Canadian plaza and beyond. Except for the summer months, auto traffic generally moves at a brisk pace. The Public Bridge Authority is trying to boost auto traffic – they've even taken to posting advertisements for Canadian gambling joints at the tollbooths – but thus far the general trend of auto traffic is downwards. The only continuing traffic problem at the Peace Bridge is the trucks getting through US Customs. And it's only the trucks that the officials anxious to expand bridge capacity really worry about. Trucks are where the big money is.

Would having more Customs inspectors and a larger Customs facility speed things up? Of course. But it will take far more than merely telling Congress that we've got a problem here and we need more staff to deal with it. Robert Trotter, assistant commissioner of the US Customs Service, pointed out recently that since 1973 the priority job for US Customs has been drug enforcement, not trade issues, and since the southern border has 90% of the drug traffic, the crossings there are getting nearly all the new Customs staff appointments. Trotter suggested that the norther border states do some of the concerted lobbying the southern border states have done so well, but it's unlikely that Congressional lobbying alone will transcend the current national hysteria over illicit drugs. Expansion of Customs staff at this border would help our problems enormously, but unless the current gaggle of presidential candidates decide it's a vote-getting issue, it's probably not going to happen in the near future.

THE PLAZA MESS

A much larger truck processing facility would help the congestion problem. The Buffalo and Fort Erie Public Bridge Authority, the agency that administers the Peace Bridge, has long been aware that the plaza is the real bottleneck in the system. Even so, the PBA and its staff have thus far avoided any specific plaza planning and deliberately separated its bridge and plaza construction projects. That was so they might avoid a full-scale environmental impact study, which would have forced them to deal with the wide range of community concerns that have surfaced over the past year.

Their idea was that they could build a bridge from the American side (where they already owned a landing site, the current plaza) to the Canadian side (the Canadians didn't care what kind of second bridge was built so long as it fed more trucks carrying Canadian goods to America and brought more gamblers carrying American money to Canada). Then, after that was done and irrevocable, they could address the plaza issue. With more bridge lanes the congestion problem would surely be far worse than it is now. If they had to eat the remaining parts of Front Park, how could the City of Buffalo say no? It was a nice end-run around the law and the community.

Bruno Freschi, whose design of a signature bridge catapulted this into a major public argument, insisted all along that it was absurd to talk about the bridge as if it were separable from the plaza. He objected strenuously when he was told that his design wouldn't be considered at all unless he made it go to the current plaza. "It's not a bridge, it's a bridge system," he said, "it lands somewhere and that's an essential part of it, not something on the side."

The PBA and its supporters never figured on the huge community response to Freschi's ideas, the focus of the New Millennium Group on the bridge and plaza issues, and the creation of the Public Consensus Review Panel by the city, the county, and the Margaret L. Wendt and Community Foundations. Last summer, the Review Panel invited the PBA, Freschi and others to make public presentations and answer questions about their design ideas. The PBA ignored the invitation. Their notion then seemed to be that they were going to build the bridge they wanted to build and no one could force them to do otherwise.

That changed when the city, the Episcopal Church home and the Olmsted Parks Conservancy filed suits that, if successful, would have blocked any new bridge construction. The PBA filed suit to force the city to let it begin construction. Judge Eugene Fahey rejected the PBA's suit, granted the city's, and kept the Episcopal and Olmsted suits open. He said he'd make up his mind what to do about them when he saw how the PBA dealt with the final conclusions of the Review Panel.

That's when the PBA decided it could no longer hang tough. If it didn't work and play well with others, Judge Fahey would keep it from doing anything at all. And that is why the Peace Bridge senior staff finally appeared before the Public Consensus Review Panel to explain what it wanted to do and why.

A GRAND PERFORMANCE

On December 16, Earl Rowe and Stephen Mayer, the two general managers of the Buffalo and Fort Erie Public Bridge Authority, presented to the Public Consensus Review Panel their plan for solving all the current problems.

Construction, Rowe said, "must begin in the summer of 2000. Yes, we need to have the shovels in the ground by next summer!" (I'm quoting from the printed version of his remarks.) Why the great hurry to break ground? Because more traffic is coming. What happens if the new bridge isn't ready for the new traffic? Then it will go elsewhere. What's so bad about it's going elsewhere? Well, it would just be bad if it went elsewhere, that's what's so bad about it.

"The purpose of your panel," Rowe said, "is to educate the public, and we are happy to have the opportunity to present the facts associated with the urgency of this project." Not so: the purpose of the panel is to pick the best bridge and plaza option and educate the public about *that*. The education component is explanatory, not primary. This is a policy-determining body, not a school or a public relations firm.

Because of a new trade initiative called Continental One, Rowe said, they expect further increases in truck traffic on the bridge. "We need to have a new bridge built by the year 2003. No other crossing can have the required additional capacity any sooner than eight to ten years." For more than a year now, critics of the PBA have pointed out that the PBA's own schedule put completion of their project at a minimum of ten years, while all the other major bridge experts who had looked at the project (such as T.Y. Lin and Eugene Figg) have insisted that a new bridge could be go through community consultation, a real environmental impact study, design, and construction in five to six years. Rowe just took the opposition's numbers and plugged them in where his own consultant's numbers used to be. Moreover, he didn't include in his projection the fact that the old bridge will have to close in 2003 for rehabilitation, and that they'll have to begin work on a new or expanded plaza about then as well. They might finish their bridge in 2003, but the complete *system* wouldn't be complete until several years later. When? He estimated six years, exactly the time estimated by Lin and Figg.

"The environmental studies have been completed," he said, "and approvals have been granted." Well, that's not quite right either: the environmental studies have not been completed – they've been avoided, sidestepped, slipped and slid around and below. That's why so many people have complained about the segmentation and it's what the lawsuits in Judge Fahey's court are all about.

The most important single new piece of information in his opening statement was that the Town of Fort Erie is making a serious move to have the bridge designated as a heritage property in Canada. A similar initiative is underway on this side of the river. I think that means they're trying to foreclose the option of removing the old bridge should the Review Panel recommend that a signature bridge be built. It's cultural blackmail, but it probably wouldn't hold up if a signature bridge were built that handled all the traffic. Would the heritage officials on either side of the border pay for the upkeep of an unused deteriorating bridge in need of major repairs? Not likely. Would the American heritage officials insist on preserving an unused bridge the very existence of which meant continued despoliation of a beautiful Olmsted park? Not likely.

Steven Mayer then discussed several large pictures on easels along the side of the room. The first showed the current Peace Bridge next to a bridge that was its twin except it had a pretty arch where the current bridge has the ugly Parker truss. Mayer said that the designers of the Peace Bridge never wanted the truss, it was forced on them by US Coast Guard height requirements. Then he showed a picture just like the first, except the old bridge had been retrofitted with the pretty arch, just like its new companion. To do that, he said, they'd have to close the old bridge for a while and replace the Parker truss with the pretty arch. They didn't have the money for that, he said, but it could be done. If someone came up with the money.

Then he showed designs for two plaza locations, one north and the other east of the current plaza. Both freed up all of the property in Front Park that the current bridge and Customs operations have consumed. Both would require rotating where the two companion or twin bridges landed and building new access roads and ramps. How would you rotate two bridges that were operating at capacity? Where would the funds come from? He didn't know about that. But it could be done if someone provided the money. How long would traffic be disrupted while that reconfiguration was in process? He didn't say.

Then Earl Rowe took over again. He warned of the dangers of considering any other design:

> We want to alert you to the fact that if a concept is presented that is a radical departure from any of these, the bridge will not be built for at least another decade. It will not be built immediately. That is a fact. Why? It will require an extensive round of environmental assessments and could lead to a super EIS that alone could take three to five years. After that, is about two years of design and several years of construction. We cannot afford it! Not only can we not afford it – we don't need to. And the costs will escalate with every time delay. We at the Peace Bridge, with the support of our users and our citizens, have addressed all the concerns raised over the past several months. Now it is the time to make an educated decision.
>
> There is tremendous merit in this plan. It has strength and viability, and meets the time frame that will help make the Peace Bridge the landmark gateway! We take this very seriously.
>
> I want to stress that we will all lose if we do not begin building this bridge by next summer. However, more positively, we all win if the bridge is begun then. Because if it is – we will be positioning the City of Buffalo and the Town of Fort Erie as urban areas whose greatest concerns are economic growth and being a wonderful place to hold a job and raise a family.

They handed out to the panel members and press packets containing pictures of the two spans, maps of the two plazas, the text of Earl Rowe's comments, and their new business cards, which carry as a logo a drawing of the bridge with the pretty span.

I thought the picture of their new bridge next to the present bridge looked klutzy, but I rather liked the look of the twin bridges. They're not beautiful like any of the signature bridge designs I've seen, but they're nice to look at in the picture. I liked the way the arches over the Black Rock Canal echoed the arches supporting the rest of the bridge. Steve Mayer said they had a certain elegance and I think he's right. They wouldn't embarrass us.

Of course, that design would be far more expensive to build and maintain, would take far longer to complete and would cause far more traffic disruption all the way than building a six-lane signature bridge and new plaza north of the present plaza. And, as Mayer said, they don't have the money to reconfigure the old bridge anyway. So it's just an idea, nothing more. They have no plans to do it.

Trust Me

It was a fine performance, really polished and integrated. Earl Rowe didn't attempt to argue any of the points raised by any of the opponents of the twin span design or schedule. He simply asserted they were all wrong and the PBA was all right. Neither he nor Steve Mayer argued the PBA's opponents' assertions that the current plaza is a disaster functionally, socially, economically, and ecologically. Their plan took no responsibility for the relocation or rehabilitation of it, only credit if someone else came up with the money to move it elsewhere later and could persuade them to rotate the landing of the bridge. They adapted as their own the time frame of their critics, and blamed the critics for the dysfunctional schedule for which they themselves had been faulted.

It was brilliant. It was at once the absurdity of *Through the Looking Glass* and the topsy-turvey Newspeak of *1984*.

When they began talking about new plaza options I thought it was maybe a divide-and-conquer gambit, that they were holding fast to their twin span idea while trying to seduce the plaza advocates away from the larger opposition. But if it was an attempted seduction, it was doomed to failure because didn't offer anything of substance. The only thing real, so far as I could tell, were the pictures – and they were photographs of a model, and one of those photographs (the one with the two new arches) was a computer manipulation of the other. The single thing they said they wanted to do and were ready to do was build the bridge they've wanted to build all along. Everything else was possibilities, things in the air.

All the alternative plans still cost far less than the PBA plan and would be completed sooner. The PBA slashed four years from its proposed construction schedule, but neither Rowe nor Mayer gave any idea where those four years went. There was no new engineering data, no new design; they just subtracted four from ten and got six. All of the other designs would have less impact on water flow and be made of more modern materials that last longer and cost far less to maintain than the present bridge or the steel suspension bridge they want to build next to it. All of them contain plaza design as part of the plan, not some later happenstance.

This is like a Warner Brothers cartoon, where the wolf, having been frustrated in earlier bullying maneuvers, now comes smiling and bearing a basket of goodies. But the goodies are full of ground glass: eat them and the wolf wins. I can't imagine the Review Panel falling for this.

One panel member said afterwards that this was the first time the PBA has been willing to be specific about anything but its companion span design. One of his colleagues was unimpressed: "The only specific thing I heard them say they wanted to do is start construction next summer. The rest is all *trust me*. Why should we trust them now? The only reason they're here is because Gene Fahey gave them no other choice."

Right he is. The PBA is in a far weaker position than it was before Judge Fahey ruled against them on three counts and held off ruling on the fourth. They cannot break ground unless the city, the Episcopal Church Home and the Olmsted Parks Conservancy say it's okay. They don't have a fallback position, they lost the easement case, they have no choice but to remain at the table. Maybe some time soon they'll decide to stop fighting the community and will join it in conversation instead.

22 PEACE BRIDGE UPDATE

The Peace Bridge Public Consensus Review Panel considering bridge and plaza options heard reports last week from its consultants on constraints on construction on the Canadian side and plaza options on the American side.

Roger Dorton, one of the Canadian consultants who joined the process late last year, told the panel that the Canadians like the present bridge more than any alternative and there are archaeological artifacts in the ground around the Canadian plaza which any new construction ideas or plans will have to take into account. Neither of those issues is new.

The archaeological issues have seemed bogus to me all along and still do. The sites and artifacts are certainly real, as are the feelings and interests of Native groups regarding them, but neither sites nor artifacts seemed to present much of a problem when the Canadians were building the huge duty free shop and truck-processing station over the same grounds just a few years ago, so it's hard to believe that's either is a real issue now. I bet that when the money for this project starts flowing ways will be found to deal with them, just like last time.

After the discussion of Canadian constraints, American-side consultants Seth Grady and Al Staufler went through the various plaza options. The idea was that they'd describe them and then the panel would ask questions, helping the consultants get rid of options that didn't work. But it took Staufler so long to get through the options that there wasn't any time for discussion.

The diagrams have all been seen and discussed previously. All the current plaza options but one are at or north of the current site. The exception is the southern site proposed long ago by architect Clint Brown and businessman Jack Cullen. Their SuperSpan design was important in opening up this public conversation, but the northern designs seem far more efficient in traffic flow and construction costs. Natalie Harder, director of regional planning for the Buffalo Niagara Partnership, made a motion to cut the Brown-Cullen plaza, but her motion was solidly rejected. So far as I know, hardly anyone on the Panel thinks the southern plaza is at all viable. One panel member said to me later, "I guess everybody wants to be nice to Jack and Clint because they got all this going." That's nice, but maybe a thank you and flourish might suffice at this point, given the paucity of time and the amount of work yet to be done. If they keep going over the same ground much more they'll reach Judge Fahey's deadline without being anywhere close to finishing the project.

Twice during his presentation Staufler said that the Episcopal Church Home would have to be moved. He said it as if it were on the order of moving the mound of salt road crews escrow in the summer for winter dispensation. Finally former state senator Joseph A. Tauriello blew up. He made an eloquent statement about the importance of that institution to the West Side, how many jobs it provided in a neighborhood where jobs were already scarce. The president of the Episcopal Church Home, Edward C. Weeks, said they very much wanted to stay in the neighborhood, but they were realists and if they had to move they would.

(A suggestion: If it does become necessary to move the Home, perhaps they could move into H.H. Richardson's Psychiatric Institute on Forest Avenue. Nobody's been able to figure out what to do with that architectural classic. It's got a lot of rooms, trees and grass, off-street parking, and no fumes or noise from trucks.)

The meeting once again made clear the foolishness of trying to isolate planning for the bridge and planning for the plaza (which was long the preference of the Public Bridge Authority and the Buffalo Niagara Partnership), and how much impact the plaza decision will have on Buffalo's West Side. At the start of this, Canadians were faulting Americans for holding up the bridge construction for mere aesthetic reasons. Now we know that it was just the other way around: the Canadian interest is merely aesthetic and the Americans are forcing the parties to focus on environmental, economic, and social issues.

What happens if the Panel recommends a signature bridge but the Canadians refuse to accept it and the Americans refuse to pour more money in the decaying old three-lane bridge presently in place? Extended deadlock, I guess, which would mean the southbound Canadian trucks wanting to cross here instead of Detroit would have to be routed through Queenston. That only adds about six miles to the trip, no big deal for the truckers and not necessarily the worst of fates for Buffalo residents worried about air and noise pollution near the plaza. Nobody's talking seriously about that possibility now, but it may very well be where things end up if the Americans opt for the more economical signature span and the Canadians refuse to give up the old bridge.

23 THE PBA'S EX-LAX INITIATIVE

It was hard to avoid the commercials on Buffalo radio and tv stations last week urging an immediate start of construction on the Peace Bridge. If you had your radio tuned to WBUF-FM 92:9 on Friday, for example, you would have heard the ads at 11:18 a.m., 4:49 p.m., and 6:18 p.m. Same thing on a half-dozen other local radio stations. If you had your tv set tuned to WKBW, WIVB or WGRZ, you'd have gotten the video equivalent of the same message in the late afternoon, early evening, mid-evening, and just before midnight. They ran day and night through much of the past week, again and again and again. They're scheduled to keep running right up to the day Judge Eugene Fahey responds to the Public Consensus Review Panel's final report.

When I first heard the radio ad I thought they were trying to sell a laxative. They kept talking about the need to *move*. Here's a full transcript with the words the speakers stressed italicized:

> Her: If we don't *move* we lose.
> Him: We've already lost over 16,000 jobs.
> Her: And 60 000 people, many of them our children and grandchildren, have left town for better opportunities.
> Him: That's why this community's top priority is new jobs and economic growth.
> Her: What's the answer? The growth of North American trade and tourism can create new jobs and economic growth.
> Him: But if we don't build a new peace bridge now we'll lose that chance to other border crossings
> Her: Along with the good paying jobs and new investments Buffalo and the Niagara region need so desperately.
> Him: It's time to move.
> Her: Business and labor leaders have teamed up to support building a new bridge *now*.
> Him: And build it with*out* taxpayer money.
> Her: After five years of study it's really quite simple.
> Him: Building a new bridge is about jobs and economic growth.
> Her: We *can* build a new peace bridge now. Without taxpayer

money.
Him: Say *yes* to starting peace bridge construction *now.*
Her: Because if we don't *move* we lose. [slight pause] Endorsed by labor and business who support jobs and economic growth.

None of the radio or tv spots identified the advertiser. Ads in which the sponsor hides are very rare. I called the radio and tv stations that aired the ads and none would tell me who bought the time. They were all apologetic about it; most said something on the order of, "We've like to tell you but we've been told not to." One sales manager said, "If this was a political ad, that information would have to be made public. But this isn't a political ad." I asked what it was, then. "A public information ad." I asked if the providers of information to the public had to identify themselves. "No," he said.

The ads were commissioned and paid for by the Buffalo and Fort Erie Public Bridge Authority, using money you and I paid to cross the Niagara River.

Here are some of the ways the ads try to deceive:

> The ads imply a direct connection between the date bridge construction begins and growth of jobs and tourism in the area. If there is any connection, it's their hurried and ill-considered plan that will cause the most harm, since their plan will take at least three years longer, cost far more for immediate construction and long-term maintenance, and cause far more disruption than any of the other proposed plans.
>
> Which labor and business organizations, exactly, endorsed the ad? The ad only says that they've got on their side the labor and business people who "support jobs and economic growth," implying that anyone who does not want to have a movement now is opposed to both.
>
> Of course taxpayer money is being used in the bridge construction project. Taxpayers pay the bridge tolls. Taxpayers pay for maintenance of I-190 and its access roads upon which the PBA's beloved truck traffic is totally dependent. Taxpayers pay for maintenance of Buffalo city streets utilized by the small portion of those trucks not headed for distant markets. Taxpayers paid

> for Front Park, which the PBA has gobbled up and destroyed for its current office and truck plaza. Taxpayers pay for the environmental effects of all those noxious truck emissions. Taxpayers are the major contributors to this project, however much the Bridge Authority board members want to deny it.
>
> When the PBA says "no taxpayer money" is being used, what they really mean is they're not taking US government grants for the bridge construction segment of their project (they will take it for the plaza construction segment five years from now). Thanks to Senator Daniel Patrick Moynihan, a huge amount of federal money is available to reduce the costs of this specific bridge construction. The PBA has refused to accept those funds because they come with a condition the PBA finds intolerable: they would have had to do a full environmental impact statement, which means they would have to consult with the community before they build. Rather than take advantage of that federal money set aside for this purpose from tax dollars we've already paid, they've elected to increase bridge tolls. That means they're making us pay *twice* for the same bridge.
>
> What, exactly, is so simple after five years of study, and to whom? Simpletons, maybe, which is what the ad takes the listeners and viewers to be. The PBA never studied the key issues; they've only issued documents about what they wanted to do. They excluded a signature span years ago and refused ever since to consider anything but its plan to twin a seventy-year-old bridge design using fifty-year-old construction technology. The only reason they're in so much legal trouble now is because they engaged in so many twists and turns to avoid the Environmental Impact Statement.

The specific purpose of the blizzard of ads on Buffalo radio and television stations is to undermine the work of the Public Consensus Review Panel and its American and Canadian technical consultants. It is also to put pressure on the mayor and Common Council so they will stop insisting that the PBA act in the public interest. It was only a few months ago that the PBA promised Judge Eugene Fahey that it would cooperate with the Review Panel's inquiry into bridge and plaza design. I guess it figures that spending a huge amount of money deceiving the public about the work of that Panel doesn't betray that promise.

This aggressive and misleading advertising campaign shows why the Public Consensus Review Panel must remain firm in its commitment to the citizens of this area. It shows why the mayor and Common Council must resist the PBA's recent burst of intense personal lobbying as well. And it shows the public how well served it was when Judge Fahey's refused to let the PBA bully its way through.

As we've pointed out in these pages before, if you call the PBA office an automated phone system tells you you've reached the "Peace Bridge Authority," which is not their name. Their name is the Buffalo and Fort Erie *Public* Bridge Authority. They don't like having the word in their title and they go to great lengths to avoid using it and even greater lengths to avoid the responsibility it implies. The Public Bridge Authority board meeting at which the decision was made to spend so much public money on this advertising campaign took place behind closed doors. The PBA doesn't think it has to serve or report to the public, it doesn't think the bridge belongs to the public. They're wrong, and no amount of misleading advertising is going to make them right.

24 JEFF BELT: "YOU HAVE TO BE AN OPTIMIST"

Jeff Belt recently returned to Buffalo after working for GM for 15 years. He is a mechanical engineer (Cornell) and has MBA from the Institute of Management Development in Lausanne, Switzerland. He became involved in the Peace Bridge question in December 1998 when he heard West Side community activist Bob Biniszkiewicz analyze the potential impact on Buffalo of a restored Front Park and a bridge plaza that was integrated with the city rather than imposed upon it. With Bill Banas, he heads the New Millennium Group's Peace Bridge Action Group and he is the New Millennium Group's representative to the Public Consensus Review Panel, the organization funded by the city, county, Wendt Foundation and Community Foundation to evaluate alternatives to the Public Bridge Authority's companion span design. I doubt that anyone knows as much as Belt about the financial, political, social, and technical aspects of the Peace Bridge issue. What follows are extracts from a conversation we had a few days ago about what he considered the key issues of the Peace Bridge affair.

STRATEGY

This is really a pretty great place for young families with children in terms of quality of life and the cost of living when you're starting out. But we aren't doing projects that are uniquely consistent with that as a strategic objective. Young families with children don't need convention centers, but they do need world-class children's hospitals. They don't need truck processing facilities in their neighborhoods but they do need parks. They need playing fields and stuff. If we can begin to resolve strategically where we're going to find our relevance in the coming twenty and fifty years, then we can begin to plan projects that are consistent.

We need to develop a regional strategy. What put Buffalo here was a grand infrastructure project called the Erie Canal and it was east-west trade. The NAFTA agreement has brought a tectonic shift in the direction of trade. Now it flows pan-North-American and soon it's going to be flowing pan-American, and it's going to be flowing north and south. We're in a very lucky position – just 90 miles beneath Toronto. I've talked with leaders in Toronto and they say that Toronto is beginning to recognize the importance of Buffalo. I think that we need to have a strategy that we're going to be relevant in the economy of North America and our relevance is going to be tied to our relationship to Toronto which is going to persist as a great North American city, really a world-class North American city. If we can begin to frame our strategic thinking like that, then we can do projects that are consistent with that strategy.

This is new ground that we're cutting. I'm very happy to be a member of the Public Consensus Review Panel and the New Millennium Group is absolutely thrilled that there *is* a panel. This is what we were asking for. The New Millennium Group suggested that I sit on the panel only because I have done the most research on the project of anybody and I have been able to commit the most amount of time, although there are others who have committed a heck of a lot of time.

I guess I just came to the forefront as spokesperson because I decided in February that I was going to commit nine months to the project. I said I'm just going to clear my calendar and I'm going to make this my full-time activity for nine months because this is my home, my home town, it's something that I think is strategically imperative. I think if we screw up the Peace Bridge that we're doomed, fundamentally. So it's worth spending nine months on it. And it's now been 12 months and I'm still committed to it because I see a light at the end of the tunnel.

DISRUPTION

The PBA and its friends say all this delay is causing us major problems, serious harm. Is there, as they insist, a real need to start construction right away?

There's a real need to improve the capacity of the Peace Bridge Gateway. I spent six years in the Tonawanda engine plant and what I worked on there was capital utilization – finding out where the bottleneck is in the sequence of part production and resolving the bottleneck. The bottleneck at the Peace Bridge is the secondary inspection facility on the US side. That problem will not be resolved until 2010 or 2011 in the Public Bridge Authority's plan of record.

Their own data demonstrates that we have enough lanes across the river to handle all the traffic for several years into the future if there were free flow at both ends. But you go by the bridge and there's always traffic backed up coming into the United States. Why? Because there's a bottleneck there and it's in the plaza. So if we go forward with their companion bridge project we're really going to do nothing but exacerbate the bottleneck. For two reasons. One, during the construction project for the companion bridge, the steel, the material, and the equipment for that construction project will be staged in the plaza and in the park, so it's just going to make a mess for people to drive around. During the connecting road project there will be detours through the Scajaquada and Delaware Park. Three thousand trucks a day through Delaware Park and down the Kensington. That will be a nightmare for commuters and people who use the park. So that's going to be very disruptive.

And then the reconstruction of the existing bridge will be very disruptive. That, again, will necessitate the staging of equipment and material in the US plaza. Then there will be the incremental reconstruction of the US plaza while six lanes of bridge of bridge capacity are funneling through it.

What we'd advocate is that the city and the Bridge Authority work together to acquire land for a new plaza, a new US plaza location, so that an all-new bridge and all-new plaza can be built in one go. Don't disrupt the existing infrastructure at all. When the new infrastructure is ready to go, just shift the traffic over with a ribbon-cutting ceremony and we're off and running.

That's the way Tampa did it, that's the way Charleston is doing it, that's the way Alton, Illinois, did it. Every other city does it that way.

That's what Bruno Freschi was always saying: there's no need to tie up the city for all that time.

He was right. And we're talking ten years with their plan. It only took two years of construction chaos to wipe out Main Street's retail when the Metrorail went in. If Metrorail could have gone in in the wink of an eye, Main Street probably would have been served by Metrorail. You'd have had trolleys and pedestrians passing down along those stores. But the stores couldn't sustain two years of construction chaos.

Our question is: can our economy, which is really sustained by the Toronto economy, sustain ten years of disconnect from that economy that drives us?

TIME

If we do a full environmental impact study, I think we're going to be able to examine the alternatives well enough to recognize that they really are doable, that we really can implement them, and that we can find the money for it. The city of Boston is spending something on the order of $100 million a week of federal money on infrastructure projects. Why? Because they asked for it. If we ask for it, I think we can find the money. Particularly if we have a strategy and a good plan that is consistent with our overarching regional strategy.

Another thing – and this is kind of a minor argument but it could become a major argument – if Toronto is successful in their effort to attract the Olympics in the year 2008, and if we go with the all new infrastructural approach, we could celebrate our participation in Toronto's Olympics with a brand new bridge and a brand new plaza and a park system. And it would all be up and running by 2008.

It could happen in seven years?

Absolutely. The Bridge Authority engineers are raising the specter of interminable delay, but the city of Tampa required five years from the time that the ocean freighter wiped out one of their twin skyway bridges until they were able to open the new Sunshine Skyway Bridge. They had no preparatory work done; they had no reason to believe that they were going to have to build a new bridge there at all. It all happened in the wink of an eye on a foggy morning. Five years later they're up and running.

And that's a huge bridge.

That's an eight-mile-long bridge. It's very doable. With the amount of work that the Bridge Authority has already done in terms of engineering and environmental impact assessments, I see no reason – particularly if they use the Public Consensus Review Panel as their body public to bounce ideas off of and get them through the EIS in an express manner – they shouldn't be able to get through that in eighteen months or so. Maybe two years. They can get the design work done in parallel with the EIS. Building a new bridge and plaza as one unified system could be completed in three or four subsequent years.

MONEY

I talked to three people in the last two weeks who heard that the PBA has been burying the Review Panel consultants with a huge amount of financial information in an attempt to convince them that the only way to go is to build the second three-lane bridge into the current plaza, that there's no money for anything else.

I wish the consultants would put more financial information on the table. They're clearly leaning toward the thesis that anything other than the PBA's plan of record is too hard. So far, however, they have not described in dollars and cents what they deem "too hard" to be. What we've got to do is look beyond the engineers' fear of moving a water main or figuring out how to fit a thruway on-ramp alongside a railway track. All of the physical features that the engineers are fearful of were put there by humans and my thesis is that if they were put there by humans they can be rearranged by humans.

All the designs I've seen for a signature bridge seem significantly cheaper, both for construction and maintenance. But the PBA and its supporters, like Andrew Rudnick, continue to insist that the twin span is cheaper. Why?

I think there was good reason to believe that a signature span would more expensive before the quotations came in for the companion bridge. I think that was a big shock to everybody. Their original estimate way back when was about $65 million for the whole project: $54 for the companion and $11 I guess for redecking the old bridge. Back then one could probably say "A signature bridge will be more expensive because all the estimates for a signature bridge have been in the range of $75 to $90 million construction costs."

But the companion bridge project as it stands right now is $110. And if they replace the Parker truss, we're probably looking at a total of $140 million for the new companion bridge and the repaired old Peace Bridge with a new through arch so that it looks more like the new companion bridge.

And it may be even more than that. During last week's meeting of the Greater Niagara Regional Transportation Council, Jake Lamb stated that the cost of the companion bridge had risen by yet another $16 million. That means the 3-lane companion bridge alone will cost about $106 million! Their total would be $156 million.

Enough for two signature bridges.

One signature bridge is all we need.

The thing is, there is no good way to estimate how much it would cost to build a companion bridge because nobody's built a bridge like that since 1927. Signature bridges are built all the time, we know what those cost. And there are lots of people who know how to build them. But nobody knows how to build a duplicate of the Peace Bridge because it hasn't been done in 75 years. So their cost estimates are probably still low.

And what about maintenance?

It's generally accepted that the maintenance of a concrete bridge is going to run one to two percent of the capital costs of the bridge. That's one to two percent of $70 to $90 million. And the maintenance on a steel bridge is going to run between three and four percent of the capital costs. So that's three to four percent of $140 million. So you're looking at maybe four times the maintenance costs.

So if we take the worst maintenance and cost estimates for the concrete bridge and the best maintenance estimate for the steel bridge that still comes to $1.8 million a year for the signature bridge and $4.23 million a year for the steel bridge, a difference of $2.43 million every single year. That's a lot of money over time.

A lot of money.

THE PLAZA

Whichever bridge is built, there's still the plaza to be dealt with. If the plaza is moved out of Front Park you're looking at land acquisition cost. The way the engineers have established the alternative plazas, land acquisition will probably run to the order of $50 million. So they simply take that $50 million and they tack it right on the whole program and they say that's the deal breaker.

I have a couple of reactions to that. Number one, if we build an all-new plaza we can make the bottleneck operation, which is the commercial vehicle inspection area, larger and more efficient, so we can actually resolve the bottleneck at the border. If you keep the plaza cramped where it is you're not really resolving the bottleneck. So for 25% more money we might be able to buy 100% more capacity. If we begin to look at it in those terms then the cost-benefit is more palatable.

The other thing that I've pointed out is, we shouldn't assume that the land that the plaza already occupies is worthless.

In the last meeting that I went to I took some paperwork with me from a report on Duluth, Minnesota. They're in the process of a $250-million highway project up there that includes Interstate 35 across their waterfront. They're going to spend $10 million capping over a section of Interstate 35 to enable waterfront access downtown. That's going to give them three acres of waterfront park that they didn't previously have. So on that basis I say, a waterfront park in Duluth, Minnesota, is worth $3.3 million an acre. If our waterfront is as valuable as Duluth's waterfront, then the land that the Public Bridge Authority plaza is occupying is worth $50 million, and we should recognize that.

Maybe our waterfront is worth more than Duluth's waterfront because the land that the Bridge Authority plaza occupies is our Presidio. It's the only elevated waterfront land on our lakeshore. It's really unique in the world. I can't think of a place where you can stand at a high point of land looking out over a majestic river and expansive body of water onto a foreign country. The closest I can come to it is Istanbul, Turkey, and there you look on to the other half of Turkey. In Istanbul they turned that Presidio into the signature park for the entire country. It's beautiful. They've got a beautiful suspension bridge across the straits. It's done properly. They recognize that what they've got is unique real estate. I think that we need to do the same, and put a dollar value on that Presidio that the Bridge Authority currently occupies.

If you get a functional Presidio – right now it's a useless area – there's going to be an immediate increase in value for all the surrounding real estate, so the benefit is not just a resurrected Olmsted park, but increased value for everything contiguous, that whole section of town. That has happened everywhere that kind of restoration has occurred.

You don't even have to make those arguments on the basis of conjecture or comparables out of town. You can merely look into our own past. When that park was contiguous, when it was functional, when we could go to that park and sit on a bench and look out over the river and the lake and Fort Erie and enjoy the view without the scent of diesel soot wafting by our nostrils, that was the most popular park in the city of Buffalo. That was what Pam Earl's whole study of Front Park and the opportunity that the Peace Bridge expansion project presents to the city is all about.

WHERE DO THE MONEY AND JOBS GO?

Their companion bridge would be built with steel that would be manufactured in Indiana and Sault St. Marie, Ontario. The steel would be fabricated in Wisconsin, then brought here and erected by B.O.T. Construction Company out of Oakville, Ontario. We don't really see where there's much in it for Buffalo in that picture. All the money is being spent out of state.

We've said all along that if we could build a concrete bridge we could make the concrete sections right here in Buffalo or in Port Colborne along the waterfront and all the bridge would be local content, local workers. But we never really knew how make that argument strong until a few weeks ago when we received a report from the Ohio Department of Transportation. They're building a signature bridge in Toledo, Ohio, over the Maumee river. One reason that they selected a concrete cable-stayed bridge instead of a steel bridge is because of the local content.

The numbers are as follows. For a $40 million steel bridge with a local construction company building it, they determined there would be $7 million of the $40 million spent for local erecting services. The rest of the money that would be spent on steel from out of state because there's no real structural steel industry left in Ohio.

For an equivalent $40 million bridge made of concrete, they determined that $38.4 million of that money would be spent in northwestern Ohio. So the cable stayed bridge won by a huge margin on local content.

A $40 million dollar bridge project in Toledo: if it's steel, $7 million into the local economy; if it's concrete, $38.4 million into the local economy. Five times as much.

I think that the Toledo Duluth stories are very relevant to our situation here in Buffalo. Both Toledo and Duluth are industrial cities on the Great Lakes and like Buffalo they've been in many ways left high and dry by trade routes that moved elsewhere. They both have lots of empty grain elevators on the waterfront. But they're high quality of life, solid practical cities, and they don't make decisions on the basis of willy-nilly imagery or whatever else. So we're not talking about comparing Buffalo to Boston and San Francisco. We're saying, look at what the people in Duluth and Toledo are doing. They're real people. They're like us.

I don't feel very confident that the construction workers in the PBA's companion bridge plan are going to be local either. B.O.T. is in Oakville, Ontario, and that's only an hour commute away. Construction guys don't mind jumping in the pickup truck and driving down the QEW for a day's work. They may all be workers in Oakville and there could be nothing in it for workers in western New York or Fort Erie. I think that needs to be part of the consideration and I advocated as much at the Public Consensus Review Panel meeting last week.

We have a box on our chart that says "economic development" and I said "Look, I think we need to be specific about this. We need to look at local labor content, local material content, we need to look at the disruption created by the construction project. And user cost as well. If we are going to tie up the bridge for ten years with construction, all the people sitting in traffic are sacrificing their time , which is worth money too. That needs to be evaluated in some way."

THE ENVIRONMENTAL IMPACT STUDY

I assume these things would all be addressed in a full EIS, which the PBA has done everything it could to avoid.

Yes, they would be. The EIS should not be hard to do because the most difficult thing in an EIS, from what I've been told, is scoping the project – deciding what the range of alternatives are, what you can do, how far you can go one way and the other and what land you can use and what you can impact, how big anything is going to be, is it going to be two lanes, three lanes, four lanes, et cetera. The second most difficult job is identifying the affected public. I think that we already have scoped the project in the Public Consensus Review Panel. I also think that we've identified the affected public on the US side because the panel includes something like 30 members representing all kinds of different organizations and communities in Buffalo and Western New York. What we need to do is add some Canadian representation – they're not sitting at the table but those representatives have been identified – and we've got it all pulled together.

So now it's just a question of doing the engineering, gathering the data, presenting it, discussing it, and deciding what the best project is in terms of economic and social benefits and minimizing the negative environmental and social externalities. I think we've done the hard parts.

It's not starting from scratch.

By no means. A lot of the data has been gathered and the people who are representing the communities where the bridge lands are all saying that everybody up there would love to make way for a better plaza. If it requires the acquisition of property that houses are sitting on or the Episcopal Church Home is sitting on, everybody is willing to be reasonable and move, and a lot of people want to move.

PUBLIC MONEY AND PRIVATE DEBT

It's my understanding that there are several sources of federal money for various parts of projects of this type available that the PBA has not availed itself of. Is that consonant with what you know?

There could be money from ISTEA – the Intermodal Surface Transportation Efficiency Act – and T-21, which is a five-year extension of that. Senator Moynihan put a clause in one of those documents to make $125 million available for bridges at international border stations. Being that he is the senator from New York and that we only need one bridge at our international border, it's specifically for the Peace Bridge project.

The Bridge Authority doesn't want to use other money because they have a lot of money. They want to get themselves decidedly back into debt. That insures their existence. That's why we need to have a reform measure to validate what it is that they do well – which is managing the gateway and maintaining the bridge, and not focus on what anybody can do. I can pay off debt. All you have to do is whip out a coupon and lick a stamp twice a month. That doesn't take a genius.

If they availed themselves of these funds wouldn't some of these financial issues that they're raising disappear?

Yes. Furthermore, there's really no telling what's too expensive. Their revenue is exploding and it's increasing. It's been outrunning all of their projections. In 1998 their total revenue was about $25 million. They're big, a very big operation. And their surplus is about $10 million a year. They imagine repainting the steel bridges every 15 years, but with a surplus of $10 million a year why bother repainting? You could just knock them down and put up new ones every ten years.

In the early '70s they came close to paying off all of their debt and as a consequence could have been vulnerable for dissolution. Every time they've come close to paying off their debt they've initiated new capital spending programs and floated additional debt to guarantee their existence.

I think that there should be a Public Bridge Authority charter revision effort that would guarantee the existence of the Public Bridge Authority in perpetuity, but guarantee their existence with a view toward them maintaining the bridge, and maybe the surrounding parkland too. Give them a portfolio of things that have to be done. There's a lot of expertise in managing and maintaining an international gateway. Recognize that expertise in perpetuity and take the guillotine of declining indenture off their neck, as it were. If somebody on the US side would come in and say, "Look, we're going to charter the Bridge Authority in perpetuity but we're going to state you don't have to be in debt. Paying your debt is not part of your mission per se. We're going to recognize your expertise in your mission." That could remove a lot of the weirdness that goes on. Because I do think that they feel threatened. They've been in a situation where they could be out of debt since 1990, so it's a long time to be out there naked.

WHAT'S GOING TO HAPPEN?

The Review Panel has a few more meetings with its consultants and then it issues its final report. What's that report going to say and what happens next?

If at the end of it all we say, "Go ahead and build a twin span in the plaza where it is, what you said you wanted to do to begin with," we will look like a bunch of idiots. We just held up this project for a year for what? And not only that, we'd then have to explain to the rest of the world why after a year of thorough study we decided that building a duplicate of a 1927 bridge was the best we could do when the rest of the world has figured out how to span rivers like the Niagara. We'd look like idiots.

I think that the Review Panel is going to say that there are reasonable alternatives to the Bridge Authority's plan of record. And on that basis I think that Judge Fahey is going to say, if there are reasonable alternatives they really must be examined thoroughly through the formula of the environmental impact study. As a consequence he'll mandate that the Bridge Authority perform a full environmental impact study. And I think that the Bridge Authority will go along with that. They'll be able to augment the capacity of the commercial inspection facility a little bit by adding a scale and an inspection booth at primary, and they may come to the community and ask for some relief in parking trucks on the tennis courts or something of Front Park while we sort this out. And if everybody's working in good faith we'll say, "Yes, do it, to get us through the rough patch here."

Do you think there is any possibility at this point that the PBA will renege on its promise to accept the Review Panel's recommendation if the Panel recommends a northern plaza, a plaza design that gets out of the Park and away from it, and a signature six-lane span?

I suppose there's that possibility. I think that the biggest concern is the city. If the city holds firm on the easements, the PBA can't renege. They can't build a bridge to the middle of the Black Rock Canal. They'd have to pick up quite a head of steam to leap that final 120 feet.

Do you have any concern that the city might weaken?

Yes. I'm very much concerned. It's difficult. The city has lots of other problems that they'd like to focus on and they don't want to have to focus on the Peace Bridge forever. The Peace Bridge is a problem that shouldn't be a problem. There's a developer up there [the PBA], a public entity, that should be responsive to the public's desires who is self-funded and fully resourced, fully capable of doing this all by themselves by simply going out and buying property and assembling it and building what they need to build. The city does not necessarily have to be involved in this and I'm sure that they'd like to turn their attention to problems that are more specific to the city agenda. I think that the biggest reason why the city remains steadfast is because we gathered signatures of close to 11,000 people during our petitioning drive during the late spring and early summer last year and most of those people who signed the petition in favor of an all-new bridge and an all-new plaza are city residents. Those are people who vote and I think that's really something that has to weigh on the city's mind because they want to do what the constituents want. It's their mandate.

And Judge Fahey?

There is an agreement between the city and the Public Bridge Authority to respect the decision of the Panel, but that agreement is not a constitutional amendment, so no matter what happens, Judge Fahey still has the hammer in his hand. The law says what you have to do in terms of environmental review. If he decides that the environmental review was adequate, there's really nothing anybody can do about it. The Bridge Authority will just go ahead. If he decides that it wasn't adequate, there's really nothing that the Bridge Authority can do about it. They have to go back and do a nonsegmented EIS. So he's the most powerful player in all of this.

I have faith that Judge Fahey is going to say "You can't do a $200-million infrastructure project in the middle of an urban community on a segmented basis. We've got to look at this holistically." If that was not the letter of the law, I think it was the intent and the spirit of the law. I think that's what he's going to say, I hope that's what he's going to say. And with that in my heart, I put my full effort into the Panel because I think that the Panel is an important group to have assembled to help the Public Bridge Authority, as an applicant, get through the EIS process.

OPTIMISM AT THE BORDER

Right now I think that the government in Fort Erie is very frustrated with the citizens of Buffalo and I think that they put my name on top of the list of most frustrating people, and maybe your name right there next to mine. Mayor Redekop told me that he wished that we would just get our act together. I think that we are getting our act together. And I think that once we do get our act together he's going to be true to his word and he's going to work with us and we're going to have a happy result of all this.

This project could be seminal. It literally could signify our ability to turn ourselves around. There's nothing that's inherently wrong with Western New York or Southern Ontario. We just haven't figured out what we want to be in the new millennium yet. We haven't reinvented ourselves since the steel plant was built in Lackawanna in 1901. And now is our chance. As a binational region – that's the way I think we should be reinventing ourselves.

I think we're going to have some good results. I really do believe that this is the beginning of a turnaround and we just need to draw some of our strategy into focus and once the strategy is in focus then the good people of Buffalo are going to be able to get a lot of things done. I'm an optimist. You have to be an optimist.

25 THE ENGINEERS ARE STEALING THE TRAIN

The Public Consensus Review Panel, the group set up by the city, county, and two local foundations to consider Peace bridge and plaza options, heard a preliminary report from its Canadian and American consulting engineers at a 4 ½-hour meeting February 17. It seemed less like a consultant's report than one of those Fidel Castro talkathons in Havana.

POWER IN POWERPOINT

The engineers talked for all but 25 minutes of the 4 ½ hour meeting. They showed 49 PowerPoint slides and talked in detail about each. They reported at length and showed drawings and paintings of 3-lane companion and 6-lane cable-stayed bridges, as they had been asked to do. They also reported at length on and showed drawings and paintings of a hugely expensive six-lane suspension bridge no one had asked them to look at and which everyone agrees never was or will be an option for this project. They told the Panel what they thought about social policy, New York State bonding law, community development and all sorts of things for which they apparently had not the slightest competence and only the most marginal and stale data, such as 10-year-old census and commerce reports and estimates of apartment occupancy based on drive-by sightings.

They passed out a 48-page booklet that contained printouts of every PowerPoint slide but the most important one, the page with the cost estimates for the bridges. Some charts had egregious multiplication and conversion errors (such as multiplying 38.47 by 200 and getting 1,480 instead of 7,694 or saying 746 meters equals 656 feet instead of 199 feet) that suggested a document assembled in haste. Everything was printed in huge letters, like those newspapers for the nearly-blind, and at the bottom of every page they printed in equally large letters the names of the two engineering firms hired by the Americans on the left, the two firms hired by the Canadians on the right, and in the middle, in much larger letters, the phrase "Bi-

National Team," as if this were some kind of jock convention.

WHAT HAPPENED TO BRUNO'S BRIDGE?

The only reason we're not suffering construction of an ugly companion span now is because Buffalo architect Clinton Brown and businessman Jack Cullen looked at the PBA's companion bridge plan and said we could and should do better. They identified the need, but the thing didn't take fire until Bruno Freschi, then dean of U.B.'s School of Architecture, and San Francisco bridge designer T.Y. Lin, came up with the idea of a single-pylon cabled-stayed segmental concrete bridge. It would, they said, not only be a breathtaking landmark, but it could be erected faster and less expensively than the PBA's steel companion bridge. Later, famed bridge-builder Eugene Figg visited and said the same thing: the community could have with far less cost and less disruption a gorgeous bridge using 21st century technology rather than a stodgy bridge using early 20th century technology.

The Review Panel's American and Canadian consulting engineers were impressed by none of that. They drew pictures of and did preliminary calculations for a straight cable-stayed bridge with two pylons, which they said was too complicated and too expensive to build. To my knowledge, no one had previously suggested a two-pylon cable-stayed bridge. That was their hypothetical design and they found it wanting. Who cares?

They did no drawings or calculations for anything like the curved single pylon cable-stayed bridge suggested by Bruno Freschi and T.Y. Lin. The reason:"Curved," they said, "has complications we don't need." Whose needs were they talking about – theirs or the community's? They didn't say and no one on the Panel asked.

They said, "The only thing going for it is some people like this idea of a curved structure. But there's nothing else going for it." I guess you could say that about any beautiful design when there's an uglier design available to take its place. If the only function of a bridge is to get trucks from one side of a river to the other, they're right. Most of us think there are other factors that deserve consideration.

Yes, a curved bridge requires more calculating and a greater variety of concrete forms than a straight bridge, and that takes time and money. They didn't say how much more expensive or how much more design work would have to be done. Are they concerned about a week or about six months, about $1 million or $20 million?

All they said was they weren't interested in it and the fact that people here liked it was of no moment whatsoever.

So the engineers, without any direction from the Panel, have flushed from consideration the single design more people prefer than any other. That's not consultation; that's hubris.

NUMBERS GAMES

The engineers considered four bridge designs. There was a spread of $15 or $20 million in three of the four bridge construction cost estimates; in each case, the Canadian team provided the high number, the American team the low number. The only bridge cost they agreed upon was the companion span advocated by the Public Bridge Authority; that's because they accepted the PBA's numbers.

The engineers also accepted the PBA's timelines, which I found stunning. For several years the PBA said its bridge-plaza plan would take ten full years. When several internationally famous bridge experts said they could design a signature bridge and plaza, have the required environmental impact study, and cut the ribbon in seven years, the PBA at first went into denial and then did a presentation for the Panel in which they subtracted three years from their timeline. There was no explanation how they shrunk their project by thirty percent, they just did it. It was on the order of "Let there be light." Well, maybe that's a little grand. Say it's on the order of that famous Dali painting with the melting watches.

But the engineers are economical: they didn't let those years disappear, they moved them elsewhere. They *added* two or three years to all the alternative plans. In their report, the signature design that everyone else says will take no more than seven years will take nine or ten years. Why? They project a three-stage design-bid-construct sequence with two years in which everything stops for the bidding. New York State construction projects are done that way, but the PBA is not subject to that law. Standard practice in the construction industry now is design and bid. There was no need to add those two-year extensions to the alternative designs. When pressed for justification the engineers had none. They just did it, just in case.

They included in all the alternative plans two years for an environmental impact study, but they did not include any time for any studies for the PBA. That seems hardly realistic, given the cases on exactly that issue now on hold in Judge Eugene Fahey's court. Judge Fahey has made it clear that he has no intention of letting the

PBA engage in a $200-million construction project without the consent of the community. The engineers ignored all of that.

The result of all this adding years to alternatives and subtracting years from the PBA plan was a prediction by the engineers that a signature span or any other design couldn't break ground for 51/2 years, and the PBA could break ground in moments. I think of my Italian buddy Teresa who says, at moments like this, "Sure, and if my grandfather had wheels he'd be a wheelbarrow."

They estimated that the cheapest plaza would be merely modifying the present one, which is no surprise since doing nothing is almost always cheaper than doing something. They estimated $50 million for acquiring property for a new plaza north of the current site, which thereby made all alternatives to the single plan the PBA is willing to fund much more expensive. Several Panel members asked them to factor in the value to the community of a restored Front Park. They refused, saying it was too difficult. Jeff Belt said they might look at what comparable cities have spent recently developing or restoring parks. The engineers continued to demur, saying it was too difficult so they just wouldn't include those numbers but the community and county could do that if it wished. Ed Cosgrove, Judy Fisher, and Mark Mitskovski were eloquent about the value of the restored park system to the community, and all said it was unreasonable to ignore it. The engineers just wouldn't go there. They were willing to estimate the cost of buying property for a new plaza but adamantly refused to look at the benefits to the city of a restored Olmsted park system.

They made other curious choices with their numbers. The included the cost of design in all the alternative bridges but did not include in the total cost for the companion span the $10 million the PBA has already spent on design. By that logic, the PBA could erect its bridge then look around and say, "Our bridge is free." Well, no, it isn't, and that $10 million is part of their costs. Neither did the engineers include in their estimates for the PBA bridge the cost of redecking the old bridge, doing all the repairs the PBA says it doesn't need but it does (the 1927 bridge is decrepit), and replacing the Parker truss with an arch so it matches the new companion bridge. They put the cost of the companion span at $110 million, the PBA's numbers. I don't know anyone else who believes those numbers. They didn't include any projections for maintenance, nor did they note that the annual cost of maintaining an anachronistic steel bridge is several times the cost of maintaining a modern concrete bridge.

Where they subtracted costs from the companion span, they added them to the cable-stayed bridge. All the outside consultants have put it at $80 or $90 million, they put it at $125-140 million. I asked one of the engineers why their numbers were 50% higher than everyone else's. Well, he said, he wasn't sure. Maybe they took other things into account. What things? Hard to tell. They'd have to look at it.

I thought that's what they'd been doing the past four months, looking at it.

When Panel members questioned their numbers they responded, "Well, these aren't our final numbers, they're just our tentative numbers." Then why was everybody there? If the numbers don't mean anything why were they reporting them and why did everyone spend four and a half hours of a Thursday afternoon listening to them and discussing them? They promised accurate numbers the following week, at the Panel's February 23 meeting.

Mostly, it seemed, they looked at the deleterious effects of things. Buffalo Commissioner of Human Resources Joe Ryan blew up at that. He asked why they didn't look at the advantages to the city of getting some people into better housing, having a restored park, a more harmonious plaza, and so forth. The engineer's answer went on for a long time but I think the sense of it was that it was easier to calculate what things cost than it was to estimate the good that would be result from the work.

The effect of all of this is, the PBA's juggled and worked-over numbers look like a great bargain and anything else looks like a great burden. It was an astonishing performance, almost surreal, and I think it left some members of the Panel not just exhausted but also stunned. They had thought the consultants were working for them, now the consultants were setting the limits, telling them what they could and couldn't have, what they should and shouldn't want.

THE PBA

The role of the PBA in all of this remains weird. Even though they agreed to take part in the review process several months ago in an attempt to get Judge Eugene Fahey to go away, they never sat at the table. At most meetings, Earl Rowe (the PBA's general manager/-corporate services) or Stephen F. Mayer (general manager/-operations) sits on the periphery. On rare occasions both are there. They rarely take part in the conversations but sometimes put on full-scale performances, as last month when they spent an hour detailing plaza options they were willing to consider if someone else paid for them. This time, Earl Rowe stood up to announce that the PBA had no comments now but would prepare a formal response for the February 23 meeting.

No one ever challenges the PBA directly at these meetings, not even when they're doing absurd things, like the time they disappeared those three years from their timeline. "We were told to treat them with kid gloves," one member of the Panel told me. He said there had been a lot of worry that if Rowe or Mayer were confronted directly they would stop participating in the process. That strikes me as unlikely, given their anxiety to impress Judge Fahey with their sincerity and good-citizenship, but the fact remains that their utterances are hardly ever challenged and the parts of the engineers' report that were obviously put in there to represent the PBA were allowed to slide by with hardly a critical glance. It will be interesting to see if the nice-nice-hands-off policy continues now that the PBA has gone public with its huge radio and tv anti-Review Panel ad campaign.

THE *BUFFALO NEWS*

The *Buffalo News* headlined its article the next day, "Early figures on bridge top out at $250 million/Authority's companion span is least costly, at less than $110 million." That article appeared on the last page of section C. Few people would read it, but nearly everyone who skimmed the paper would read the headline, which was probably the intention. You had to read carefully and deep to figure out that the $250 million bridge was not the signature span, and you wouldn't find out at all from the article that the PBA bridge would in fact cost a good deal more than $110 million.

"The cheapest alternative is that proposed by the Peace Bridge Authority," wrote *News* reporter Tom Ernst. Sure it is – but only if you assume it has no maintenance costs and the old bridge needs no rehabilitation and the alternative bridges cost half again as much as experts in the field say they'll cost. "The most costly would be $356 million," he wrote. He didn't say that this most costly alternative was a design suggested by no one on the Panel or in the community, that it had been thought up and tossed in at the last minute by the engineers.

One item in the article was worthy of note: "The authority in

1988 selected a companion design to add three lanes of traffic." I didn't know that the PBA had selected the companion span twelve years ago, nearly a decade before the design competition sponsored in part by the *Buffalo News*. Did the *News* know when it put the public through that exercise that the PBA had locked it all down in 1988?

WHAT NOW?

The February 22 meeting at which the engineers were supposed to submit more reliable numbers and the PBA and the Episcopal Church Home were going to make detailed responses to the February 17 presentation was canceled without explanation early in the week. I would guess the meeting was canceled because the engineers weren't ready with reliable numbers and didn't want to be embarrassed again or members of the Panel decided it was time to take the steering wheel back from the engineers and were trying to figure out how to do it. These are some of the still-unanswered questions:

> What are the real costs of the various bridge and plaza designs?
> Will the engineers start answering the questions they're being paid to answer or will they continue dancing to the PBA's tunes?
> Will the huge amount of money the PBA is spending in its anti-Consensus-Panel-start-construction-now advertising campaign have any effect on the panelists or politicians?
> If the engineers come in with a lousy decision will the Panel reject it and send something else on to the PBA?

We'll start getting the answers out at six p.m. on Tuesday, February 29, when the engineers present their draft bridge system recommendation in a public meeting at the Channel 17 studios. They told me that their report will consist of one bridge and plaza system, no alternatives. (That presentation will be broadcast from 6-8 p.m. on WNEQ channel 23 and WNED AM 970. It will be rebroadcast on both Sunday March 5, from 3-5 p.m.)

The key public meeting will be the following Tuesday and Wednesday, March 7 and March 8, beginning at 5 pm both evenings. At that meeting the public will be able to comment and ask questions, presumably of both the engineers and the panelists. (Those sessions will be broadcast live on WNED-AM.).

Soon after that, the Public Consensus Review Panel will makes its recommendation to the Public Bridge Authority. That is when we'll learn the answer to the key question: Is the Buffalo and Fort Erie Public Bridge Authority capable of respecting any opinions or needs other than its own?

26 THE ENGINEERS SPEAK: BUYING THE PBA PARTY LINE

The American and Canadian engineering firms hired by the Public Consensus Review Panel and the Buffalo and Fort Erie Public Bridge Authority presented their preliminary recommendation for a bridge and plaza system at a public meeting Tuesday night at the WNED studios in downtown Buffalo. (That two-hour presentation will be rebroadcast on WNED/17 on Sunday at 3:00 p.m.)

The good news is that they recommended that the bridge operation move out of its present location into a new plaza slightly to the north, which would return to the community Front Park and Fort Porter.

The further good news is that the some of the engineers' numbers are looking more realistic. Two weeks ago they said the PBA would get its bridge and plaza done by 2007; now they say it's 2010. The engineers are also saying that the PBA's twin-arch design at the new plaza will cost $10 million more than a cable-stayed bridge at another new plaza. Which is to say, they've decided that the PBA's plan is neither faster nor cheaper.

The rest of the news is dreadful.

– After considering several bridge designs of their own devising, the engineers decided that the PBA's companion span is the best way to go, modified so the truss on the old bridge matches the truss on the new bridge. That is, they recommend a twin span. They said that "a twin arch will provide signature-quality aesthetics," which tells you something about their sense of signature, quality and aesthetics.

– They dismissed without any serious consideration at all the curved concrete cable-stayed bridge proposed two years ago by Bruno Freschi and T.Y. Lin. They never thought it worthwhile to talk with Freschi or Lin or to examine any of their detailed supporting documents. They refused to include in their report the single bridge design the PBA feared most of all.

– They recommend an environmental impact study for a new plaza (which at this point couldn't possibly be avoided), but not for the bridge, thereby endorsing the PBA's fiction that the bridge and the plaza are totally separate projects. That fiction has only one function: to let the PBA avoid submitting its plans to the close scrutiny and community involvement required by US law.

It is, all in all, a very detailed and very well presented and very dreadful piece of work. The Public Consensus Review Panel, which paid for part of it, should be enraged. The Buffalo and Fort Erie Public Bridge Authority, which I suspect paid for a larger part of it, is no doubt delighted.

From 5:00-7:00 p.m. on Tuesday and Wednesday, March 7 and 8, the Panel will listen to public responses to those draft recommendations. On March 10, the consultants will deliver to the Panel their revised recommendation. On March 14, the Panel will ratify or reject that recommendation and on the next day it will send its own final recommendation to the Public Bridge Authority.

THE BEST PUBLIC OPINION THAT MONEY CAN BUY

In the weeks just before Tuesday's consultants' report, the Public Bridge Authority spent a huge amount of money trying to crank up public hysteria about the need to start construction immediately. The campaign was managed by the Buffalo firm of Collins & Co, which drenched the city's TV and radio stations with commercials implying that the area would suffer great economic harm if any more time were spent analyzing data or engaging in an Environmental Impact Study or even thinking about what might be best for the community. Last week, the PBA helped organize a meeting in Fort Erie at which only friendly speakers were invited to perform. The day after the WNED presentation, they got the Buffalo Building and Construction Trades Council to put on a dog-and-pony show focusing on the need for jobs now, jobs now, jobs now.

I remember a fascinating interview Bill Moyers did in 1980 with Fritz Hippler, who had been in charge of film production for Adolf Hitler. Fritz Hippler was responsible for what is probably the most notorious hate film of all time, *Der Ewige Jew* (*The Eternal Jew*, 1940). Hippler looked back on his work with more than a little pride. The great secret of propaganda, he told Moyers, is "simplify and repeat, simplify and repeat." All the great propagandists since have learned that lesson. Simplify and repeat. Get rid of the complicating details and say it again and again and again and after a while, if you've done it well enough, most people will forget there ever were complicating details or other ways to think about things.

The purpose of the current PBA public relations campaign – which began with a $100,000 media buy three weeks ago and may by now have more than doubled that – is to put pressure on the Panel, the Episcopal Church Home, the Olmsted Parks Conservancy, and Judge Eugene Fahey. The PBA wants all those agencies and people, each of whom has a voice in the final bridge and plaza design choice, to back off and leave them alone. For most of their history they've been left alone and they and their friends have done just fine that way. The current attention is new to them, and very distasteful. They don't like us meddling in what they think is their business.

Simplify-and-repeat advertising campaigns go on all the time in politics: candidates and parties spend massive amounts of money and orchestrate events designed to give the impression that truth and God are on their side. I don't remember ever seeing this kind of massive and relentless campaign waged by a public agency trying to influence public opinion on a public works construction project, especially while it is ostensibly taking part in an objective review of exactly that project. That suggests that the stakes are probably bigger than most of us thought. I wish I could find out how big the stakes really are and who will really make the big money out of the steel companion bridge project, but I can't because the PBA keeps the details of its financial activities secret.

FOREPLAY

On Saturday February 26, the Public Consensus Review Panel conducted a five-hour meeting at which the consulting engineers hired to evaluate various bridge options were asked to listen. Lately, the engineers seemed to be setting their own goals for the research and the Panel wanted to remind them what this was all about. It was also an opportunity to introduce some new information and listen to the responses of the engineers to recent questions about its data and analyses.

Members of the Consensus panel spoke, as did representatives from the city of Buffalo, the Olmsted Parks Conservancy, and the Episcopal Church Home. The engineers responded and talked about their own concerns. Mostly, it seemed, they wanted to focus on cost, and nearly everyone else wanted to focus on the human consequences of various bridge and plaza designs. The only member of the panel who consistently pushed for giving cost priority was Natalie Harder, the representative of the Buffalo Niagara Partnership, which has long been the only community organization to endorse the steel companion span.

(I have often faulted the Buffalo Niagara Partnership for its uncivic posture in the Peace Bridge affair, but now I wonder if it isn't a little unfair of me to expect from it a measure of civic concern and responsibility it doesn't claim for itself. The Buffalo Niagara Partnership isn't a community *service* organization; its purpose is to help its members make money. That mission is expressed unambiguously in a box next to Partnership President Andrew Rudnick's message in the 1998 membership directory: "Like other employer organizations, the Partnership represents the private sector and advances the 'bottom line' interest of its members.")

If you made cost and quick profits the primary desiderata and you accepted the numbers the engineers were offering, then you probably had to accept the notion that the companion span was the only way to go. But attorney Joe Crangle, former chairman of the Erie County Democratic party, gave what I thought a reasonable response to Harder's plaint: "Let us come out of this with a plan that makes sense economically and environmentally and let other decision-makers find the money." Jim Kane, Senator Moynihan's representative on the panel, suggested that the whole discussion about cost was compromised and distorted by the PBA's refusal to accept "funds for this particular project allocated by the federal government."

The engineers took many of the Buffalo concerns to heart. In their presentation Tuesday night, they gave far more attention to social issues and far less attention to dollars than in the past. Even so, they managed to deliver the PBA's agenda.

NUMBERS GAMES

I have difficulty trusting a lot of the numbers offered as evidence by the consulting engineers. The most obvious weirdness this week had to do with bridge maintenance.

The PBA says it currently spends about $4 million a year on maintenance for its two plazas and one bridge. Bridge General Manager Earl Rowe phoned the Larry Hunter show in December and said they were spending $400,000 to $500,000 on maintaining the bridge alone (that leaves $3.5 million per year spent on the two plazas, a huge amount of money). Rowe said the half-million dollars didn't include major paintjobs because those were so expensive they were capitalized and tucked in another part of their financial statements.

The consulting engineers seem totally unaware of any of that. They project the 75-year maintenance costs for the companion bridges with twin arches at $11.3 million. That comes out to $75,333 per year per bridge, and in that amount they include the 15-year paintjobs the PBA excludes from its maintenance budget. They're off by a factor of 660% – without paint.

What this tells us is, the engineers didn't even look at what is really going on at this bridge at this time. They pulled their numbers from some abstract industry standard or they made them up on the fly. If the engineers didn't check obvious numbers like these, how can we trust their projections into the future on things that are only imagined?

WHO'S PAYING FOR WHAT?

Panel member Brenda McDuffie, of the Buffalo Urban League, tried to get Roger Dorton, who works for Buckland & Taylor, one of the Canadian engineering consultants, to understand that a whole range of community values were important in the discussion. He kept going back to a short list someone developed seven months ago. McDuffie explained that the Panel had learned a great deal in all this time and had expanded and rearranged that initial list of topics considerably.

Dorton was resistant and seemed far more angry than her comments warranted. McDuffie suggested that the choice of topics should rest with the Panel, since they had hired the consultants. "You work for us," she said.

Dorton's face got redder and he told McDuffie in a sharp voice, "I don't work for YOU! We have a partnership here!" Ah, yes, something most of us had forgotten because of all those documents with "Bi-national Team" in big letters across the bottom: Dorton and the other Canadian engineers don't work for the Public Consensus Review Panel. They were hired and are being paid by the Public Bridge Authority. The report the Public Consensus Review Panel will judge after next week's public hearings are being provided by a group of American engineers hired to explore all the options and a much larger group of Canadian engineers hired by an organization that wants to destroy all the options but one. The PBA has refused to sit at the table and discuss any of these issues but perhaps it has figured out a way to trump people willing to speak openly.

Most of the Saturday events were watched in silence by Bridge managers Earl Rowe and Stephen Mayer. They were bracketed by two press agents and by their attorney, Arnold Gardner. Rowe spoke only once, late in the day and in answer to a question. The other four never spoke to the room at all, though they exchanged comments amongst themselves all morning and afternoon.

OH, CANADA

The Canadian position has been constant from the beginning of this: stonewall, stonewall, stonewall. They have refused to participate, they have refused to talk, they have refused to share. All they have done is issue ultimata. From their point of view that makes sense – Fort Erie is a town largely supported by the Bridge, so the more bridge activity the better, even unnecessary maintenance work occasioned by sticking with an anachronistic technology. Officials and business executives in more distant parts of Canada don't care about quality of life in Fort Erie and surely not in Buffalo; they care only about their profits from truck traffic. As far as they're concerned, Buffalo is a bottleneck for their trucks, nothing more.

Judy Fischer, a member of the Panel and the Buffalo Common Council, pointed out that over the years Buffalo has given up a great deal for this bridge and the Canadians have given up nothing. We've lost a landmark park and endured a huge amount of pollution from trucks idling at the plaza (asthma rates are much higher in that area than anywhere else on the West Side) and the town of Fort Erie has gotten jobs. The Canadians have a lot of demands, all designed so they can make money out of American markets, but they offer nothing in return. Not one Canadian official has agreed to take part in the panel. Only one has deigned to visit, Wayne Redekop, mayor of Fort Erie, and he basically said they would consider nothing but the PBA plan.

Joe Crangle said, "We hear about this economic benefit from this Peace Bridge international crossing. And it undoubtedly is true. But true to whom? True to the 300,000 people of Buffalo? I wonder. I really wonder.. . . We don't mind being the funnel, we don't mind being the catalyst for this, but hell, we deserve something more than just a twin span and 17 acres of plaza and no guarantee how it's going to benefit the people of this city." Panel member Andres Garcia, who represents the West Side Hispanic Community, said, "NAFTA is doing nothing for people walking down Niagara Street."

As we've pointed out here before, there's been no evidence offered that all this increase in truck traffic will do this area the least bit of good. It will provide more jobs at the Peace Bridge and maybe a few more brokerage houses will open up, but that will be more than offset by damage to the infrastructure and air and noise pollution. They have consistently made claims that the trucks will benefit us and have consistently refused to produce any data showing how that is so. They raise the red flag of diverted truck traffic: if we don't increase lanes here immediately then trucks will go to Detroit or Queenston. Nonsense. What trucker with a load for New York or Philadelphia is going to drive five hours out of his way and buy all that extra diesel fuel to cross at Detroit because of an extra half-hour or hour delay here? And what if a truck crossed at Queenston instead of Buffalo? The Queenston bridge now handles a million trucks a year; it could triple that with little difficulty. It's not like the toll money is coming into the local economy. The PBA is in competition with the three Niagara Falls bridges, but the rest of us aren't. Those are all bogus issues.

THE CONSULTANT'S COMPROMISE

The engineering consultants equated Buffalo's desire for a plaza that did more good than harm with Fort Erie's fondness for "Heritage aspects of existing arch bridge." So they compromised: they gave Buffalo a decent plaza that would free up the park and they gave Fort Erie another bridge to hang out with the one it has now.

But those aren't equivalent demands. No one in Fort Erie was carrying on about the Peace Bridge as historical object until the PBA's twin span plan was challenged. All the attempts to have the current bridge declared an historical monument were manufactured after people in Buffalo started agitating for a better design. The Fort Erie folks have made demands, but their only presence in the process has been in that one visit to the Panel by Mayor Redekop and last week's performance from which all opposition to the PBA plan was excluded.

The engineers' compromise is no compromise at all. Fort Erie gives up nothing in this deal; it just gets more jobs, more people hired to paint the anachronistic steel cables. Why should Buffalo have to look at two ugly bridges to underwrite that?

The PBA is spending a fortune in an attempt to get out from under the lawsuits of the Olmsted and Episcopal Church Home. If that works, then the only thing between a twin span and us is the city of Buffalo refusing to give the PBA permits to drive construction trucks across city streets. If Olmsted and the Episcopal Church Home are out, if Judge Fahey is out, then our only hope for a decent bridge rests in Buffalo's city hall. With all the money the PBA has to lobby and pressure, how long can the local pols hold up? The PBA started out clumsy, but of late they've gotten very good, and they've bought very expensive legal and public relations counsel. They've spent a lot of money trying to buy what they couldn't get any other way.

Will the engineers stick to their disproportionate compromise in their final draft? Will the Review Panel have the backbone to recommend something more rational? Will the PBA pour more money into this, muddying the process still further? I can't imagine any resolution in which the key players – the Public Consensus Review Panel and the Buffalo and Fort Erie Public Bridge Authority – will both get their money's worth out of this exhausting and bitter process.

PROMISES, PROMISES

The Public Bridge Authority promised Judge Eugene Fahey it would accept the Panel's recommendation, but within days one member of its board told a Fort Erie reporter that accepting a recommendation and acting on it weren't at all the same thing. The PBA's current campaign to turn public opinion against the Panel suggests that, should the Consensus Panel issue a report different from what the engineers recommended the PBA may very well continue its pattern of avoidance, delay, misrepresentation, and denial until it has no other choice.

The Review Panel has taken the position that public examination of key facts will lead to a decent conclusion. The PBA has taken the position that you're better off buying the conclusion you want, you're better off with simplify and repeat, simplify and repeat. The consulting engineers seem caught in the middle. Their report Tuesday night seemed nice and neat but it isn't; it's a conceptual mess. It offers to return to us a beautiful park that should never have been destroyed in the first place and the price they suggest is the imposition of an ugly and costly anachronistic bridge. That may make engineering sense but it doesn't make any other kind of sense, and that's why the Public Consensus Review Panel will almost certainly tell them that this just isn't good enough. It just isn't good enough.

27 PBA's TV Ads Pulled by All Buffalo Stations

All four Buffalo commercial television stations and all local cable stations on which the Buffalo and Fort Erie Public Bridge Administration placed spot ads designed to put pressure on the Public Consensus Review Panel and Judge Eugene Fahey decided last week they would no longer air those ads.

When the first anonymous ads advocating an immediate start of bridge construction appeared on local television stations, *Artvoice* called the stations and asked who had paid for them. We were variously told that they had been instructed not to reveal the sponsor or that these were educational spots so there was no need to identify a sponsor.

After *Artvoice* questioned that policy in print ("The PBA's Ex-Lax Initiative", February 17), at least one station seems to have queried the FCC or the network legal department. We have received a copy of a fax indicating that the PBA's refusal to allow its name to be used in connection with the ads it had purchased put the stations out of compliance with section 317 of the Communications Act of 1934. According to the memorandum, section 317 specifies, "whenever you are paid to broadcast any material other than product advertising, you must announce that the broadcast is paid for and who paid for it....For advertisements that are political or that have the discussion of a controversial issue of public importance, you also must place in your public file the names of the chief executive officers or members of the board of directors of the organization responsible for the advertisement....If there is evidence that someone other than the named sponsor is the real sponsor, the station must inform the named sponsor that it will not broadcast the message."

So, rather than risk their FCC licenses, the stations pulled the ads. Local radio stations seem to be doing the same thing.

The PBA could continue to air its advocacy commercials if it would agree to put its name on them, just as candidates for office or lobbyists for legislation do. The continued absence of the ads from Buffalo tv and radio stations tells us that the PBA continues to be unwilling to admit that it was the sponsor.

When the Peace Bridge question is finally resolved it might be a good idea for our elected representatives – the attorney general among them--to look into this. Is it legal for a public agency to secretly spend a fortune disseminating misleading information as part of an attempt to influence public policy?

28 THE RIGHT THING TO DO

(Statement prepared for the Public Consensus Review Panel public meeting at WNED studios, 7 March 2000)

Thank you for allowing me to speak in response to the consulting engineers' report. Before I begin, I would like to identify myself and explain why I am speaking to you this evening.

My name is Bruce Jackson. I've lived in Buffalo for 33 years. I'm State University of New York Distinguished Professor and Samuel Capen Professor of American Culture at the University at Buffalo. My wife, Diane Christian, is a State University of New York Distinguished Teaching Professor. We've raised our children here, and we love this community.

Except when I've had class or been out of town, I've attended all public meetings of this Review Panel and all meetings of the Bridge Task Force organized by Common Council President Jim Pitts. I've read everything I could find – engineers' reports, annual reports, legislation, memos, financial statements, even all the minutes of the Buffalo and Fort Erie Public Bridge Authority going back three-quarters of a century. I've gone to board meetings of the Buffalo and Fort Erie Public Bridge Authority – but since they go into closed executive session whenever they talk about anything important regarding the new bridge, I gave up on that. I've talked to scores of people involved in the process or likely to be affected by the results. I've written 27 articles on this subject for *Artvoice*, Buffalo's weekly newspaper, and one editorial for the *Buffalo News*.

I did all of that because I think this process and what comes out of it will have profound consequences for this community for decades to come. It really matters which bridge and plaza we get and what happens to the community in the getting of them. It is equally important that the public have a voice in what is done to us. These things should not be decided in rooms from which the public is excluded on the basis of private interests forever kept secret.

By profession I am a teacher, and the largest part of my research has been in criminal justice matters and in analyzing spoken and written documents. I've written about the use and abuse of statistics and the difficulties inherent in evaluating data provided by interested parties. I have worked as a consultant for many organizations, among them the Arthur D. Little Company of Cambridge, Massachusetts, and the Library of Congress. That training and experience helped prepare me to read and analyze PBA statistics and consultants' reports. I should also mention that my academic training began in engineering and physics. Numbers are not mystical to me and I've never thought that people who argue on the basis of them automatically deserve any more belief or trust than anyone else.

I will direct my remarks to four topics: the issue of compromise, the possibility of decent design, the question of time, and what is to be done now.

1. The Issue of Compromise

The consulting engineers, several Canadian politicians, and even a few politicians on this side of the border, have urged compromise on us. They say that we're getting our parkland back, so we should compromise by giving the Canadians the companion bridge they hope to start building this summer. Even though all the most reliable data show that a six-lane concrete bridge and plaza system will cost less to build and maintain and will be completed faster, they say "Compromise. Agree to let them start building a twin span now."

But there is no compromise there. It is little more than what the PBA told us it was going to do two years ago. Letting us build a decent plaza in exchange for immediate construction of a twin span is like selling us our own car and then charging us rent for the garage in which it is to be parked. The logic is faulty and the results would be wretched.

The bridge and plaza are physically linked and they are linked in terms of their impact on the environment and the community. The full environmental impact study on a non-segmented bridge and plaza project that the PBA has so long opposed must be undertaken before a spade cuts the ground. The consulting engineers didn't understand the importance of that, but their concern was engineering, not human or long-term environmental issues.

No matter. None of that has anything to do with Canada. Those social and environmental issues are entirely on this side of the border. What we do once the bridge lands here is no more the business of the Canadians than what they do with it there is ours. This is Buffalo territory, not international territory. If they want something beautiful or ugly in their plaza, something that blends with the environment or desecrates it, that's their business, something they have to work out with the PBA. And we have to work out with the PBA what kind of plaza we will have. That's what those lawsuits are all about.

But the bridge is truly international. If the PBA or the Canadians want a bridge we don't like, we don't have to let them land it in our country. If we want a bridge the PBA or the Canadians don't like, they don't have to let us land it in their country.

The only issue involving both us and Canada in all of this is the bridge itself: what shall it look like and when shall it be built and what inconveniences shall we endure to let that come about? Those are issues for discussion and perhaps compromise – but I'll remind you that thus far no Canadian official has been willing even to talk about it with us. We've received only ultimata and comments about what bullies we Americans are. It's untrue: the Americans are not the bullies. We're the ones who keep saying let's talk, let's see what we can work out and they're the ones who keep saying, "The only thing we can work out is what the PBA and Ottawa tell us is the best thing."

Why have the American board members of the PBA represented only Canadian interests in this? I don't know. I wish I did. I wish they were taking part in this process. I wish they thought enough of the American public to be willing to talk with us. They don't, so we can only speculate, and in the silence take whatever steps are necessary to protect our interests.

2. THE POSSIBILITY OF DECENT DESIGN

The issue isn't about the Freschi-Lin Bridge, which the consulting engineers refused to consider at all, versus the Twin Span. Nor is it the choice between the ungainly designs offered by those engineers who were obviously committed to the twin span idea.

It is about whether the PBA is willing to consider any design for a bridge and plaza system other than the one it came up with long before it held any public sessions more than three years ago. It is about whether the PBA is willing to consider honestly what is to be gained by building a six-lane bridge that is cheaper to build and maintain and which will be constructed faster and with less disruption to traffic and the community than anything it has proposed. It is about whether the PBA is willing to suspend its private interests and act as if it really were a public agency.

There are plenty of first-rate highly-experienced engineers and designers out there who can design bridges far more beautiful than anything those consultants showed you last week and who have experience in building them that those consultants lack. When your engineers said "This can't be done because it's too difficult," I heard them saying, "We don't know how to do this." There are people who *do* know how to do this. Gene Figg has built hundreds of such bridges and so have T.Y. Lin and Santiago Calatrava. There's no reason an honest design competition couldn't be held right now. There is more than enough time to get it right.

3. The Question of Time

In the past month, the PBA drenched the community with ads saying we had to get started on construction now else dire consequences would result.

That is not true. Their design, if fully implemented, will take longer than any other northern plaza design suggested by anyone. I cannot believe that every bridge designer who has looked into this project is wrong and only the engineers hired by the Public Bridge authority are right.

And there is a further complicating factor their engineers did not discuss: the old bridge is sick, very sick. They can repave it and tune it and tweak its wires, but will not last as long as the new companion bridge. It is almost 75 years old and even with expensive repairs it will soon be at the end of its life. So not very long after all this proposed construction is completed – ten years, fifteen years, twenty-five years – that entire old bridge will have to be replaced in its entirety, and the community will have to endure another four or five years of one inadequate bridge ferrying traffic, four or five years of construction vehicles and commercial trucks pounding and polluting city streets. There's no need for the community to go through this twice. If we build a new six-lane concrete bridge it won't have to.

4. What to do?

This has been a long and difficult process. The PBA has spent a fortune in recent weeks trying to buy this, they've lobbied many of you, they've taken ads trying to get the public to turn against you.

The PBA has so much money, so many people at its disposal, so many lawyers, so many press agents, they have so many documents to pour on your tables. I know you've all spent hundreds of hours on this task and at times it no doubt has seemed thankless and pointless.

But this is not the time to quit in weariness. When this is over, they go back to running their bridge and we have to live with the results. You, me, our children, our children's children, all of us.

Restoring our destroyed Front Park is not enough. From the beginning they tried to separate the bridge and plaza and they're trying to do it now. Divide and conquer. But why shouldn't we have a viable plaza *and* a beautiful bridge?

It is not now, and it never has been, an either/or situation for us. They keep saying that again and again, but that doesn't make it any more true. It isn't either/or. Of course we can have a plaza that restores to our community the park that was wrenched from us, one that integrates in a useful way with city streets. And of course we can have a bridge that is not an affront to the eye, a bridge that is as beautiful as it is functional. The people who say "You can't have both" are people who have other interests at heart.

Few politicians are capable of thinking beyond the next election or two. They try, but only rarely are they capable of saying to the people who poured large chunks of money into their recent campaign or who they hope will pour more money in the next one, "I'd like to help you but I've got to think of the public interest ten or twenty years down the road." No: they compromise. That's what the lawyers almost universally recommend. Compromise. "You've got a position, he's got a position, hey, let's find someplace in the middle we can all live with." There is no truth for such people; there is only something "we can all live with." That's maybe the easy way to go, but it's not always the right way to go.

And in this instance, with this bridge and plaza, it is surely not the right way to go. The American members of the Buffalo and Fort Erie Public Bridge Authority have not chosen to represent us in this. We don't know why and we probably never shall. We don't know how long we can count on our elected politicians standing tall in this.

It all comes down to you. This entire issue is in your hands and your hearts. Please don't let them wear you down. Don't let them convince you that the bridge no longer matters or that Buffalo can be bought off with a promise to design a new plaza. We can and should have *both* bridge and plaza and you have the power and the responsibility to see that we get them. Do what you know is right.

29 Fighting All the Way

All sides in the Peace Bridge Border War were going pedal-to-the-metal for the past week:

> The team of Canadian and American consultants hired by the Public Consensus Review Panel and the Buffalo and Fort Erie Public Bridge Authority turned out not to have been asking the questions the Public Consensus Review Panel thought they were asking.
>
> 85 people told the Review Panel what they thought of the consultants' preliminary report.
>
> The PBA and the Canadian government continued their vigorous attempts to pressure the PCRP to vote for a steel twin span.
>
> after its misleading commercials were cancelled by all Buffalo tv stations the PBA went back to doing it in print.
>
> Erie County Executive Joel Giambra and PBA chairman Vic Martucci tried to broker a last-minute compromise but the other nine members of the PBA turned it down flat.
>
> The *Buffalo News* continued shilling for the PBA, and
>
> The consultant's final report still hadn't arrived five days after the deadline and nobody was saying why.

What the Engineers Missed

On Tuesday morning, March 7, the Public Review Consensus Panel met with its Canadian and American consulting engineers to discuss problems in the engineers' preliminary report (which had been made public at a session televised by WNED on February 29). The meeting revealed major differences in what the engineers thought they'd been hired to do and what the Panel thought they were paying to get. Some members said they that after reading the details of the report they couldn't figure out how the engineers managed to recommend the twin span with the rebuilt Parker Truss, since the body of the report seemed to say that was the most expensive and time-consuming of the viable options.

The engineers said that the most important factor was when construction would start, and since all the options other than the twin span would require an Environmental Impact Study, the twin span therefore had a great advantage. There was, as I recall, a stunned silence, followed by several vigorous responses. Former Erie County district attorney Edward Cosgrove, representing the Common Council on the Panel, said, "When did you start accepting the schedule of the PBA?" Mark Mitskovski, vice president of the Peace Bridge Columbus Park Association, said, "The time that matters isn't when you start. It's when you finish, and what you finish with." Review Panel co-chair Gail Johnstone suggested that when the engineers set to work on their final report they do it with the end rather than the beginning of the process in mind. The engineers said they would take that into consideration.

VOX POPULI

That night and the next, a studio audience of as many as 800 listened for more than 7 hours as 85 people gave the panel their opinions of the preliminary report and the PBA's plans.

Nine speakers (5 of them Canadian officials) endorsed the twin span, 18 didn't address the bridge question at all (some focused on plaza issues, some talked about general public policy matters, one urged the panel to do whatever was best, one thanked everyone involved she could think of by name, one talked about exercise and how much he liked going back and forth across the bridge in nice weather, and one discussed at length the kind of light show she thought would work very well atop whatever bridge was in place when this was over).

The other 58 speakers (some of them Canadians) all endorsed a signature bridge. Some of those simply expressed opinions, some detailed their own experiences dealing with the PBA, and some presented hard data. The last group included Senator Moynihan's representative Tony Bullock, who informed the panel that there was a great deal of Federal money specifically available for this project should the PBA want to avail itself of it; UB professor John Mander, who discussed the economics of pre-stressed concrete cable-stayed bridges; and Jay Rohleder of the Figg Engineering Group who talked about the specific costs and timelines of such a project here.

John Mander had one of the best lines of the evening about the twin span: "It is a travesty of technology where we end up with a double mediocrity."

JOE CAMEL

Nearly all of the supporters of the twin span – the Canadians, PBA chairman Victor Martucci, a woman who represented Ciminelli Construction, and county legislator Al DeBenedetti – reminded me of those tobacco company executives who testified before Congress a few years ago, the ones who said that if anyone could show them any real evidence that cigarettes caused harm they'd of course stop peddling the deadly poison, but they'd never seen any such evidence or there wasn't any such evidence, so they'd continue with what they were doing, thank you. (One person who urged the panel to vote for the twin span was West Side real estate developer Bob Biniszkiewicz, who didn't support the twin span so much as he feared that the PBA wouldn't develop the plaza if the panel didn't recommend a twin span. His concern seemed to be solely with property values.)

Martucci's first sentence was misleading. He said he was chairman of the "Peace Bridge Authority." There is no Peace Bridge Authority. The organization he chairs is the "Buffalo and Fort Erie Public Bridge Authority." They avoid using that word – *public* – whenever they can, all of them, the people who put the messages on the answering machines, the general managers, and now the new chairman. In his eight-minute statement Martucci said "Peace Bridge Authority" six times; he never said "Public Bridge Authority" once.

Not one of the twin span proponents engaged the very specific testimony from John Mander or Jay Rohleder, both of whom provided exactly the kind of data they said was missing. Neither did I ever hear any of the engineers on the binational consulting panel ever deal with their very specific information.

Which suggests that those of us who have tried using facts and reason to get the PBA and the Canadians to change their minds are probably answering the wrong question or arguing the wrong argument. We keep trying to figure out why they prefer the steel bridge when cable stayed segmented concrete is better and faster and cheaper, and we keep looking for arguments to show them that the signature span isn't just pretty, but practical.

There's got to be a compelling reason why what seem like solid facts to so many of us are of no interest to the PBA or its supporters. Surely something more than mere truck traffic is behind the arrogant presentation by the representative of the Toronto city council who told the Panel that the Toronto council had voted about what we should do, something more than the number of lanes on the bridge is behind the barrage of letters to the Panel from a half-dozen Canadian provincial and national politicians demanding or urging an immediate start on construction. If they were really interested in the problem of truck delays they'd be politicking the US government to add staff to the real bottleneck, the truck processing operation. So we probably haven't seen the real reasons for all of this. It's difficult enough finding the political payoffs on this side of the border; I don't know where to start looking on the northern side. But I haven't a doubt that they are there and they are compelling.

WHAT I THINK THE MAYOR SAID

Mayor Masiello was expansive and clear about the proposed new plaza. When it came to the bridge choice, he seemed to be pushing hard for the twin span but didn't come out and say it.

He thanked everybody and every thing who has or had anything to do with the argument now in process: the funders of the panel, the panel members, the PBA, Bruno Freschi, WNED. He said the sort of good-sounding but empty things mayors say at events like this, such as, "I pledge to you that whatever new bridge system is finally recommended, it will be the capstone of a revitalization of the west side of Buffalo" (the underlining is in the text his staff handed out).

He said he agreed with the consultants that a new plaza was better than the old plaza and the restoration of Front Park and reclamation of Fort Porter were good for the city. He went on at length at how much he supported a "signature gateway" and how he'd do all he could to see that it worked. That's good. His support will help that move along, if the PBA doesn't find a way to weasel out of it, as many of its critics expect it will.

The Mayor didn't get to the bridge for a long time. This is what he said:

> As representatives of the community, if you feel the 'signature' bridge issue is most important, and the recommended bridge does not pass the 'signature' litmus test, then you must not ratify the final recommendations.
>
> On the other hand, if you feel economic development, a signature gateway, the restoration of Front Park and Fort Porter, the revitalization of the West Side community are most important, and this recommendation provides a unique opportunity for economic development dollars and jobs to flow into the region as soon as possible, then you must ratify the final recommendations conditioned upon the building of the northern plaza

When he was done several people said, "There's Tony, waffling again as usual." They were wrong: he wasn't waffling at all. He just wasn't owning up to what he was saying and he was setting himself up for a good speech later: however things go he'll be able to claim he was on the winning side.

Of *course* he was pushing the PBA/Canadian agenda.

The worst part was how he espoused the PBA's duplicitous either/or gambit: you're either for aesthetics or you're for everything else. In that equation how could *anyone* choose aesthetics? Aesthetics *or* economic development, a new plaza, community development, and restoration of the Park? You'd have to be nuts to choose aesthetics, a silly aesthete to choose aesthetics, a person with no sense of responsibility to choose aesthetics. It's like saying, "The choice is yours: you can dress your children in pretty clothes *or* you can give them food and send them to school and give them health care. The choice is yours and I'm not taking any position on it."Sure you're not.

Four people worked on the mayor's speech. You'd think that one of them would have pointed out to him that no such choice is necessary: choosing a signature bridge doesn't mean rejecting the plaza. You really *can* choose two from column A, if you want to.

THE PROPAGANDA CAMPAIGN

All week long, the PBA kept up its assault on the Review Panel. Denied local tv time because it wouldn't put its name on its commercials, it shifted to direct political pressure. There were letters from

New York and Canadian politicians who didn't represent anyone in any of the affected districts on this side of the river and who hadn't attended any of the hearings urging the Panel to accept the Canadian/PBA plan. The PBA even hired an Albany lobbyist. At a press conference Sunday afternoon, Mark Mitskovski said, "The PBA signed an agreement with the Public Consensus Review Panel to work towards a public consensus, not to undermine it with hired lobbyist and hundreds of thousands of dollars in advertising expenditures."

The PBA is continuing in print the same kind of misinformation it had in the commercials that got kicked off the air. This question and answer from an 8-page glossy insert they published in last week's *Business First* should give you an idea of the quality of information they're putting out:

> Can a single-span bridge be built in the same amount of time as a companion span?
>
> No. Any new plan or option will require a new Environmental Impact Statement (EIS) and only then can construction begin. This will take eight to 10 years.

Eight to ten years for an EIS? Even their own engineers it at two years max, and that's a worst-case scenario. And referring to "a new Environmental Impact Statement" implies that there is an old one, which there isn't. The PBA has spent a fortune avoiding one. An EIS is the public works equivalent of truth-in-lending, and the PBA will do nearly anything to avoid that.

THE PBA'S DAILY

On Thursday March 9, the day after the public hearings, the *Buffalo News* ran an editorial saying that the twin-arch design (which would rebuild the old bridge so it looked like the new bridge after the new bridge was up) was a great improvement over the simple companion span design (which would have left the ugly Parker truss in place). That is certainly true, but it surely isn't anything new: the PBA offered it as an option years ago and several times said it would consider such a design – if someone else came up with the money. Bridge general manager Stephen F. Mayer said that when he talked to the Review Panel about alternative options last December.

The editorial went on to say that there was a far better option "a majestic cable-stay bridge that connects to a new northern plaza combines the aesthetics this community deserves with the practicality necessary to capture the economic benefits of increased truck traffic. And it does that while restoring Front Park and freeing the lower West Side from the intrusive effects of smoke-belching rigs rumbling through the neighborhood." Right on. The editorial says such a bridge would be $10 million cheaper. It even argues the thus far unexamined claim that there would be problems connecting such a bridge on the Canadian side: "But that's why God created human ingenuity. Given the boost such a bridge would give to the community, it's worth trying to solve those problems."

Great stuff, huh? The *Buffalo News* finally coming down on the side of reason, finally considering the community's interests in this key issue, finally getting the point.

No. It's just sucker-bait. They say all those good things, only to wind up once again in the pocket of the PBA: "The Peace Bridge Authority remains anxious to get this project done. The community wants it done right. Despite our preference for the cable-stayed option, we fully recognize that the graceful twin-arch design meets both those objectives."

How does "that graceful twin-arch design" meet the objective of getting "it done right?" That plan will result in enormous disruption of life on the west side of Buffalo. It will send thousands of trucks lumbering through Delaware Park every day for six years. It will cost two or three times as much to maintain as a signature span and before very long the old bridge will have to be replaced in its entirety, once again subjecting the city to this dirt and disruption. The editorial acknowledges none of the evidence easily available indicating that the PBA's numbers on cost, time, and disruption are all seriously defective. (The *News* has never applied any of its vast investigative resources to an examination of any of the PBA's activities and claims.)

If it were just lousy journalism we could mark this, say "What's new?" and go on to something else, but the editorial is far more pernicious than that. The logical inference from its final paragraph, which the PBA will surely point out, is this: Since the *Buffalo News* says either choice is just fine, why not start construction in June as we've planned all along? Why wait another moment?

Would the *News* have been so cavalier with the city's interests and so much in the service of the PBA's if there were another daily newspaper that could call them on the hypocrisy immediately? When Warren Buffett killed the *Courier-Express* he did this community as much harm as when the PBA destroyed Front Park.

A COMPROMISE DENIED

The engineers were supposed to get their revised report to the Panel office last Friday but they didn't deliver. In the absence of the report, County Executive Joel Giambra made a brave attempt to broker a compromise and PBA chairman Victor Martucci, who had seemed adamantly opposed to anything but the twin span in his March 7 appearance at the public meeting, seemed willing to give it a try. They met with Jeff Belt, the New Millennium Group representative to the Review Panel, to talk about the possibilities.

All the proposals for a 6-lane bridge going into a northern plaza had focused on a bridge north of the current bridge. In his statement to the Panel Tuesday night, Tony Bullock, Senator Moynihan's chief of staff, had pointed out that the bi-national engineering team never considered the possibility of a 6-lane concrete bridge just *south* of the current bridge, in the same location the PBA is planning on putting its second span. If that could work – and there seemed little reason to suppose it wouldn't – then all the Canadian permits the PBA already has would still be valid and there would be none of the problems some Canadians have claimed would arise with archaeological sites if any other location were chosen (I've always thought those claims enormously overinflated, but this would obviate them anyway.)

Such a bridge would land at the new northern plaza, just as the twin span would, but once it was up the old Peace Bridge would no longer be needed. The Canadians who claim undying affection for the old bridge could maintain it as a pedestrian and bike bridge or as an antique – -if they wanted to pay for the upkeep. In all likelihood, their affection would have faded fairly soon after the new bridge became operational and the old bridge would have been torn down.

"The hope was that Martucci could gain the PBA's approval of a two- week extension of the engineers' deadline so that this alternative could be studied," Belt said. "In fact, all indicators were that the 6-lane concrete option would be about $15 million cheaper than the twin span plan, it would create more local construction jobs and construction could begin this summer if the PBA was permitted to keep the bridge and plaza segmented in the environmental review. Giambra's plan was to introduce this solution to the bridge impasse while coming out in strong support of the northern plaza and restoration of Front Park."

There were problems with this idea. If construction was to begin this summer, then it wouldn't be the soaring cable-stayed bridge that had caught everyone's imagination. The PBA would start work on the cofferdams and there would be a quick design process. There would be no environmental impact study. And since it would be landing further to the south than any of the north-plaza signature designs, it would keep the northernmost portion of Front Park, an area Olmsted called "The Bank," from being restored.

The compromise the PBA, the consulting engineers, and the Canadians had proposed – that they build the bridge they wanted to build all along and we build the plaza over which they have no say – was no compromise. It was just a suggestion that we roll over and wait to see how inventive they'd get. This *would* be a compromise; it would require some give on both sides. Was it worth doing? That's what the two weeks' examination by the consultants might reveal.

Belt thought it was at least worth studying and talking about and so did Martucci. But the PBA board – all nine of them according to Belt – refused to accept the two-week delay the study would require. They were confident they controlled enough votes on the Review Panel to force a recommendation for the twin span now and they saw no reason to look at any alternative.

I asked Belt if he was certain that the PBA vote against consideration of a possible compromise was unanimous. He said that, so far as he knew, it was. If he's right, that means Barbara Kavanaugh, who represents New York Attorney General Elliot Spitzer on the PBA, voted with the Canadians against the Buffalo community on this issue. This might explain why Assistant Attorney General Kavanaugh hasn't returned one of the last four telephone calls from *Artvoice* asking questions about PBA policy and why Spitzer's office has been so quiet about the Peace Bridge issue for so long.

CONSULTANT'S REPORT? WHAT CONSULTANT'S REPORT?

The report that the Canadian government officials, the PBA, lower West Side real estate developer Bob Biniszkiewicz, and a small group of New York officials are urging the Panel to accept was only a preliminary report. No one has seen the final report. It was due last Friday, March 10, but it didn't arrive. It hadn't arrived by the afternoon of Tuesday March 14, when the Review Panel had been scheduled to vote on it. It hadn't arrived by noon of Wednesday March 15, when *Artvoice* went to press.

Review Panel co-chair Gail Johnstone wrote a letter to Victor A. Martucci reminding him that the 12 November and 2 December agreements between the Authority and the Steering Committee of the Panel had "provided that the PCRP has 14 days from the time the Bi-National Project Review Team submitted its Final Report to issue its determination. Since the PCRP has not as of this date received the Bi-National Project Review Team Report, be advised, on behalf of the steering Committee, that the PCRP will issue its decision no later than 14 days from receipt of the Final Report from the Bi-National Team by the co-chairs."

Which is to say, the co-chairs weren't going to ask their panelists to vote on a report that didn't exist, no matter how heavy the political pressure from Albany and Ottawa to do so, and if and when the report did arrive, the co-chairs had every intention of not rushing to judgement on it. It is now possible that the Review Panel will not issue its final report until after the March 27 date Judge Eugene Fahey has set for his final ruling on the Episcopal Church Home and Olmsted Parks Conservancy lawsuits against the PBA.

What if the engineers take seriously the data produced by Mander and Rohleder? What if they think the concerns of the people who live here matter? What if they start looking at when the construction ends and what we end up with? What if they take into account the huge amount of federal funds Tony Bullock said would be available to a project that really was open and above board? Will they change their preliminary conclusion? The PBA is confident that they won't, that they will ignore all information offered in response to their preliminary report and Gail Johnstone's suggestion that they don't base their decision on the PBA's timetable. Is the PBA hoping or does it know things the rest of us don't?

I don't know what's going on under the table, but so far as I know there are no deals presently *on* the table. The PBA said in December that it would consider a northern plaza, but only if the city would let it reroute Niagara Street and if someone else would find the money to build it; thus far, it has committed itself to nothing and neither has the city. The attorney for the Episcopal Church Home and Olmsted Parks Conservancy told the Review panel how much it would cost the PBA if it elected to expand its current plaza or if it decided to build a new plaza; the PBA hasn't responded to any of that. All that exists is a huge amount of political pressure from

Albany and Canada for the Review Panel to ratify a consultant's report that doesn't even exist.

It's gotten mean and nasty, but it isn't over yet.

30 WHAT THEY SAID

These are from statements given at the Public Consensus Review Panel's sessions on March 7 and 8, 2000 at the WNED studios. Except for the selections from PBA chairman Victor Martucci's 's remarks and the final paragraph of Tony Bullock's remarks , which were transcribed from a tape, the passages that follow are selected from printed or handwritten texts provided by the speakers.

TONY BULLOCK (chief of staff to Senator Daniel Patrick Moynihan)

...In general, the consulting team - under the direction of this panel - has provided an invaluable service to the community. They have clearly shown that the plaza design as originally proposed by the Public Bridge Authority is wholly inadequate and would not serve the needs of Canadian and American users of this facility into the 21st century.

It has been well-established that the delays and inefficiencies plaguing the current operation at the Peace Bridge are primarily due to its plaza operations. Bridge capacity itself is largely irrelevant to this ongoing problem. Providing emphasis on plaza design was the right thing to do. By endorsing the engineering team's recommendation for either plaza option B or plaza option E, the review panel will acknowledge the clear superiority of such alternatives and help spare us all from the disastrous consequences that would otherwise result from the original PBA proposal.

The consultants – I would argue – got it half right. They are right on target in their treatment of the plaza. But they are dead wrong in their conclusion on the bridge span itself. I would further argue that the analysis is not, as yet, sufficient to make such a determination. The tortured logic that has resulted in a recommendation to proceed with the PBA's original twin span – or an enhanced twin span that replaces the Parker Truss with a cloned arch – should be discarded.

Using the cost estimates provided by the engineering team on page 55 of their presentation, the enhanced twin arch plan (advanced by the PBA) will cost more than the single-span, cable-stayed option as proposed for plaza B ($164.4 million v. $163.7 million). Yet the conclusion of the engineering team objecting to a single-span alternative identifies additional funding as the primary reason why such an alternative should be rejected. The conclusion makes no sense whatsoever and is not supported by its own data.

More significant is the glaring deficiency in the examination of alternatives to the twin span proposal. For the moment, let's rule out a full suspension bridge. There is no constituency for such a concept. Let's also put aside the use of plaza option D at LaSalle Park for much the same reason. The reality here is that the only viable option to the twin span proposal is a single-span, six-lane bridge that would utilize more or less the same alignment as the existing span.

As such, why didn't the engineering team provide an analysis on an apples to apples basis of a single span cable-stayed design (or any other type of six-lane bridge) in the exact alignment of the proposed companion span that would utilize plaza sites B or E?

Even if one accepts the cost estimates provided by the engineering team on page 55 (and I do not) the cost differential between the two bridge construction estimates is negligible – roughly $13 million. But when you add to the companion bridge proposal the cost of rehabilitation of the existing bridge and the proposed new decorative truss, the cost of the twin span far exceeds the single span alternative.

Why then should we be willing to accept the conclusion that we cannot afford the single span alternative? Why was such an obvious combination of span design and plaza location – one that would have cast the single span option in a favorable light – not provided by the engineering team?

There are further aspects of their analysis that warrant concern. The greatest shortcoming in their report lies in its treatment of the existing span. There is no examination of the $17.9 million figure assigned to the rehabilitation of the existing span. Who came up with this number? Where is the detail that would support such a number? The greatest uncertainty in this entire debate is what will it really cost to rehabilitate this old bridge? What will be discovered when the deck is removed? What is the structural integrity and general condition of the piers? Is there truth to the rumors that some of the piers no longer make contact with the river bottom lands? Why does the Federal Highway Administration give the bridge such low marks in its most recent inspection? Why is it that the superstructure of this bridge gets an FHA rating of poor condition and the substructure gets a rating of serious condition? And why is this of no apparent concern to the engineering team? I am sure that the members of this panel will recognize the obvious need to have a full and candid assessment of the condition of this bridge before making any final recommendation on which span alternative is best for Buffalo and Ft. Erie.

The PBA twin span proposal rests entirely on a terribly shaky premise: that this 75 year old bridge is ready for another 75 years of service. But what if that's not so? The PBA's economic argument for a twin span falls apart completely if it turns out that significant structural renovation and expensive time-consuming repairs to the existing span are needed.

To gloss over this critical premise would be tantamount to malpractice. Anyone who has been involved in renovation work on a home, commercial building or structure of any kind, knows that estimating such costs can be very difficult. As a rule, however, you don't really know what your dealing with until you get into it and, generally, it always costs more and takes twice as much time as you thought it would. Considering the PBA constant emphasis on time and money, it is somewhat astonishing that they have not provided greater detail on this crucial component of their own proposal.

What will we do if it turns out that the existing bridge needs $40 or $50 million in structural and maintenance repairs? What if that work adds two or three years to the time-line?

One thing is for sure, the old bridge will never get any wider. It's current lane widths are well below current engineering standards at 11 feet and always will be. There are no breakdown lanes on the old bridge and there never will be. Why was safety not more of a consideration in the analysis of the engineering team?

And what of maintenance costs? The table depicting comparative maintenance costs on page 56 of the engineering report utilizes a 75 year life-cycle analysis. The standard in the industry is to use a 100 year life cycle. Could it be that the life expectancy of the existing bridge will not permit the use of this industry standard as a basis of comparison? Have the engineers done any sort of cost-benefit analysis on this proposal that would justify – or at the very least identify – the investment that would truly be required to keep this bridge in service for another 75 years? As Senator Moynihan pointed out in his letter to Coast Guard Bridge Engineer Nick E. Mpras on March 12, 1999, the FHA's overall structural sufficiency rating is a pitiful 17 out 100. Not a passing grade by any measure. It is apparent that the cost estimates for the construction and maintenance of a single-span, six-lane bridge have been overstated while the same estimates for the twin plan have been understated. Industry standards show that steel bridges require three to four times more annual maintenance than do concrete bridges. Yet the figures do not reflect the usual margins. In the case of an 75-year-old bridge in obvious need of major repair, this maintenance figure on an annualized basis could balloon out of control. Again, where is the cost-benefit analysis that would be essential to this discussion?

As has been said before, a six-lane, modern bridge provides a superior alternative in every measurable way to the twin span proposal. We don't need a decorative arch, we need adequate lane width. We don't need a new lift of asphalt to pretty up a structurally deficient, 75-year-old bridge designed with 19th Century technology that will create an ungodly maintenance burden on the toll payers and future users of this crossing. We need a modern, safe, low-maintenance bridge option that satisfies the cost constraints, maintenance concerns and design objectives that the people of this region deserve and have every right to expect.

Bad ideas don't get better with time. The companion plan is bad engineering, bad economics, and bad planning. It should be dismissed not on the basis of its aesthetic shortcomings but rather on its obvious economic and engineering inferiority to options that are indeed readily available.

Buried a bit within the engineering report is an option that makes a great deal of sense. Accept the recommendation of plaza option E but couple it with a six-lane, single-span bridge in precisely the same location as the proposed companion span. Utilize its alignment, utilize its footprint, utilize whatever permits may exist for its construction and we are on our way to an end result that satisfies nearly everyone's concerns.

Archeologically-sensitive sites are avoided. Maintenance worries over the old span are no longer. Engineering and permitting progress on the existing application can be conveyed to the new design. Alignment, approach roads and other cost efficiencies are retained. An improved plaza design can be realized. Safety issues are resolved and a new design can inspire communities on both sides of the Niagara.

As you conclude your work, I urge you to be swayed not by public relations campaigns but by the merits of the arguments raised.

If the argument for a twin span was convincing in and of itself, it would not need so much advertising to sell itself. The essence of a fine recommendation is contained within the framework of the engineering report. It just got lost on its way to the conclusion. Help the report redeem itself. Connect the right bridge to the recommended plaza and let's get this thing started.

One last point people have raised is, what kind of money might be available from the federal government? The best answer I can give you to that question is: a lot. We just passed under the leadership of Senator Moynihan and former Senator D'Amato, the largest transportation bill in the history of the universe: $214 billion, approximately $14 billion of which comes to the state of New York in various programs and various categories of the DOT hierarchy. There is money to be applied to this project but only when the community decides what it wants to do and the people in charge think that makes some sense.

VICTOR MARTUCCI (chairman, Buffalo and Fort Erie Public Bridge Authority)

...The agreement was simply this: to commission a binational team of engineers to evaluate the Peace Bridge authority's bridge and plaza plans against alternatives recommended by the PCRP. The binational team of engineers were asked to make a recommendation for a new bridge system based on established criteria by the PCRP. The PCRP would vote yes or no on the engineers' recommendation. The Peace Bridge Authority, the mayor, and the county executive agreed to pursue the recommendation of the PCRP if they voted to ratify the recommendation of the engineers. The binational team of engineers has made its draft recommendation. We are here tonight and tomorrow night to publicly comment on that draft recommendation. On March 14 the PCRP will vote to ratify the recommendation. If I'm stating the obvious or reciting the facts that we all know I apologize, but I believe it's important to clearly define for the public the process that we all agreed to follow. For the end result to be legitimate in the eyes of the public we must not deviate from the process.

The binational team of engineers has recommended a companion span bridge and a plaza to be located north of the existing plaza. The bridge plan also calls for removing the Parker truss on the existing Peace Bridge and replacing it with a twin Black Rock arch identical to the arch proposed for the new bridge. The recommendation of the engineers is not the system proposed by the Peace Bridge

Authority. The recommended system will require the PBA to seek out financial partners for the project.

What's critical, however, is that the recommended system allows this community to achieve one important goal: and that is to begin construction this summer.

...You see this is not just an issue about steel and concrete, superspan or twin span, aesthetics or functionality. It's about jobs. It's about the economic future of this region. It's about securing a future where our children and our grandchildren won't have to leave town to find a good paying job. In my opinion this is the beauty of the draft recommendation submitted by the binational team of engineers. This recommendation acknowledges the importance of aesthetic design, functionality and cost, but most importantly it gives the region a chance to move forward today, to capitalize on the incredible opportunities that cross-border trade between the US and Canada offer. And the opportunities are real. As we speak a coalition of government, business and community leaders from Toronto to Miami are working to establish a direct trade corridor linking these two cities and every marked in between. Dubbed Continental One, this trade corridor will pass directly through the Niagara region via the Peace Bridge. Buffalo and the Niagara region will be in a natural location for distribution and logistics related business, providing good-paying jobs for our community...Ladies and gentlemen, this issue is about jobs...

SAM HOYT (New York State Assembly)

The engineers' recommendation which was presented to all of you last Tuesday is unacceptable and does not accurately reflect what is desired by our community nor what is best for our community. I suggest that they go back to the drawing board and examine how to give our community what we want – a signature bridge and a new plaza. However, the engineers' recommendation is just that – a recommendation. You have the power to accept or reject the engineers' recommendation. The PBA agreed to accept your decision when they joined the process.

At a meeting in Albany with the western New York Delegation, the PBA said that they would explore removing the Parker Truss – when and if they have the funds, and if it is feasible. In fact, the presentation which they delivered last week also confirmed this – they recommend that studies to replace the arch should be pursued. I seriously doubt their intention to build a northern plaza as well. Every indication is that they will build south of the existing plaza.

Their own documents show that the bridges turn SOUTH not north.

I understand that they have suggested that a Northern plaza is agreeable to them, and that Mayor Masiello has endorsed this plan. I agree that this is a viable option, but I am concerned that without a complete plan once construction begins on the bridge, when it comes time to build the plaza, they will change plans and state that they can't go in that direction. I'd like to believe them, but frankly, I don't trust any promises from this Authority – they have a long history of ignoring the public and undermining the public trust. They had to be dragged into this process kicking and screaming, and only joined at the last minute under duress to forestall the judge's ruling on the lawsuits. Then, despite joining the process they engaged in an advertising campaign designed to undermine the process and didn't even identify themselves as the sponsor of the commercials.

This is why I encourage you to approve nothing less than a complete plan. If we allow them to begin construction now, without a commitment for the balance of the project, we may tie our hands later. I will not stake our community's future on empty promises and uncertain possibilities.

As a public official who represents more than 120,000 constituents, it is my duty to listen to and respond to the public that I represent. The Public Bridge Authority, as a public entity, is required to do the same. Yet, everything they do ignores this fact. As early as 1996, I sent a letter to the PBA, which I have provided to you, regarding my concerns over their intention to conduct a segmented EIS. I felt then, and I feel now, that this project cannot move forward until a full, non-segmented EIS has been undertaken. If the PBA had listened then, instead of fighting public input and behaving in their typical unresponsive arrogant manner, we might not be here and could have already commenced construction.

The PBA's ad campaign stresses the fact that we need to start building now – but the critical fact is not when we start, it's when we finish. The PBA's plan for the companion span will create 10 years of delays and detours at the border, while a signature bridge and an all new plaza can be built more quickly and with no disruption at the border. Furthermore, Buffalo is no longer a steel town – the PBA's steel companion span will pump more than $100 million into another city's economy, a significant portion of this for repairs to a decrepit 80-year-old bridge. We heard last night from Figg engineers that a pre-stressed concrete bridge can be built using more than 80% local materials. This is the sort of economic boost we need. Additionally, using local materials makes sense because I want to use local labor.

The PBA doesn't have the money to construct this bridge – as they indicated Tuesday night, they need an increase in their bond cap to build – and to get a bond increase from the state legislature, they need to go through me. I can not in good conscience, and will not, support a bond increase for the companion span which will wreak economic havoc on our region for 10 years. We cannot afford this type of devastating impact on our community. If the PBA expects the State Legislature to listen to them, then they better begin listening to the public that we legislators represent.

I have introduced legislation, which I have also provided to you, which would make the operations of the PBA more open to public accountability, and begin to bring the PBA more in line with other public authorities. Another piece of legislation would allow the cities of Buffalo and Fort Erie a direct appointment to the PBA, adding more community input to the PBA's activities as well as requiring Buffalo and Fort Erie's common council's approval of all major capital projects....

There has been a lot of focus on the plaza recently, and this is good - the plaza is critical, but it is not JUST the plaza that matters – there are more than 12,000 signatures on petitions, most from city residents, saying they want a signature bridge. The advocates of the companion span will likely attempt to portray my position as holding up progress – Nothing could be further from the truth. I recognize as well as anybody the potential for job creation and economic development, and the great asset a restored Front Park would be to our city. I also recognize the potential for the revitalization of the West Side.

Other speakers have implied that unless we begin a companion span now that we will jeopardize all of these important objectives. I disagree. We can have all of these things and a great bridge. Buffalo is a great city – we all know that. However, I fear that Buffalonians have accepted mediocrity for far too long. We have an opportunity to break away from that and show the world that we can do something exceptional: something exceptional that can be accomplished in the same time frame or less; something exceptional that can be done within a similar budget and without the negative impact at our border during construction.

We have the occasion to create a legacy by building an iconic signature bridge. We are not going to get another chance to do this right, let's not squander this once in a century opportunity.

JACK CULLEN (businessman)

This is clearly an outmoded relic of another age, originally a grand design, but a design defiled at the very outset when it was built without the final arch on the US side. When did it suddenly become a revered historical icon, a treasured landmark? Only very recently, when supporters of the Twin Span scheme sought to elevate it to that status for their own self-serving purposes.

RON RIENAS (representing the Town of Fort Erie)

I can share with you that the Fort Erie Council has tremendous respect for these matters and the Peace Bridge's Canadian portion does have a heritage designation applied to it. Any effort to remove this designation and to demolish the Bridge would be vigorously opposed by the Town of Fort Erie…

For Erie does not want a dramatic, ostentatious presence. The difference in the Fort Erie view and that of some in Buffalo is as pronounced as the two-land, 30mph scenic Niagara Parkway on the Canadian side of the River and the four-lane, 55mph expressway on the American side. Fort Erie's view is that the Peace Bridge should not be dominating and overpowering, but rather, should compliment and respect the River, the Niagara Parks lands, the old Fort, the Native Archeological sites and the community itself. The existing Peace Bride, at least the Canadian portion, does this quite well and the companion span should do likewise. Let us not forget that the Niagara River is not even a half mile wide, it is not San Francisco Bay or Tampa Bay with its seven mile long Sunshine Skyway.

If it's a sense of arrival you're after, which we in Fort Erie are, then the place to do that is in and around the respective plazas. That is why the Town of Fort Erie has embarked on a Gateway Master Plan in partnership with the Niagara Parks Commission and the Bridge Authority. I encourage the City of Buffalo to do likewise with its Plaza.

ANTHONY M. MASIELLO (mayor, City of Buffalo)

As long as I am Mayor, this new bridge system is what we are collectively going to work on. This is a symbol of our renaissance! This is a symbol of our rebirth! The significance of the symbol is the community effort we put forth; what we do together towards reach our goal, should be what we are remembered for in making it happen.

STEVEN LANA (pediatrician, Buffalo)

…I have more questions now than ever before. Here are just 15 of them. Help me to understand.

1) Whose bridge is this anyway? The PBA acts as if Buffalo is merely a thoroughfare for truck traffic, 80% of which is destined for far away profit centers and special interests. Our streets, parks, air quality, safety and very identity are at best of secondary importance to the movement of trucks. Who really benefits from this?

2) T.Y. Lin, world-famous bridge builder, and Bruno Freschi, a Canadian citizen, recipient of his country's highest civilian honor for his work in the field of architecture, produced a stunning design that promised a five-year completion span for less money. Why was that plan eliminated so casually?

3)UB Professor of Engineering John Mander and noted bridge designer Eugene Figg both stated that a single six-lane bridge and new American plaza could be completed in 5 years. Are they wrong? ...

5) The PBA's original cost estimates were off by 40%. Did the consultants conduct an independent accounting of cost or simply accept the PBA's latest figures?

6) The consultants propose moving the plaza to a site NE of the existing one but want to start construction of the twin before plans for the new plaza are developed. Will they merely point the new bridge somewhere toward the American side? ...

9) What is the dollar value of the new park land that will be gained for Western New York by the restoration of Front Park and Fort Porters? And was this factored into total cost?...

12) If we can build a bridge and plaza simultaneously without disrupting existing traffic patterns and routing thousands of heavy trucks through our city streets – why choose the chaos of a multiple-step construction project?

13) Time and again we've heard the bottleneck to traffic flow is the US customs plaza, yet your plan solves this problem last! Shouldn't this be given top priority? ...

15) Why does the PBA refuse to do a full EIS as required by law for projects of this magnitude and significance? With all the work that's already been don an EIS could be completed in ten to twelve months. To paraphrase an old saying, "It's not when you start, it's when you finish that matters.

I'm tired of hearing "just do it." Most of us were raised to believe to "do it right or not at all."

ROSS ROBINSON (businessman, Niagara Falls, Canada)

...Since this debate began, the perception has been that all Canadians want to "just get on with it, and build a twin or companion span."

The "American vs. Canadian" angle has been very cleverly created and perpetuated by very clever and I trust well-paid public relations consultants. It makes great press. It is also just not true.

...The public Bridge Authority and their public relations consultants have been ruthless in their determination to "just get on with it," and to build their plan of record...

For two years, and again last night, we have heard that the Peace Bridge has received Canadian heritage designation. The truth, confirmed with the Town of Fort Erie and the PBA this morning, is that no such Canadian, Ontario or even local municipal designation exists. In reality, Fort Erie's local architectural conservancy board are now preparing to apply to the Town of Fort Erie to have the Peace Bridge designated...

For this once-in-a-century opportunity, I respectfully urge panel members, engineers and politicians to think less like politicians and more like leaders and statesmen.

JOHN MALONEY (Member of Parliament for the riding/district of Erie-Lincoln, which includes the town of Fort Erie)

...A new bridge is needed now as an economic catalyst for our respective regions, to exploit the trade and commercial boom between Canada and the United States and the jobs that both create. We wish to prevent the forfeiture of progress by default.

Thus we can see that this issue is more than a Buffalo – Fort Erie issue or a Western New York – Southern Ontario issue. The Peace Bridge is an integral link in the trade, tourism, and social infrastructure of our two countries....

I agree with David Collenette the Canadian Minister of Transport's assessment in a letter to you that the review team exercised due diligence in reaching their decision which was based on agreed--upon criteria, including functionality of traffic and plaza operators, local and regional economic impacts, environmental impact, aesthetics, completion schedule and cost.

The preferred alternative advanced by the engineering team is the twin span companion bridge, replacing the Parker Truss with twin arches thereby providing signature quality aesthetics while ensuring the heritage aspect of the existing bridge.

Perhaps not everyone here agrees with the decision, but I do. I urge you to join the consensus behind the preferred alternative. I am supported in my position by many of my colleagues in the House of Commons as confirmed by their letters to you....

The greatness or mediocrity of aesthetics of a bridge often differ only in the eyes of the beholder. My eyes find our historic Bridge the majestic, strong and gracious monument to Peace that it is....

DOROTHY NYSTROM (lifelong resident on Columbus Parkway, Buffalo)

...In my association with the PBA for the past ten years, I have tried to cooperate with them, but have found them to be arrogant, deceitful and trying to run rough shod over everyone. I had hoped that through the hard work of the Review Panel my opinion of them would change but so far it remains the same. I hope that somehow the men in the PBA will change their tactics. Expect a miracle – we may just get it.

I have attended almost all of the so-called public meetings conduced by the PBA. We tried to cooperate with them but they would not even consider any alternative possibilities, other than their own plans. We were not there for input – we were there to be told what was going to be.

The PBA quietly bought up half a city block under the pretense that they needed to widen the present plaza and we were to receive a buffer zone between their property and ours. Once they had acquired the property they told us they were building a four story parking ramp and an administration building that would abut the rear of our property. When asked how they could do this they told us that since these buildings were off-site and not on the plaza they could do whatever they wanted to. We offered to sell them our properties so they would have a buffer zone and eliminate the problem of a common sewer system. Construction there in the past has caused some neighbors to have their basements filled with feces.

Their reason for not buying our property was that they didn't have the money. I question that. It is my understanding that the PBA has spent millions of dollars building public buildings for the town of Fort Erie. Some of these buildings are state of the art aesthetically beautiful ice skating rink and a new town hall for the public officials. With all due respect to our Canadian friends of which I have many, I have to ask how can we believe the opinions of these public officials under these circumstances? Would you come to a meeting such as this and bite the hand that feeds you? I think not.

And why is it that Canada gets all the social and economic benefits from the PBA and we are treated with absolutely no respect or consideration? And also where have all the American representative on the PBA been all of this time?

Because of the PBA actions no one in their right ind would want to buy my property at the present time – unless I give it away, which I don't intend to do. I have had years of exposure to the noise, fumes, dirt, pollution and psychological abuse from the PBA. Therefore, I favor plaza Plan B – in order to get away from this overbearing neighbor.

I don't trust the PBA in any way, shape or form, so I want some stipulations written in stone.

1. The property currently owned by the PBA on Busti Avenue be taken away from them. Until this happens, I want them held responsible for the upkeep of these properties so they don't continue to deteriorate and will not need to be condemned.

2. They are never allowed to do to others what they have done to us by putting off plaza buildings wherever they choose.

3. Since no environmental studies have been done I have other concerns. If Plaza plan E is adapted as it is, it would place my home in such close proximity to all the noise and pollution we would be adversely affected for the next 8-10 years or more. Plus our property values would be nil. We would have the demolition of the Episcopal home, the building of the bridge and plaza on our right, the tearing down of the present plaza and restoration of Fort Porter behind us. We can't open our windows now. What will it be like then?

Mayor Masiello said that we should be willing to compromise. I suggested that since our small group at the edge of the plaza near Rhode Island would be forced to live under these adverse conditions we should be compensated for the taxes we have to pay, loss of property value , repainting our houses, and even medical care, or else new properties be purchased for green space between the plaza and residential homes.

If our voices and concerns are not heard loud and clear then we are not taking any more abuse, the only recourse left us is to pursue the legal means that are open to us.

I realize what a monumental task you have undertaken and I thank you for it. Now I hope that you have the stamina and courage to do what is right for all of us.

BRIAN MCALONIE (Buffalo resident)

...Eiffel Tower, Statue of Liberty, George Washington Monument, Great Pyramids, The Leaning tower of Pisa, The Acropolis, The Coliseum, England's Big Ben, Lincoln Monument, George Washington Bridge, CN Tower, Sears Tower, London Tower Bridge, St. Patrick's Cathedral, St. Paul's Cathedral, St. Peter's Basilica, The Louvre, The Chrysler Building, Empire State Building, Woolworth

Building, Falling Water, Guggenheim, Getty Museum, Golden Gate Bridge, Sydney Opera House, Taj Mahal, United Nations Building, World Trade Center, NYC's Center Park, The Capitol Building ,The White House, The Canadian Parliament Building, Mt. Rushmore, St. Louis Arch, Guggenheim at Bilbao, Arc de Triumph, Notre Dame Cathedral, Grand Central Station, The Brooklyn Bridge, The Parthenon, The Roman Aqueducts, Rock 'n' Roll Hall of Fame, Lincoln Center, The Pantheon, Museum of Modern Art, Museum of Natural History, Taliesen, City of Baltimore, City of Pittsburgh, City of Cleveland, City of Toronto, Buffalo Psychiatric Center, The Botanical Gardens, The Butterfly Conservatory, Buffalo & Erie County Historical Society, Buffalo City Hall, The Pan Am Exposition, Our Lady of Victory Basilica, Albright-Knox Art Gallery, Guarantee Building, Ellicott Square Building, Niagara Mohawk Building, Central Terminal, The Olmsted Park System of Buffalo, Liberty Building, Darwin Martin House, The Larkin Building...a Signature Peace Bridge!

March 23, 2000

31 TURNING TRICKS ON THE NIAGARA FRONTIER

THE CONSULTING ENGINEERS' FINAL REPORT

The four engineering consulting firms hired by the Public Consensus Review Panel and the Buffalo and Fort Erie Public Bridge Authority to evaluate ways of dealing with Peace Bridge capacity and condition delivered their final report and recommendations on Thursday, March 16, six days late.

Two of the engineering firms – Ammann & Whitney and The Louis Berger Group – were hired by the Review Panel and were charged with representing the community. The other two – Buckland & Taylor and Parsons Brinkerhoff – were hired by the Buffalo and Fort Erie Public Bridge Authority and were charged with representing the Authority. It was a misalliance doomed from the start and objectivity never had a chance. How could a group, half of it charged with representing the public good and the other half with protecting a narrow economic franchise, be expected to do anything but cut out a deal?

And cut a deal they did. The panel recommended a companion span (which would allow the Public Bridge Authority to continue its present operation without having to submit to the inspection of any outside agency) and a new toll plaza for Buffalo. Two clients, two construction projects. Questions about what might best serve the city or region and the commercial industries using the bridge system were secondary to that division of interest. At the end, data didn't drive them to an inevitable conclusion. Rather, an inevitable conclusion determined which data they chose to consider at all.

When I read their report I thought of Herman Melville's comment on consultants in "Billy Budd," his great story about legalism in conflict with justice: "There is nothing namable but that some men will undertake to do it for pay."

Why was the report almost a week late when the engineers knew that the Review Panel was operating under pressure of a pending court hearing with Judge Eugene Fahey? I heard that one or two of the four wanted to recommend a signature span and the others did not. They had agreed from the beginning that they would deliver a single recommendation: there would be no split vote and no minority report. So they wrangled and negotiated until the signature span holdout or holdouts finally gave in.

I asked one of the four principal engineers if that story were true, and, if it were not, what *was* the reason for the delay. He wouldn't answer, so I can only speculate – not that it much matters at this point. If they all signed on to the report, *when* they decided to give their name to it is of far less import than *what* they gave their name to.

Five Recommendations They Were Willing to Provide

These are the five things the four engineering firms said we ought to do:

> 1. Set up a legally-binding partnership between Buffalo, Fort Erie, and the Public Bridge Administration.
>
> 2. Move the plaza to the northern location designated in their various documents as plaza E, the one advocated by the PBA.
>
> 3. Begin construction of a companion bridge this year.
>
> 4. Commission design studies to replace the Parker truss on the old bridge.
>
> 5. Pursue funding for additional customs & INS staffing at the border.

The first is political and, as Reverend Ivery Williams pointed out in the final meeting of the engineers and the Review Panel on the morning of March 7, none of their concern. They were asked to make technical evaluations of bridge systems, not to design international organizations. Since there already is an organization that makes legally-binding agreements between Buffalo and Fort Erie – the Buffalo and Fort Erie Public Bridge Authority – this recommendation is also redundant. The present mess arose because that organization wouldn't act as a responsible citizen. Why would creating a new organization consisting of exactly the same components be any more useful or ethical?

The second recommendation – moving the new plaza to the site suggested by the PBA – is unobjectionable, unless you object to the PBA unnecessarily chewing up a big hunk of Niagara street.

The third – build the twin span and do it soon – is a shame, scandal and abomination, and shows you why some people say you hire a consultant for one, but not both, of two reasons: to find out what you didn't know ***or*** to convince somebody else that what you've already decided is right, no matter what the facts or truths are.

The fourth – design something to replace the Parker truss – seems reasonable enough, since it would be hard to replace the Parker truss without designing the replacement. This recommendation would be unnecessary had they done a responsible job with recommendation number three.

And the fifth – fix the mess at the border processing – is important, given that the present congestion on the bridge is not the result of too few lanes on the bridge but of too few inspectors available to move trucks through Customs and INS once they are off the bridge.

(Each section of the consultants' report was prefaced by a blue page with pictures of several bridges and, in large type, the word "Section" followed by the appropriate numeral. The text on the blue page for the recommendations says "Section 8." The engineers probably didn't mean anything special by it, but to anybody who spent time in the US military that phrase has a very specific meaning. A "Section 8" is the kind of discharge they gave you when they decided you were totally looney.)

FOUR QUESTIONS THEY AVOIDED

These are four of the key questions raised by the Panel and by speakers at the March 7 and 8 hearings they chose not to address:

> Would clearing up the border processing mess obviate the need for a new bridge entirely? That is, if the bridge simply spilled into clear highways would it have to be replaced at all? (Its overall decrepit condition notwithstanding.) One of the Canadian representatives at last week's hearings suggested moving the whole US processing operation to the Canadian side where, he said, there's plenty of room for it. We already do that sort of thing at the Toronto airport and it works perfectly well.
>
> If the only reason for a new bridge is to handle increased truck traffic, wouldn't a better ***regional*** solution be to

> route them through Lewiston-Queenston, which can easily double or triple its present truck load, or convert the hardly-used lower railroad bridge at Whirlpool Falls to a truck-only route (that bridge is in far better condition than the current Peace Bridge, is strong enough to handle the load, and could be converted at a fraction of the cost of building a new bridge here). Both those locations already have or can easily construct links to I-190 on this side of the border and the QEW on the other.
>
> How soon will the old Peace Bridge, the one that they recommend be duplicated, have to be totally replaced: ten years, twenty years, thirty years? And how long will that second bridge construction at this location take? How much will it cost? How much disruption will it impose on the Buffalo community and on cross-border truck flow?
>
> Wouldn't all the involved communities and industries be better served by a full environmental impact study that included the bridge and plaza as a complete system, and wouldn't that be a far more responsible way to approach this whole issue from a professional engineering point of view?

WHY PBA DIDN'T WANT THE ENGINEERS TO ASK THOSE QUESTIONS

The PBA has fought very hard to keep this from ever becoming a regional discussion. They are in fierce competition with the three-bridge system in Niagara County. Except for the town of Fort Erie, which profits handsomely from the present situation, that competition does not serve any of the legitimate stakeholders well in any regard. But the PBA's isolation does ensure the continued prosperity of the small group of well-paid contractors it hires to do hugely expensive maintenance on its decaying steel bridge and, if its plans are successful, on the new high-maintenance companion to it.

One experienced local politician said to me this week, "When you people talk about the concrete bridge being far cheaper to maintain you think you're bringing up an advantage. To them it's a disadvantage. They want a bridge that costs as much as possible to maintain. Think how many jobs they have to pass around. That's what politics is all about."

Another said the same thing another way: "The reason the PBA is fighting so hard to keep that old bridge is simple: if they go to a six-lane bridge other agencies will be involved and those agencies will demand to look at the books and the contracts and their sweet franchise will be jeopardized."

WHAT THE ENGINEERS DELIVERED

All in all, the consulting engineers delivered a thoroughly contemptuous and contemptible report. Contemptuous because it took no serious cognizance of any of the community issues raised over the past year, and contemptible because, at great cost, it merely provided a rationale for the conclusion the PBA has been trying to impose on the community all along. Much of it was predicated on numbers provided by the PBA itself, which is on a par with advising someone to buy a used-car on the basis of the salesman's description of its internal condition.

They ignored nearly everything said in response to their preliminary report by members of the public, members of the Review Panel, and by people with far more experience in this kind of bridge construction than any one of them has. If you doubt that the substance of this report was determined before any of the recent hearings, notice how many times in the 100 pages (many of them charts, maps, and drawings) they refer to the "heritage" status of the present bridge. The present bridge doesn't *have* any heritage status, a point made at the hearings by signature span advocate Ross Robinson (a Canadian) and by a Fort Erie official who had asserted the existence of such status on Tuesday, only to retract the claim in a letter to the Panel's steering committee on Wednesday. Clearly, the consulting engineers were not going to have their conclusions obfuscated by data.

Those four firms are out of this now sorry affair but in case anyone is taking nominations for the Engineering Consultants' Hall of Shame let me list their names one final time:

Ammann & Whitney Consulting Engineers
The Louis Berger Group, Inc.
Parsons-Brinkerhoff Quade & Douglas, Inc.
Buckland & Taylor Ltd.

THE CALENDAR

The Panel planned to discuss and vote on the engineers' report immediately after it arrived Friday, March 10. That would have allowed two full weeks before the March 27 date on which Judge Eugene Fahey had said he would announce his decision on the lawsuits filed against the PBA by the Episcopal Church Home, the Buffalo Olmsted Parks Conservancy, and the City of Buffalo. The report kept not arriving and the engineers kept silent about the problems they were having and when those problems might be resolved. The steering committee of the Review Panel began to worry that the judge would deliver his decision before anybody saw the engineers' report, before the Panel voted, and before the PBA responded. That might have made the entire Consensus Review process a waste of time, which surely would have delighted the PBA.

During the week, Gail Johnstone, co-chair of the steering committee, informed Victor Martucci, chairman of the Buffalo and Fort Erie Public Bridge Authority, that the report had not been delivered on time and that once it arrived the Panel intended to take up to the full two weeks it had been promised to consider it in their agreements last December and January. Her message may have been simply informative. It may also have had a strong implication: we're not going to take the heat if the engineering team the PBA hired in a foot-dragging process that put this project on hold for two months, is now trying to muddle the process again with another delaying tactic.

Two members the Panel told me that Martucci, in response, demanded a vote from the Panel in four days, but Judge Fahey would have none of that. On Thursday, March 16, while everybody was still waiting for the engineers' report to arrive, Judge Fahey revised his schedule so the delay wouldn't cause unnecessary problems for any of the parties to the lawsuits. This is the new calendar:

> *Thursday, March 30:* the Review Panel votes on the engineers' recommendation
> *Tuesday, April 4*: Judge Fahey, the litigants and their attorneys meet to hear the PBA's reaction to the Panel's vote
> *Wednesday, April 5*: the PBA delivers its official response to judge
> *Friday, l 7*: Unless something that happens on April 4 or 5 gives him reason to put things on hold one more time, Judge Fahey renders his decisions.

CROSS-BORDER CON-MEN

What would you say if I said, "Let's be equal partners. You pay half a million bucks and I pay nothing"?

Yeah. Me, too.

That's exactly the deal Canada and the US have in regard to bridge plaza facilities. The PBA and its Canadian supporters have been waving the flag of "new bridge, no tax dollars" at us for a long time now. It turns out what that really means is, no Canadian tax dollars, but lots of American tax dollars.

Here's how it works. The United States government rents from the Buffalo and Fort Erie Public Bridge Administration the space it needs for Immigration and Naturalization Service, the Customs Service and the Food and Drug Administration. It also pays rent on parking spaces. For that space, the PBA bills the US government $439,239.91 annually. Since 1984, the United States government has paid the PBA $5,316,170 in rent. The rent goes up or there is a lump sum payment whenever alterations are made.

(That information was provided to Senator Daniel Patrick Moynihan in a letter dated March 16 from Thomas J. Ryan, Regional Administrator of the US General Services Administration. Ryan's letter was written because the PBA kept stonewalling Joe Crangle, representative of the Community Foundation on the Review Panel. On three separate occasions Crangle had asked the PBA for information about plaza costs and they wouldn't tell him anything, so he asked Senator Moynihan's office for help.)

The Canadian government has never paid anything for any of the far larger facilities on the Canadian side of this or any other border crossing with a toll. Nor has the Ontario government. Nor has the Fort Erie government. Nothing. Zero.

That is because the Canada Customs Act specifically makes it a free ride for the Canadians:

> The owners or operator of (a) any international bridge or tunnel, for the use of which a toll or other charges is payable...shall provide, equip and maintain free of charge to Her Majesty at or near the bridge....adequate buildings, accommodation or other facilities for the proper detention and examination of imported goods or for the proper search of persons by customs officers.. .. The owner or operation of an international bridge or tunnel is liable for all reasonable costs incurred by the Minister....

In ordinary English that means every year Americans pay almost a half million dollars at this bridge for services that are provided free to Canadians. It means when the Buffalo and Fort Erie Public Bridge Authority brags about not using tax dollars it is bragging only about *Canadian* tax dollars, since American tax dollars are underwriting its plaza operations on this side of the border and everybody's toll payments are paying the entire cost of the Canadian plaza.

That half-million is chickenfeed compared to what we'll be paying once the new plaza is built. The rental rates Thomas J. Ryan reported to Senator Moynihan are for the old plaza and are specified in a lease that is 16 years old. The rent for the new plaza will increase to cover all the costs of all buildings and parking spaces used by any US government agency. We don't yet know what the total will be, but it certainly will be more than the new Rainbow Bridge plaza, where the GSA is currently paying $1.9 million per year, with increases of $200,000 in the annual payments every five years over the 30 year lease period. There, too, the Canadian government pays nothing for the space, or parking or renovations: the entire $40 million for renovations to the Canadian plaza came from tolls.

The Buffalo and Fort Erie Public Bridge Authority is an international body – five Canadians, five Americans. Why are the five American members of that board willing to give all that American money to the Canadians in exchange for nothing, or nothing they are willing to tell us about? Why do they keep telling us that they run their bridge with no tax dollars when they know perfectly their operation is underwritten by *American* tax dollars?

We can't blame the Canadian government for screwing us in this deal (or the town for Fort Erie for doing everything it can to keep the cash cow healthy). Every country has every right and even the obligation to get as good a deal as it can for its citizens. In this case the Canadians got and are getting a very good deal indeed.

The problem is with the five Americans on the PBA, who are party to this, and the several members of the New York state assembly and senate who give them blanket endorsements in everything they do and say. Not one of them ever told us about this gross imbalance and they have lied consistently to keep us from knowing about it.

You learn things like that and you have ask three questions:

> Who are they really representing?
> What do they get for it?
> What can be done to make them accountable to the public?

32 THE VERNAL PEACE BRIDGE Q&A

WHERE THINGS ARE

The day this issue of *Artvoice* reaches the stands marks the end of the long investigative process begun last summer by the City of Buffalo, Erie County, the Community Foundation, and the Margaret L. Wendt Foundation. That process created the Public Consensus Review Panel, which hired engineers to evaluate several bridge and plaza options. The Panel met this morning and voted to reject the engineers' recommendation

The Buffalo and Fort Erie Public Bridge Authority boycotted the panel until it seemed likely that Supreme Court Judge Eugene Fahey was going to rule against them in lawsuits filed by the City of Buffalo, the Episcopal Church Home and the Olmsted Parks Conservancy. The PBA never did join in the Panel's meetings or encounters with the public (though Panel member Natalie Harder of the Buffalo Niagara Partnership was ever sensitive to and articulate about the PBA's interests), but they did hire two engineering firms to represent them in the technical evaluations.

The bi-national team of consulting engineers, as they came to be called, recommended moving the American plaza to the north of its present site and allowing the PBA to go ahead with the twin span bridge design it has advocated all along. The team also recommended an environmental impact study on the new plaza construction but not on the bridge itself. All the citizens' groups opposing the PBA's plans have urged a full environmental impact study, not the segmented version the PBA wants and the engineers recommended.

Some Panel members seemed worried that if the recommendation were rejected the PBA would keep its plaza where it is and would even try to consume more of Front Park. Other members pointed out that if the Panel rejected the recommendation, the City was still in a position to refuse to issue the construction permits and could continue withholding them until the PBA agreed to act like a responsible corporate citizen. If it didn't go ahead with the new northern plaza plan its own engineers recommended, then it couldn't begin construction on any bridge at all. The only member of the Review Panel to support the engineers' recommendation for the twin span was Natalie Harder.

In a closed session in Judge Fahey's chambers next Tuesday, the PBA is scheduled to tell the three litigants its response to the Panel's vote on the engineers' recommendation. Judge Fahey is scheduled to rule on all the pending lawsuits three days later. He might do nothing for the present or he might order the PBA to do a full nonsegmented environmental impact study before any ground is broken on any new projects this side of the river. The Episcopal Church Home and the Olmsted Parks Conservancy might settle their suits against the PBA and the City might lose its will or nerve and back down. A big election is coming up, the PBA's attorney is a major Democratic Party fundraiser, and we know no more now about the backroom aspects of this than we did a year ago.

So we're at a key junction where things might go in any of several directions. It may be some time before anything is resolved.

SOME QUESTIONS, SOME ANSWERS

And many questions continue to be asked, some of them answerable, some not. Last June, *Artvoice* published "The Great Summer Peace Bridge Q&A" in which we responded to 12 frequently asked questions about the Peace Bridge affair (e.g.: "Does the Peace Bridge make money?" "Who uses the Peace Bridge and what do they pay?" "Why are the lawsuits necessary?" "Why is the PBA so adamant about the companion span?" and so forth).

In the intervening months, we've learned more about the answers to some of those questions and we've heard many new ones. So, at this pivotal moment in the long decision-making process, we're going to do another Q&A. Every question that follows was sent to us by email, US mail, or was asked over the phone or in a face-to-face conversation.

What does "signature bridge" mean?

A bridge that is not only functional but beautiful enough that it would be identified with the community it serves.

What is the PBA and what does it do?

The organization's full name is the Buffalo and Fort Erie Public Bridge Authority, though its agents and directors and the *Buffalo News* refer to it as "The Peace Bridge Authority" whenever possible.

Most people refer to it as "the PBA," which works equally well for those who want the "public" left in or kept out. The PBA was created in 1933 to operate the Peace Bridge after the initial owners got into financial difficulties when traffic dropped off because of the onset of the Depression and the end of Prohibition. It is a public benefit corporation created by the US government, the Canadian government, the Province of Ontario and the State of New York. No one I've talked to really knows what a public benefit corporation is, which is one reason the PBA gets away with so much mischief. The mission of the Authority is to oversee maintenance of the bridge and to set tolls at a level that will permit proper maintenance and service of bond debt. If it stopped there we would not be in the present mess.

What does the typical board member of the PBA do?

PBA board members decide who gets the maintenance contracts and how much they will be paid for that work. They set wages for bridge employees, amounts of pensions, terms of paid and unpaid leaves, and levels of staffing. They hire the principal operating officers – presently Stephen Mayer and Earl Rowe – and they tell those officers what to do. They decide when to float bonds and how much interest to pay and which brokers will get the very lucrative commissions on those bonds. They decide on construction necessary for the bridge operation and who will get the construction contracts. They select and hire consultants for various things. They make contracts with the people who run the duty free shops. They decide whether or not any of the bridge's spending will go out for bid. Sometimes they make public relations or land deals they think or claim are in the bridge's interest, such as spending more than $20 million to provide a skating rink and city hall (on Rte 3) and courthouse (at the corner of Jarvis and Central) for the Town of Fort Erie.

They also decide what kind of audits to have. They have an annual financial audit, which says whether or not accountants agree with their assertions about how they spent their money and what they have in the bank. They do not (so far as I know) permit or authorize management audits, which might indicate what people in their employ do or whether they do anything at all or whether they do it at all well.

The PBA's ten board members also decide to commission and how much to pay for advertising, such as the recent television and radio campaign devoted to influencing the Public Consensus Review Panel and New York Supreme Court Judge Eugene Fahey. They also decide on special hires, such as the Albany lobbyist they recently set to work on behalf of the twin span.

Much of what they've done in recent years is secret, or at least what they've talked about is secret and how they've managed certain things are secret. They go into executive session a great deal now. Some of their recent minutes are nothing more than a call to order, a motion to go into executive session, a notice that they came out of it, and a motion to adjourn. Before this twin span issue came up they hardly ever went into executive session.

How do you get on the PBA?

All five Canadians are appointed by the Canadian Minister of Transport in Ottawa, and he tells them how to vote on all major issues. (None of them has any expertise in any aspect of bridge construction, operation or management, or social or ecological issues: two are lawyers, two are accountants and one is an eye surgeon.)

Two of the Americans are appointed by the governor of New York State, the other three serve *ex officio* – New York's Attorney General and Director of Transportation, and the director of Niagara Frontier Transportation Authority. The Transportation seat is currently occupied by a political appointee, not by anyone in the Transportation department. Traditionally, the Attorney General's seat is occupied by the director of the Attorney General's Buffalo office, currently Barbra Kavanaugh. Her predecessor, appointed by Attorney General Dennis Vacco, a Republican, was the owner of a steel mill.

The American members of the Board used to be paid for their service but now they do it for free, or on time they're being paid for by the state anyway. All five Canadian members get $100 per day (Canadian).

What happens to the PBA when the bond debt is paid off?

It dissolves and ownership and management of the bridge is transferred to the same transportation agencies in Ontario and New York that take care of other roads and most other bridges. There seems little danger of that happening. The PBA has regularly sought new bonding and has kept the debt large. With the construction of a new bridge and plaza there will be enough debt to keep the PBA going for the remainder of this century.

Why does Fort Erie do everything the PBA tells it to do?

A lifetime resident of Fort Erie said: "This is a small town. You throw $100,000 around and it's a big deal. These people are throwing millions. The courthouse. The town hall. The double-surface rink. That's $20 million in a town the total real estate assessment of which is maybe $58 million. If you were mayor what would you do when they told you they wanted something?" Another said, "The Bridge contributes to everything over here. Whatever Fort Erie wants, they get. And whatever the PBA wants, Fort Erie gives."

What problem does this bridge and plaza construction project try to fix?

Auto traffic on the Peace Bridge has declined in recent years, truck traffic has nearly doubled and it will continue increasing – all as a result of NAFTA, the North American Free Trade Agreement. The current bridge and plaza system was not prepared to handle all that truck traffic. The problem has been partly alleviated by construction of a processing center on the Canadian side that helps truckers get their paperwork in order so they aren't held up on this side for merely technical reasons. Even so, some trucks are inspected and that takes time and they're big so they take up a lot of space.

Will adding three lanes – in the form of a second three-lane bridge or a new six-lane bridge – solve the problem of congestion on the American side?

No, and no one thinks they will. The primary delay is from the truck processing, not the number of vehicles passing through. If there were no processing, traffic would flow smoothly and there wouldn't be any backups. Would that continue if truck traffic keeps growing at its present rate? Probably not. What happens if a car or truck breaks down even now? A mess. Much of the short-term problem would be solved if we could get more Customs and INS inspectors here, but that's almost impossible. Most of those resources go to the southern border. The primary mission of US Customs Service has nothing to do with trade; it is the interdiction of drugs, and drugs mostly come in from the southern approaches. It's unfortunate that our government prefers to pour resources into an intractable social problem perhaps best solved by other means and ignores a very tractable logistics problem easily solved by available means, but that's how it is and – until all the northern border governors, senators and representatives start screaming and standing together on this – the situation won't change.

Is a new bridge at the current site the only option?

No. There are at least two viable alternative solutions to the truck-flow problem.

One is to route the trucks across the Lewiston-Queenston bridge. The customs brokers would not lose any jobs; they'd either keep their operations here or they'd move them up to those two bridges. The negative impact on Buffalo's economy would be minimal, and the positive impact on its environment would be significant.

Another option was suggested by the operators of Detroit's Ambassador Bridge last November: build a new bridge adjacent to the railroad bridge a mile north of the Peace Bridge. There would be an easy connection to the I-190 on the American side. On the Canadian side there is a wide swath of former railroad bed that provides a straight shot to the QEW, so constructing a connecting road there would require no displacement of residences and no meddling with archaeological ruins. The space is there, it is already cleared, it is unused.

The Ambassador group proposed a full-service bridge at that location, but what might make even more sense is to put a trucks-only bridge there. That would leave the present Peace Bridge functioning, until its condition deteriorated so much it had to be replaced. When the Peace Bridge was being repaired or replaced, automobile traffic could be diverted to the truck bridge; if that caused serious congestion problems, some of the truck traffic could be temporarily diverted to Lewiston-Queenston. Free of trucks, the Peace Bridge could shift its far smaller plaza into a northern location that would not require rerouting Niagara street and which would permit full restoration of Front Park and Fort Porter.

Buffalo wouldn't get a signature bridge out of that kind of project, but neither would it have to endure ever increasing amounts of noxious fumes and traffic problems caused by the trucks landing in a residential area. I'm not recommending it – I'd like to see a signature bridge here and see no reason we couldn't do one. But it's important that we don't get locked into the kind of either/or pattern the PBA keeps telling us is our only option. There are many options, and the full EIS might very well provide an opportunity to explore them honestly, which has not yet been attempted.

Was PBA Chairman Victor Martucci correct when he told the Review Panel and the Buffalo News that building the twin span now is "about jobs."

Yes, he was. But hardly any of those jobs will go to workers in Erie County. Nearly all of them will go out of the state or out of the country.

The consulting engineers told the Review Panel that the old bridge would last another 75 years. Upon what data did they make that prediction and are they right?

They used data provided by the Public Bridge Authority and it is unlikely they are even close. "It is irresponsible simply to assume that the bridge is ready for another 75 or 100 years of use," Senator Daniel P. Moynihan said in a letter to Review Panel co-chair Gail Johnstone earlier this week.

In the lead paragraph of its March 28 front-page story on the Bridge's condition, the *Buffalo News* reported that "no one has taken an underwater look at its piers since 1996." That's all it was: a look. Someone on top stuck a camera under the surface. There has not been an real underwater inspection of the piers since a barge rammed the bridge more than 15 years ago. There hasn't been an underwater inspection extensive enough to be included in the federal reporting since before 1981, though the condition of the piers is key to estimating maintenance costs and life span of the bridge. If those piers are deteriorating or are unstable – as some engineers insist – the maintenance costs would be enormous, the bridge would be out of commission a great deal of the time, and, like a tooth that looks okay on the surface but is rotted on the inside, would not be salvageable for the long term anyway. If you look at the piers, you can see deterioration at the waterline; there is almost certainly far more damage from the swift current beneath that.

The PBA's engineers have predicted the cost of redecking the old bridge, but most bridge engineers say there is no way to predict those costs because no one knows what structural problems they will find once the current deck is peeled away. This bridge was built with the same kind of high-sulphur steel used to build the *Titanic*. It does not take welds well, which is why the entire Peace Bridge is held together with nuts and bolts. Repairs to such high-sulphur steel structures are far more difficult than to the easily-weldable steel used in recent decades.

The simple fact is this: all the numbers they've giving us are guesses at best-case minimums, and some of those are not believed by anyone but the PBA, if indeed even they believe their own numbers. No one knows how much the rehabilitation and maintenance cost will really be, or how long the bridge will survive.

But isn't it true that the Federal Highway Administration rating for the Peace Bridge improved this year?

Yes. It went from a sufficiency rating of 17 (on a scale going from 1-100) in 1997 to 58.9 in 1998, a huge jump.

The change was based on ratings of three factors, each evaluated on a scale of 0-9): superstructure from 4 to 7, substructure from 3 to 6, and structural evaluation from 3 to 6. A 3 for substructure means "serious condition – loss of section, deterioration...local failures are possible" while a 6 is "satisfactory condition – structural elements show some minor deterioration." A 3 in the structural evaluation is "basically intolerable requiring high priority of corrective action," while a 6 is "equal to present minimum criteria."

How did the bridge improve so radically in a single year? It didn't; the only thing that changed from 1997 to 1998 were the numbers the PBA submitted to the FHA. The FHA ratings are based on self-reports; they don't go out and do any inspections, they trust the bridge managers to tell them the truth and to be consistent in their reporting. The PBA improved its reported numbers by 100%.

This sudden improvement in reported condition, with no improvement in fact, reminds me of last December when Stephen F. Mayer, the PBA's general operations manager, told the Review Panel that the construction plan they had said would take ten years would only take seven years. When asked where the three years went, Mayer said that they had recalculated the schedule.

The PBA says it doesn't need any public money to build their bridge. Does it really have enough money to build the bridge and new plaza?

No. They are admittedly counting on US tax dollars to build the plaza and new access roads. Those no-tax-dollar claims have to do only with the bridge itself, and to do that they would need a large increase in their New York bond limit. State Assemblyman Sam Hoyt says he is going to block any such limit increase until the PBA starts acting like a responsible corporate citizen. Federal funds are available for this bridge construction project, but thus far the PBA has refused to apply for them because they would obligate the Authority to open its books and contract-letting practices to outside inspection.

The PBA has sufficient cash reserves to start building a twin span, but not to finish one, so there is serious question whether it could at present even get the bonds permissible under the current limit. Absent the assurance that the PBA can finish, those financiers won't float the bonds, and absent that, all the PBA can do is start the bridge and then hope to blackmail the state into lifting the bond limit to get a half-finished eyesore done with.

So if they get the bond level raised, they could do the bridge without tax money?

No, not even then. That claim is a legal fiction understood by the manager of any non-profit institution with a federal or state tax exemption.

In real terms, tax exempt bonds are a government subsidy designed to help museums, schools and other public entities compete in the financial market. The tax exemption lets those organizations pay lower rates of interest on their bonds because the bond owners don't have to pay taxes on the income. This is using tax money, whether or not a cent of public money ever changes hands. Neither are they paying property taxes on their bridge, buildings, or land. If they are not paying the same property taxes you and I pay, that means our tax dollars are underwriting their operation. They use tax money every day, and will continue to do so. It's all a matter of where and how that support shows up in the books.

What will be the effect on Buffalo's West Side if the PBA is allowed to proceed with its twin span plan?

A mess. According to the most recent traffic plans, for as few as four and as many as eight years, trucks will be rerouted along Niagara street, through Delaware Park on the Scajaquada Expressway , and out to the Thruway on the Kensington Expressway. That will not only clog traffic on those routes – neither of which was designed to be a major truck thoroughfare – but will also have detrimental effects on Delaware Park (if the joggers on the Ring Road are bothered by car exhausts now they're going to need oxygen tanks once the trucks start slogging through the park), on housing values in the top-dollar locations bordering the Park, on access to the Albright-Knox, Burchfield Gallery, Historical Society, and Buffalo State College, and more. In general, the quality of life in that part of the West Side and the safety of road use on those two highways will be severely degraded.

The PBA says going to an environmental impact study now would set the project back two years. Is that fair?

Fair to whom? To the Buffalo community? Sure it is. To the American taxpayers, who have been getting royally screwed by Her Majesty's bridge laws on the other side? Certainly. (The current arrangement specifies that all expenses for all Canadian government operations on bridge property be paid out of tolls, and all expenses for all American government operations be paid out of American taxes.) Had the PBA been willing to take part in a nonsegmented EIS three years ago when it was first urged on them, this would have been long over. It's their evading tactics, not the community's, that have caused the delay. We're not the bad guys in this, no matter how much their smarmy television ads implied otherwise.

Furthermore, an EIS won't take two years. It probably won't even take a full year. They project that number only to argue that going to a six-lane signature bridge would delay things even more. Much of the work has already been done. If they'd stop stalling and evading, the rest could be done fairly quickly.

What's wrong with the Parker truss?

The truss – the big green boxy structure atop the part of the bridge crossing the Black Rock Canal – was not part of the original design. That section was supposed to be like all the others with graceful curved support elements beneath the roadway. Then the builders were informed that the bridge had to clear the canal by 100 feet all the way across to provide clearance for the masts of larger sailing ships which then plied the Great Lakes. That meant the section had to be supported from above, rather than below, and it had to rise higher than had previously been thought necessary. The builders hated the truss and always thought it detracted from the beautiful design of repeating arcs crossing the river and canal. Washington officials are insisting that any new bridge continue to clear the canal by 100 feet even though no boats in operation nowadays need anything near that clearance. Their reason: "It might be needed in the future." If it weren't for that anachronistic height requirement, the bridge could be flatter and the design issues would be far simpler.

But whatever happens, the PBA will build the northern plaza and if they don't build a signature span they'll at least replace the Parker truss with something to match the companion span, right?

Wrong. If a court forces them to do it they will, but otherwise it is unlikely they will do anything but build a companion span.

Last week Mayer told WEBR reporter Mike Desmond what he had told the Review Panel last December: that the PBA would consider building a northern plaza when and if someone else gave the PBA all the money it needed to do the entire project. He added that someone would also have to get that plaza though an EIS, otherwise the PBA has no intention of doing anything other than building its twin span.

Mayer also told Desmond there would be no determination about replacing the Parker truss on the old bridge until there was a satisfactory determination of its historical significance. What historical significance? The truss is an eyesore, a bureaucratic emendation to the original design. No one defended it in the past or defends it now, but all of a sudden the PBA is saying its historic value must be investigated. The only reason for that bit of flummery is obfuscation and delay. The fact that Mayer would say such nonsense to a reporter tells us that the PBA is still planning on building the bridge it has wanted to build all along, and that it plans to stall and stall and stall when it comes to doing anything else.

Which bridge system would be cheaper to build, become fully operational earlier, be less expensive to maintain over the next century, cause less disruption, and look better – twin span or six-lane prestressed concrete?

Six-lane prestressed concrete on all counts.

If it's true that the companion span is the least efficient option, why does the Canadian Minister of Transport in Ottawa instruct his five delegates to the PBA to insist on it, and why do the five American delegates to the PBA follow their lead?

In part it's because the politicians don't want to renege on the deals that were made for who would get the lucrative steel construction and maintenance contracts. Part is an unadmitted but not insignificant measure of that anti-Americanism makes the Canadians unduly resistant to alternatives coming from this side of the border. Part is that the Canadian members of the PBA arrive at their meetings knowing exactly how they're going to vote that day and the Americans almost never do.

And then there is saving face and blockheadedness.

Never underestimate how far people will go to preserve face. Nearly a decade ago, the PBA assumed that the old bridge was of historical value and could not be torn down. They were wrong. Because of that assumption, they never gave any serious consideration to a six-lane bridge; they even excluded it when they first told their designers what to think about. By the time serious opposition arose, they were deeply committed to and identified with their design and feared they would look stupid if they acknowledged they had missed the obvious at the onset. The higher Canadian officials who had backed them because their plan seemed reasonable then, now backed them because they didn't want to look foolish and incompetent either. They didn't want to admit they had appointed a bunch of people with no competence to make a decision of this magnitude and had exercised no intelligent oversight when they locked the process down. A sad fact of politics is that politicians don't just occupy their predecessors' offices; all too often they choose to occupy their predecessors' errors as well.

Why do they think they can get away with this?

> Because they've gotten away with so much so often in the past.
> Because they think we're stupid.
> Because they think if they stonewall long enough we'll get tired and will give up.
> Because they've got a lot of money to buy what they want and they think they'll be able to buy the opinion they want here just as they bought it across the river.

They have so much money and they are so committed to their plan. Can they be stopped?

Yes. In the short term, it is important that the City of Buffalo not give up in weariness or because of other compelling issues or because of frustration.

In the long term, a solution may come from the courts. The courts may give the PBA and the Canadian government the excuse they need to work with this community rather than in opposition to it. For more than two decades a succession of Democrat and Republican governors refused to settle the Attica civil rights case, though any one of them could have done so, could have blamed it on Rockefeller, and could have pointed out that the settlement was far cheaper than the millions being spent on attorneys and staff defending the suit. Not one was willing to bite the bullet until a Federal appellate court said it was time to end this, whereupon the Pataki administration shrugged its shoulders, said 'okay,' and settled. 'You can't blame us for the Attica settlement,' Pataki's team seemed to say, 'the court is making us do it.' No loss of face there.

And no loss of face here if a judge tells them they have to take part in a full EIS and if in the course of that process they realize that the six-lane signature bridge really is better and faster and cheaper. They can do the right thing, if they can find an excuse for doing it.

Here are nine questions we haven't been able to answer. We'd very much like to hear from anyone who has specific information that would help us with any of them:

What, if any, US interests does Congressman John LaFalce represent in this?

What business or political interests caused the Buffalo Niagara-Partnership's president Andrew Rudnick and development director Natalie Harder to consistently represent Canadian interests in the Peace Bridge affair?

Why did the directors of the Buffalo Niagara-Partnership allow or encourage Andrew Rudnick to pursue with such vigor and for so long a policy that would cause this area to lose many jobs that would otherwise have come here?

Why have New York Attorney General Spitzer and the head of his Buffalo office Barbra Kavanaugh, both of them signature span advocates a year ago, gone silent on the bridge issue?

Was it legal for the PBA *to spend hundreds of thousands of dollars in an attempt to pressure the Public Consensus Review Panel and the litigants in Judge Eugene Fahey's court?*

To whose drumbeat are the five American members marching and – given Governor Pataki's and Attorney General Spitzer's expressed support of the signature span – why are they doing it?

Why would the mayor of Buffalo fence-sit about a project that thus far has promised only huge harm to his city?

Why would anyone in a position of public responsibility want to route geometrically-expanding truck traffic into an urban area when there are alternatives available?

Given all the questions and concern and the dark veil of secrecy that in recent years has covered much of the PBA*'s operations, why has the* Buffalo News *never used its large investigative reporting staff and great financial resources to look at any aspect of the Public Bridge Authority operation?*

33 Humpty Dumpty Time at Peace Bridge Plaza

"When I use a word," Humpty Dumpty said, in a rather scornful tone, "it means just what I choose it to mean. Neither more nor less."

"The question is," said Alice, "whether you CAN make words mean so many different things."

"The question is," said Humpty Dumpty, "which is to be master. That's all."

What the Consensus Panel Learned

The penultimate motion offered at the final meeting of the year-old Public Consensus Review Panel, which took place at D'Youville College the morning of March 30, came from Natalie Harder, spokesperson for Andrew Rudnick and the Buffalo Niagara Partnership. Harder moved that the Panel accept the report of the engineering team hired by the Panel and the PBA to consider what might be done about truck congestion problems at the Peace Bridge. The engineers had recommended a plaza north of the current site, immediate construction of the companion span the PBA has advocated all along, continued avoidance of an environmental impact study for the bridge project, and a few other things.

No one seconded Harder's motion.

I asked one Panel member later why Harder's motion couldn't even find a second. "They gave us something you can't vote yes or no on," he said. "It's a bunch of weasel words. They were all tired and wanted to get out of this crazy town and figured if they did that we'd let them go. The fact that the PBA came along in October and tried to queer the process is no reason for us to go along with them. Our public mandate was to listen to the public consensus and tell them all we've learned in the previous eleven months. And that's what we did."

Panel co-chair Randy Marks thanked Harder for her motion. Then Edward Cosgrove, former Erie County DA and the representative of the Buffalo Common Council on the Panel, moved that the Panel accept its own report, which recommended construction of a northern plaza, a six-lane concrete bridge, a full environmental impact study, and establishment of an oversight commission to ensure that the work is done correctly. It urged federal elected officials to pursue the wide range of funding opportunities the PBA has thus far ignored and to explore ways of righting the imbalance in Canadian and American investments and benefits. It urged state officials to consider raising the bonding limits that would make all this possible. The Review Panel report also analyzed in lucid detail ways in which the engineers' report had been too vague or ambiguous to permit rational action and ways in which the engineers' report was right on the money. That is, it was a report built around the engineers' report, one that tried to capitalize on its valuable parts and show how its defects might be corrected and, most important, brought the final recommendations into harmony with the engineers' own data.

Randy Marks called for a voice vote. The yeas seemed loud and clear. Then he called for the nays. No one said anything but Natalie Harder raised her hand from her elbow to her shoulder. Marks and the other co-chair Sister Denise Roche, president of D'Youville, could see her vote, as could steering committee members Robert Kresse and Gail Johnstone and the few other people at the front table. But hardly anyone else in the room could see it. I sent her an email later to be sure that she had voted nay and she wrote back that she had.

Hers was the only vote against the Review Panel's report.

Waffles for dinner at City Hall

A few hours after the meeting, Mayor Anthony Masiello and County Executive Joel Giambra issued a joint statement thanking the Panel for its work, endorsing the recommendation for a northern plaza, urging construction of a six-lane signature bridge, telling the PBA to ignore everything else in the Panels' report and apologizing because the Panel hadn't behaved as the PBA had thought it ought.

The letter was patronizing, insulting, lacking in both statesmanship and or grace. It was an astonishing document, a grumbly thanks to two dozen citizens who had given a huge amount of time in the service of this city and county. They deserved better, every one of them.

I've been told that Giambra's signature is probably a result of his weakness after his recent surgery and the fact that the bridge doesn't loom large on the screen of his chief of staff Bruce Fischer.

The mayor's participation in this sorry document seems like another manifestation of what his critics cite as his main fault: a tendency to respond to whoever most recently talked to him forcefully on any particular issue. In this case the forceful talkers were Victor Martucci (chairman of the Buffalo and Fort Erie Public Bridge Authority), Andrew Rudnick (president and CEO of the Buffalo Niagara Partnership), and Larry Rubin (attorney at Kavinoky and Cook, who has worked with Arnold Gardner representing the PBA and who was Commissioner of Community Development in the Griffin administration).

All three had met the previous evening in the mayor's office with Masiello, Giambra, Fischer, Joe Ryan (Rubin's successor in Community Development), and Rich Stanton of the city law department.

Martucci, Rudnick and Rubin brought with them to that Wednesday meeting the outline of the discussion points the Review Panel had agreed upon Tuesday evening after a straw poll showed a clear majority favored the six-lane bridge. The PBA/Partnership damage control team knew it couldn't prevent the vote scheduled for the next morning but, as the Giambra-Masiello letter showed, they were still capable of some last-minute mischief.

JUKE AND JIVE AND A MOMENT OF TRUTH AT PEACE BRIDGE PLAZA

The following morning, Friday March 31, the Buffalo and Fort Erie Public Bridge Authority called a press conference at its Front Park headquarters. While twin-span advocate John LaFalce and PBA attorney Arnold Gardner stood against a far wall watching everything and talking to hardly anyone, PBA chairman Vincent Martucci introduced himself, bridge general managers Earl Rowe and Stephen Mayer, and the PBA board (Buffalo's Barbra Kavanaugh wasn't there, he didn't say why).

The first thing he said was untrue: "I'm Victor Martucci, chairman of the Peace Bridge Authority." As we've frequently pointed out in these pages, there is no *Peace* Bridge Authority. There is a Buffalo and Fort Erie *Public* Bridge Authority. That entire team goes to great lengths to avoid reminding us that they are answerable to the public.

Then he repeated what has of late become almost a mantra for him: the Review Panel, he said, had no right to do anything except vote the engineers' report up or down. The agreement between the PBA and the PCRP last October precluded the PCRP from saying anything about anything. "Pursuant to the agreement, the engineers would then make a recommendation to the PCRP. Now, here is the important point: the PCRP would vote yes or no to ratify the engineer's final report...nothing more and nothing less. This was the agreement that the PBA and all the other parties entered into in good faith."

Well, no, it wasn't. Nothing in the agreement specifies anything like this. The PBA's attorneys may wish they had put it there and Martucci may wish it were there, but they didn't and it isn't and no amount of repetition is going to make it so.

"As a partner in the process," he said, "the PBA agreed to vigorously pursue in good faith, the recommendation of the bi-national team of engineers."

Not so. They agreed to pursue the recommendation of the *panel's steering committee.* This is the passage from their agreement signed last October:

> As part of this agreement, the Peace Bridge Authority commits to pursuing the selection recommended by the Steering Committee, while Mayor Masiello commits to using all powers granted to the Administrative Branch by the City Charter to move the necessary easements forward.

That's pretty simple English; it doesn't take a lawyer's mumbo-jumbo for us to get through that sentence.

Then he said something that was true but grounded in a deception, one that has skewed this entire process:

> The recommendation of the bi-national team of engineers is the same conclusion reached by three previous studies. In 1996, the *Buffalo News* conducted a design competition. In 1996, the PBA concluded its review. And in 1998, the Buffalo Niagara Partnership made its recommendation based on an extensive review. All four of these reviews reached the same conclusion...a companion span best meets the criteria of time, money, environmental and economic impacts.

Those groups did recommend as Martucci says they did, but he doesn't mention a key fact that only came out later in the question and answer period: the Public Bridge Authority had excluded from every one of those studies any possibility of a six-lane bridge. The fix was in from day one. I asked Martucci about that and he asked Bridge general manager for operations Stephen Mayer, who has been on staff for a long time, to answer. Mayer said that in 1994 "the decision was made, which the Authority supports, that the bridge is frankly in very good condition. So it really made no sense to tear down the bridge. For a lot of reasons: there's archeological, there's historical, there's also functional reasons. So then the issue became what was the best way to add capacity."

The decision was made? I've been a teacher and editor for 33 years. The passive voice is a red flag for me. People use it when they don't want to take the rap for something, they use it when they want to pretend nobody did anything anybody has to get concerned about. So I asked Mayer, "Did you or did you not exclude from serious consideration a six-lane alternative?"

"We excluded it because it did not make any sense to take down the Peace Bridge," he said. To my knowledge, that is the first time anyone at the PBA said something all of us who have been examining this issue closely have long suspected: no design other than the twin span has ever stood a chance in any of these design conferences, charettes, charades, or whatever else they called them.

One final quotation from Victor Martucci's prepared statement: "We stand ready to honor our commitment to pursue the bridge and plaza system as finally recommended by the bi-national engineering team."

The bridge and plaza system recommended by the engineers was what they *wanted*. What they made a *commitment* to was what the Review Panel Steering Committee recommended. And that they do *not* accept, which is to say, they have chosen to renege on the contract they themselves initiated.

In civilized society there is only one place to deal with organizations and people who renege on contracts: Court. Which is where the action now moves.

This Friday Judge Eugene Fahey delivers his long-awaited ruling on the suits filed against the PBA by the Episcopal Church Home, the Olmsted Parks Conservancy and the City of Buffalo. No one ever knows what a judge will do until the judge does it, but smart money is saying that Judge Fahey will order the PBA to do a full environmental impact study on the bridge and the plaza projects.

If the PBA decides to make nice and utilize all the data accumulated by the Public Consensus Review Panel, an EIS can be completed in less than a year. If, as they suggested at last Friday's press conference, they choose to appeal, then the process could drag on for years. Only this time they'd be hard put to convince their masters that it's American civic foolishness impeding the march of progress. This time everyone would see very clearly who has muddled this process all along. And it's not us. And it never was.

34 PARTNERSHIP RETHINKS THE BRIDGE ISSUE

The strong coalition of Buffalo and Niagara Public Bridge Authority twin-span supporters is unraveling.

Two weeks ago, the *Buffalo News* gave signals that it would begin asking hard questions of PBA members instead of rephrasing PBA press handouts and calling it news.

Last week the board of directors of the Buffalo Niagara Partnership confronted the Partnership's increasingly alienated position from the community on Bridge issue.

One board member pointed out that the PBA's insistence that its twin span was the fastest way to go had been mooted out by Judge Eugene Fahey's decision forcing the PBA to give up its artificial segmentation of the bridge and plaza projects and mandating an environmental impact study. Another said the consulting engineers' report submitted to the Public Consensus Review Panel indicated that the choices were perhaps not as clear-cut as they had previously seemed. Another said, "We're holding a losing hand here." Another said, "We've got to convince the PBA that appealing Judge Fahey's ruling won't accomplish anything and it will just waste two more years." And another said, "The problem is the Canadians."

He was probably referring to the monolithic Canadian opposition to anything other than a companion span. The mayor of Fort Erie, the premier of Ontario, the Canadian premier, and the Canadian Minister of Transport all recently importuned Governor Pataki to intercede on behalf of the twin span. They also urged the PBA to appeal Judge Eugene Fahey's decision, which would force the PBA to obey New York State environmental law. All five Canadian members of the PBA are appointed by the Minister of Transport in Ottawa, so they will probably do as they've been told.

"We didn't take a vote," one Buffalo Niagara Partnership board member told me, "but I think the clear consensus was that we should build a six-lane concrete bridge and a new northern plaza. And if we have to go to the Canadians on our knees, let's go there on our knees. We've got to do something to get them to pay attention to all this new information."

Canadian Consul-General Mark Romoff, a member of the Partnership board, said nothing in response to the discussion. He has been a key behind-the-scenes player in this from the beginning. For some time observers thought that Romoff's dedication to the twin span reflected
Canadian national policy formulated in Ottawa, but lately some say that the statements by Canadian federal officials were based on recommendations and data from Romoff. "You think Prime Minister Chrétien in Ottawa spends his time worrying about bridge design in Fort Erie?" a Canadian businessman said. "Of course not. That's what he's got a consul-general in Buffalo for."

As always, everything we learn about this affair leads to new questions. Here are some that come to mind:

> Mark Romoff will be key in solving this border problem. Is he willing and able to play that critical role?
>
> What will Ottawa do when Romoff reports that the PBA's principal base of support has reversed its position and now wants a reasoned and open examination of alternatives to the twin span?
>
> Will the PBA's obstinacy and delaying tactics force Canadian and US policymakers to consider another route for Continental One, the proposed Toronto-to-Miami trade corridor? If so, the huge number of jobs supporters insist will come to this region will go elsewhere.
>
> It seems clear that Continental One will benefit construction and engineering companies and manufacturing and distributing companies. Will any of its supporters now step forward to tell us what benefits will accrue to this region from Continental One and if those benefits will outweigh the environmental damage it will surely cause?
>
> How will the five American members of the PBA vote at their meeting Friday if a motion is made to appeal Judge Fahey's ruling? Barbara Kavanaugh was not at the PBA's last meeting and we've been told that she has not taken part in Peace Bridge issues because the New York Attorney General was involved in the litigation and she was avoiding a conflict of interest. But what public interest is

> really being served if that avoidance of conflict of interest means the Americans are always one vote short at PBA meetings?

More to come.

May 24, 2000

35 The Peace Bridge War: A Report from the Front

Where things are

Things are quiet in the Peace Bridge War right now. We won't know until later whether this is merely a lull in the battle or the attenuation of noise that signals a real end to hostilities.

There was a lot of bluster and frothing from the Buffalo and Fort Erie Public Bridge Authority immediately after the Public Consensus Review Panel's March 30 vote, which accepted the consulting engineers' recommendation for a new plaza north of the current plaza, but rejected the engineers' recommendation that a three-lane steel companion bridge was preferable to a six-lane concrete bridge.

The engineers tried to satisfy two political entities – New York and Fort Erie. They gave Buffalo a decent plaza and Fort Erie a second bridge. Several members of the Review Panel pointed out that the engineers' recommendation for a twin span was contradicted by the engineers' own data earlier in the same report. That data said the twin span and necessary rehabilitation of the current bridge would take longer and cost more than building an entirely new six-lane bridge, and that in the long run the six-lane bridge would be far cheaper to maintain. Panelists also pointed out that the consulting engineers had been asked for a technical, not a political, recommendation.

At a press conference the following day, PBA chairman Victor Martucci said the Review Panel didn't have the authority to make its own recommendation because it had signed an agreement with the PBA that it would do nothing but vote on the engineers' report. The Review Panel had signed no such statement.

A week later, on April 7, New York Supreme Court Justice Eugene Fahey ruled that New York environmental law required a full environmental impact study of any major bridge and plaza construction. The PBA's claim that it could commence and finish construction of a twin span faster than anyone could commence and finish construction of a six-lane bridge died right there. If they had to obey the law, then all bridge designs had the same starting line and all the hoopla about getting more trucks across the border faster was meaningless. (It was never true anyway, since all their rhetoric was about the *start* of construction, never the *completion* of it.)

PBA attorneys said they were unhappy with the decision and the PBA would decide at its next meeting whether or not to appeal. Several senior Canadian officials urged the PBA to appeal immediately; none of those statements acknowledged the existence of New York environmental law or the fact that the PBA had as many American as Canadian members.

One prominent Buffalo corporate attorney said to me, "They have to file an appeal just to save face. But if they have any brains they won't pursue the appeal. Fahey's a scholarly judge and that's a very well-grounded opinion. If they actually pursue an appeal they'll just waste two more years and they'll lose anyway."

"So why bother filing at all?" I asked.

"So they have some leverage at the next stage."

"To get what?"

"I haven't any idea. I don't know if they do either."

A week later, the executive board of the Buffalo Niagara Partnership, an organization that had long supported the twin span idea, decided that Judge Fahey's ruling meant it was time to alter course and to support a six-lane concrete bridge plan. Several members of the Partnership board later said they thought filing an appeal of Judge Fahey's ruling would be pointless and time-wasting. "We sent the PBA a message saying that," one board member said to me.

To date, the PBA hasn't done much of anything in response to the Review Panel recommendation and the judge's decision – at least nothing admitted in public. There have been vague threats that there will be no new bridge at all, that the truck traffic will be routed elsewhere (oh, punish us, punish us!). A truck driver phoned *Artvoice* on his cell phone a few weeks ago to say that the traffic jams on the bridge were worse and he was convinced it was a deliberate slowdown to put pressure on the Americans to give in. One lane is currently closed for some heretofore unannounced cleanup work and large signs on the Canadian side proclaim "EXPECT DELAYS." Are those delays really necessary or are they just more of what that truck driver was suggesting? One sad result of all the indirection and misdirection in the past is, it is difficult for most of us to take anything that happens there at face value any more.

WAS THE WAR OF ANY USE?

Yes.

First, we probably won't have that ugly twin span to look at, we won't have huge trucks rumbling through Delaware Park for six years, and we'll probably attain full six-lane functionality far more quickly and cheaply.

Second, we learned important lessons.

One of the first things County Executive Joel Giambra did when he took office was put county involvement in any new convention center on hold until completion of a full environmental impact study – which includes citizen involvement. He might have done that anyway even if he hadn't seen the unnecessary anger, frustration, expense, and delay caused by an agency trying to impose an unwanted project under false pretenses. But having seen those things he wasn't about to stumble into the same swamp.

Planners, engineers, and developers in other cities have taken the message as well. (It's like after the 1971 bloodbath at Attica: no prison disturbance since then has been settled by slaughter and torture.) I doubt that anyone in the construction industry is unaware of what happened at the Peace Bridge, and I doubt that any competent planner doesn't think about ways of avoiding this kind of mess.

We don't have to go far for examples. The New York State Department of Transportation is doing major construction work on route I-490 in Rochester. Part of that project consists of replacement of the Troup-Howell Bridge, which crosses the Genessee River in downtown Rochester. The city told the engineers it wanted a "signature style" bridge. Howard Reffel, the project design engineer and an employee of DOT, said they developed several models and figured which might work within physical and budgetary constraints. These were presented to the public. The Rochester newspaper paid little attention during this phase of the process (sound familiar?) so DOT worked hard getting out word of its public hearings and meetings through other channels. They set up an Aesthetics Committee composed of representatives of local government, artists, the landmark society, and other community groups. They listened to input, developed alternatives in a design workshop and presented them. There still was almost nothing in the Rochester paper about the project, and no editorial commentary at all (sound familiar?). The main thing was letting people know what was going on, giving various constituencies a real chance to provide information that was taken seriously.

After Reffel described all those steps to involve the community in the project I asked him if he had been aware of what had happened at the Peace Bridge. "Of course," he said. I asked if what happened here had influenced what they did there. "Sure," he said, "but what we did was the right thing to do anyway." So it is.

WHAT HAPPENS NOW?

I haven't been able to find out if the PBA is sulking, licking its wounds and resting up, lying low in the hope that the New Millennium Group Bridge Action Committee will get bored and will go on to something else, or is in stasis because it's experiencing a stalemate between the five American and five Canadian members. I suppose someday soon we'll see a puff of white smoke over their office at what used to be part of Front Park and someone will come out and tell us.

We'll have to wait and see what they do. Whatever they do, there are some long-range change we should begin thinking about.

LET THE SUNSHINE IN

This mess might not have happened if the PBA weren't allowed to operate in secret. There's no reason for it to operate in secret. Its job is to maintain a bridge, not national security.

The PBA avoids the sunshine laws and freedom of information laws that make most actions of nearly all public agencies accessible to the public. Their rationale is, the PBA is an international agency, not a state agency, so it is therefore not subject to New York State's public access laws. I don't buy that argument. The PBA is in part a creature of the New York legislature, it is tax exempt and can issue bonds because of acts of the New York legislature, so why can't it be subject to rules made by the New York legislature?

The usual reply to that is, "Because the New York legislature cannot make rules that bind Canadians," We don't have to bind Canadians. Our rules will just bind the American members of the PBA. The New York legislature should pass a law saying that the American members of the PBA can no longer operate in secret. If the Canadians don't like it, they can stay home from those meetings. There is no reason American citizens should be penalized because the Canadian government doesn't give its citizens the same access to information and protection from governmental abuse that our laws provide. (And if they do have the same protections, then let them agree to the change as well.)

Another argument against introducing sunshine is, the US can't unilaterally set up conditions on this binational agency. That's also nonsense: the Canadian government set up a condition without consulting us when it decided that all expenses for housing the offices of all Canadian officials connected with border bridges and tunnels had to come out of toll revenues, not taxes. American officials occupying similar space on the American side are forced to pay rent. Every year, American taxpayers pay millions and millions of dollars for services Canadian taxpayers get for free. Did any American legislative body ever get consulted on that ripoff? Of course not. If the Canadians can rip us off for millions of dollars every year, what prevents us from making sure that our representatives to border agencies obey US law?

REPAIR THE PBA'S DEFECTIVE COMPOSITION

The Peace Bridge was officially opened August 7, 1927. It was built by a private corporation headed by Frank B. Baird. The Buffalo and Fort Erie Public Bridge Commission, a public benefit corporation created by the United States, Ontario, and New York governments, took over in 1933 when the Depression and the end of Prohibition drove the private corporation to the brink of bankruptcy. The original membership was 6 Americans and 3 Canadians, I think to reflect the much greater American investment in construction of the bridge. In 1957, New York created the Niagara Frontier Port Authority and tried to make the Bridge Commission part of it. The primary reason for that was to grab the large amount of cash the bridge had in the bank. The bi-national Bridge Commission refused to be eaten by a State agency. After some legal wrangling, there were four changes:

> the membership of the board was increased to ten, five from each country
> the bridge would be tied to no other agency, it would be fully independent
> the two countries would divide excess revenues equally
> the sunset, the date when all bridge assets would be turned over to the two governments to be run as part of ordinary state and province facilities, was extended to 1992.

The turnover never happened because the PBA kept getting more debt and it was guaranteed existence as long as it had debt. The revenues were never divided equally because of the Canadian rental ripoff I described above. The PBA has remained independent of other agencies, but it has not remained free of politics.

Two of the American delegates are appointed by the governor

and three serve ex officio: the head of the NFTA, the director of the New York Department of Transportation, and the New York Attorney General. The three ex officio board members frequently are represented by designees: the Attorney General usually appoints the head of the Buffalo office, for example. Sometimes the governor gobbles up both the Attorney General and Transportation slots and gives them to political friends. All five Canadian members of the PBA are appointed by a single federal official: the Minister of Transport in Ottawa.

Some observers are convinced that the most compelling reason for the unified Canadian support for the steel companion bridge in Canada has only marginally to do with cross-border traffic and aesthetics, but is primarily grounded in political patronage: Canadian steel companies were promised that job six years ago and they're not letting go. Maintenance of a steel bridge is three to five times maintenance of a concrete bridge and the people who have or expect those contracts are probably in the political process as well.

What to do about that?

We'll never get the PBA fully out of politics. Politicians and cash cows like that are bonded body and soul with Crazy Glue. And we can't do anything about the control of all five Canadian members by a single political party and officer. But we can expand participation on the American side.

I wouldn't mind those people having their sweetheart contracts and whatever else they get to rake off if they just wouldn't screw us in the process. We need at least one person on that board who will in all likelihood have some interest in the city of Buffalo. At present, we don't have that, or at least we have no evidence of it.

So make two changes in the legislation:

> Make sure that the seats of the Attorney General and Director of Transportation aren't assigned to political cronies having no legitimate position in either agency.
>
> More important, take one of the governor's appointees away and let that seat go to someone appointed by a committee composed of representatives of Buffalo, Erie County, the Olmsted Parks Conservancy, the EPA and whoever else has a point of view and legitimate concern not represented now.

If those two simple things were done – let the sunshine in and give the people a seat at the table – then some of the abuses and abominations we've seen and suffered will be less likely in the future. That doesn't mean everything will go perfectly. Everything hardly ever goes perfectly. But at least we'll have a shot at decency next time around.

36 IN YOUR FACE ON MEMORIAL DAY

Except for a flurry of press releases from several Canadian politicians who seem desperate to have a steel bridge built here (no matter that it takes longer to build and costs more to build and maintain) and what seems to have been a carefully-organized but embarrassingly-obvious letter-writing campaign to the *Buffalo News*, the Buffalo and Fort Erie Public Bridge Authority and its cronies have been fairly quiet since Judge Eugene Fahey said in early April that the Authority had to obey New York environmental law.

They have been so quiet that Senator Charles Schumer was moved last week to bring Detroit's Ambassador Bridge development team back to town, the group that was here last year saying they'd like to put up a bridge with their own money and operate it as a private enterprise. In a two-hour meeting with interested citizens at the Adam's Mark last Friday, they said they could deliver far better service than the PBA with the current bridge facility and they could design and develop a bridge really suited to the area's traffic needs and the community's environmental and ecological needs. All they needed was for the governments involved to pass legislation privatizing the binational operation.

They said if we gave them the franchise, they could get double the present number of trucks going across the river here within a few years, which would be great for the Canadian and American economies. They waffled when they were pressed to say what, if anything, the truck traffic would do for the economies of Buffalo and Fort Erie (other than that their Detroit experience showed more bridge traffic required more bridge workers). But, then, everyone else who has talked about the riches attendant upon increased cross-border truck traffic has waffled on that one too.

They said they'd made their pitch last year, then had stayed away while the Public Consensus Review Panel and Judge Fahey did their work. Now that the Panel found the PBA's twin span idea defective and the Judge said it was illegal, they thought this was a good time for them to reintroduce themselves into the process. They several times alluded to mismanagement and missed opportunities by the PBA and its staff. Peace Bridge general operations manager Earl Rowe sat through this quietly. He never said anything and his face was neutral.

Schumer said he wasn't a supporter of the Detroit plan so much as he was anxious to get things moving again. He several times talked about the PBA's inability or refusal to do anything at all. He said he was a supporter of a signature span, but at this point he was starting to feel that design wasn't as much of a problem as the PBA's inability to get anything done.

Three days later, on Memorial Day afternoon, one of the heaviest US-bound auto traffic days of the year, the Peace Bridge staff showed that it was indeed capable of action. They shut down one of the bridge's three lanes so four workers could do some minor maintenance. The lane remained closed to traffic even after the workers quit for the day. Cars were backed up for miles and many drivers experienced two-hour delays getting across.

Not one of the irate drivers who called me that afternoon to describe what was going on – or not going on – believed that the timing of the maintenance work was accidental. "They're sticking it to us because they didn't get their twin span," one caller said.

This isn't the first time in recent weeks a bridge lane has been closed down at peak hours for work that could as easily have been done at night, but since Monday was a major traffic day the work was certain to inconvenience a far greater number of people. The fact that all three lanes had been open earlier in the weekend indicates this was elective, not emergency, maintenance. All that traffic was backed up because someone in the Peace Bridge team wanted it backed up or didn't care about the hundreds or thousands of drivers and passengers trapped in the cars on a hot holiday afternoon.

Most large cities around the country schedule elective maintenance work on roadways for late night on ordinary weekdays, hours when the reduced lanes will cause the least inconvenience to drivers.

Was this deliberate or accidental? If accidental, something should be done about bridge management because they don't know what they're doing. And if deliberate, something should be done about bridge management because they know exactly what they're doing.

Maybe Senator Schumer is right: if the PBA can't stop stalling, and if some people over there are making things worse just to punish us for refusing to let them have their way with the anachronistic and expensive steel twin span, maybe we should find a way to erase the blackboard and start over from scratch.

37 Language Lessons

The Peace Bridge affair is slowly creaking toward sanity. It has squandered a huge amount of financial and human resources that could and should have been applied to more useful things:

> the money the county, city, Wendt and Community Foundations put up to operate the Public Consensus Review Panel;
> the money the city, the Olmsted Parks Conservancy, and the Episcopal Church Home put up in their lawsuit forcing the Buffalo and Fort Erie Public Bridge Authority to obey New York Environmental law;
> the money the Buffalo and Fort Erie Public Bridge Authority spent trying to avoid that obligation;
> the money the Buffalo and Fort Erie Public Bridge Authority spent designing a bridge that will never be built;
> the time of everyone involved.

What a bloody waste, and all so a few people could make money building and maintaining an anachronistic steel bridge. How much better off we all would have been had the PBA just said early on, "Here's a list of the people whom we or the people who control us promised will make a lot of money on this. If we can pay them off we can do this decently." No, we had to go through all this politicking and nit-picking and lawsuiting and squandering. Politics and power. Ugh. Bah.

Legal Matters

Judge Fahey issued his final order two weeks ago, The order resolved the lawsuits brought against the Public Bridge Authority and New York State Department of Environmental Conservation. The bottom line is, no new bridge will be built unless and until the Public Bridge Authority undertakes the full environmental impact study it fought long and hard to avoid.

The delay was so the attorneys could complain about fine points of wording or timing (their meters running all the time). The delay doesn't seem to have accomplished much: the final order was almost exactly what Judge Fahey delivered in early April

Shortly thereafter, Peace Bridge general manager Earl Rowe announced that the PBA was filing a notice of appeal."The Peace Bridge Authority has taken the first step in its effort to overturn a judge's ruling that it must complete a more thorough environmental review before building a bridge," wrote *Buffalo News* reporter Patrick Lakamp in the first sentence of his June 8 article on Rowe's announcement.

The sentence was wrong on two counts. First and most obviously, there is no "Peace Bridge Authority" and never has been. The organization is the Buffalo and Fort Erie Public Bridge Authority – the Public Bridge Authority or PBA for short. The PBA avoids using the word "public" whenever possible because if it hadn't been for the public insisting on its rights they would have gotten away with this anachronistic-but-profitable-twin-bridge eat-up-more-of-Front-Park scam. Why every reporter and editor at the *Buffalo News* except Donn Esmonde continues to support them in that misnomer remains one of the Secrets of Modern Journalism.

More important is what Rowe was really saying and doing. This wasn't the beginning of an effort to overturn the lawsuit; it's the end of it. The PBA hasn't filed an appeal, it has only filed a notice of appeal, which is like getting a paddle at the auction. The auction paddle gives you the option of bidding and the notice of appeal gives you the option of appealing; neither one commits you to doing anything or limits the behavior of anyone else. Actually appealing Judge Fahey's decision would put things on hold for years, and the guys who control the strings of the five Canadian PBA board members won't stand for that. They are fairly drooling for the increased profits from those Canadian trucks coming down here full and going back up carrying nothing but bags of American money.

Rowe's language was conciliatory, full of phrases about working out solutions good for everybody and meetings of interested parties and finding a satisfactory way to get moving again. He hasn't spoken this way before and I take the diction as data. There are changes in progress at Peace Bridge Plaza.

I think Senator Schumer's invitation to the Detroit bridge developers to come to town to say how they'd handle things, and the far warmer reception those developers got this time than the last time they came and made their pitch, had exactly the effect Schumer wanted: the PBA realized that just about everyone is fed up with their delays, their attempts to feed Canadian steel and construction companies sweetheart contracts that screw the public, and their arrogant disregard for the public on the American side of the river.

Here's what I predict: a few more months of attenuating blather and puffery and posturing. While that goes on in public, more and more rational negotiation in private. I bet you a shiny Buffalo nickle that this time next year they'll be deep into the planning for a six-lane segmented concrete bridge across the Niagara River.

RUDNICK KVETCHES

Buffalo Niagara Partnership president and CEO Andrew Rudnick sent Senator Daniel Patrick Moynihan a letter complaining about the deportment of Jim Kane, who heads Moynihan's Buffalo office. Kane, Rudnick said, was "thug-like," "intimidating," "negatively irresponsible" [as opposed, I presume, to "positively irresponsible"], and given to engaging in "brutish tirades." Rudnick told the Senator that Kane "singlehandedly has eroded the respect and admiration this community has for you."

Rudnick set forth his credentials as someone of social worth and seriousness of purpose in the very first sentence of his letter: "It has been about 30 years since I sat in Sander's [sic – there is no apostrophe in Sanders] Theater listening to your lecture." That is Ivy League code for: "I went to Harvard so I'm swell and I went to listen to you back when you were just a professor so you owe me." It's a measure of how out of touch Rudnick is to think that Pat Moynihan would turn on a trusted and reliable aide merely because Rudnick belongs to the Club and begins with a bit of soft brown-nosing.

The letter seemed like the petulant grousing of a man who'd been too long on the wrong side of the Peace Bridge issue, was perhaps in a snit because the Board he served was in the process of changing its position, and so he was now just striking out in frustration. The directors of the Partnership must have cringed when they saw it quoted in the *News*.

Why write a letter attacking a public servant nearly everybody likes to a boss who holds that public servant in high regard over an issue that's over? That's just personal vindictiveness. What good does personal vindictiveness do the Buffalo Niagara Partnership?

In my experience with him, Jim Kane has been a consistent source of straightforward and reliable information on the Peace Bridge affair. If he doesn't know something he says it and if he thinks someone is lying he'll say that too, which is probably the source of Rudnick's animus toward him. When the PBA signed the agreement with the city a year ago saying it would take part in the Public Consensus Review Panel if the lawsuits would be put on hold, Kane was the one who called it for what it was: A sham and a stall. At the end, Kane said, the PBA would renege on its promised cooperation and the lawsuits would have to go on anyway, and the delay while the PBA picked engineers and brought them up to speed would only sap energy, waste time, and squander money. Kane was right on every count.

Jim Kane is a first-rate public servant and I hope he stays in government after Senator Moynihan retires next year. Maybe the Buffalo Niagara Partnership will give him Andy Rudnick's job so that organization will stop endorsing this pettifogging and personal smearing and will regain some self-respect.

EVERYTHING IN ITS PLACE AT THE *BUFFALO NEWS*

Years ago, a friend who was studying Soviet government policy told me that the import of a story in Pravda, the Communist Party newspaper that was published daily in scores of languages, was often determined more by where the story was placed than by what the words said. With all those languages, there was fear that subtleties of political change would get lost in translation, so the Party handled it visually instead. Party insiders around the world were kept up to date on the meaning of locations. If you were reading Pravda and you saw an article praising General X and it was in the lower left corner of the page, say, it meant General X was about to disappear from the scene. The same article upper right meant General X was in the inner corridors of power.

I had to think of that when I noticed the placement in the *Buffalo News* of the two Peace Bridge stories I've discussed here: Andy Rudnick's grouchy letter to Senator Moynihan was on page B7, below a larger article on skateboarding. The article about the PBA's notice of appeal was at the beginning of the Classifieds. The physical location of these two stories may indicate a shift in the Favored Son status the PBA has long enjoyed at the *News*. They're still calling it by the wrong name, but front page coverage now goes to the proposed move of Children's Hospital and the bizarre saga of the Commercial Slip. Hooray. It's about time we got this Peace Bridge issue settled and moved on.

June 22, 2000

38 MUGGERS & LEADERS

OBSTRUCTIONISTS

You know why construction hasn't started on the twin span or the new convention center? Because Buffalo suffers a lack of leadership and an overabundance of obstructionists. That's the current buzz being circulated by the people who want to build a steel twin span now, a new convention center now, move Children's Hospital now, and bury the Commercial Slip forever now.

Those two themes turned up again and again in the series of pro-steel-twin-span letters organized by the Buffalo and Fort Erie Public Bridge Authority's flacks and published in the *Buffalo News* over the past two months. It's become the buzz on the street among people whose experience of these issues is headlines and sound-bites.

A PhD psychologist at a party in one of the southtowns said, "All these good things could be happening in Buffalo if it weren't for the small special interest groups obstructing progress and if there were some decent leadership." Someone asked her, "What good stuff, exactly?" "Moving Children's Hospital to High Street, increasing capacity of the Peace Bridge, getting a better convention center."

If there were some leadership, she said, then the obstructionists wouldn't be able to obstruct and Buffalo would move forward. Half the people at the table said "Yeah, right," or some variation thereupon. The other half looked at one another silently, not wanting to spoil the otherwise pleasant lunch, all of them probably thinking what I was thinking: "This is an intelligent and thoughtful professional person. She's bought the partyline. Where do you start with someone like this?"

PARALYSIS

Consider these selections from a recent *Business First* article by James Fink, "Decisions Paralysis: Big projects slip away as debates rage on":

> Western New York's leadership can't make a decision.

> Want proof?
>
> Look at the raging debates concerning the Peace Bridge and the proposed downtown convention center. Throw in questions and concerns about the Buffalo Zoo and the Inner Harbor project....
>
> In each instance, a proposal was presented and championed only to have the respective projects bogged down in a series of debates that sometimes bordered on a war of words where personal attacks take precedence over the big picture: moving Western New York forward....
>
> "To me, I don't care who gets the credit for projects like the Peace Bridge, convention center or Inner Harbor," said Robert Bennett, president of the United Way of Buffalo & Erie County. "I think people forget to look at the big picture. Once a commitment is made to something, I would hope everyone would jump on the train and support it, instead of worrying about who gets the credit...."
>
> The decision paralysis frustration is peaking right now, largely due to the community's prolonged haggling about the Peace Bridge. Arguments about the bridge's design and environmental impact have gone from public hearings to court....
>
> Meanwhile, Canadian interests who actively endorse the twin span proposal are perplexed over Buffalo's battles.
>
> "The most important issue should be how we create jobs and job opportunities for today and tomorrow in both the Niagara Region and Western New York," said Tim Hudak, a member of Ontario Provincial Parliament whose district includes Fort Erie. "Sometimes I'm not sure people in Buffalo really understand the long-term negative ramifications they're causing by these delays and indecision."

The unstated underlying assumption of Fink's entire article is that every one of those projects was worth doing as they were designed and public interrogation of those designs was interference with – with, well, what? Making money quickly? It surely was bad, bad, bad.

The "big picture" he writes about is nothing other than the proposal itself. People ask about six years of truck traffic through Delaware Park because of the PBA's plan? That's screwing with the big picture. They mention the huge difference in childhood lung disease around the Peace Bridge Plaza and the rest of the city? That's screwing with the big picture. They point out that a six-lane concrete bridge will go up faster, cost less to build, cost less to maintain, and will free up Front Park? That's screwing with the big picture. They point out that a six-lane concrete bridge will bring a huge number of jobs to this area and a three-lane steel bridge will bring almost none? That's screwing with the big picture.

He quotes United Way's Robert Bennett saying that "once a commitment is made to something, I would hope everyone would jump on the train and support it..." But whose commitment and whose train are we talking about? In the case of the Peace Bridge, it's five Americans and five Canadians in a closed room, most of whom have no interest at all in the fate of Buffalo. Are those of us who live here bound to jump on whatever train they care to cannonball through our town? That's madness.

And then he says that the decision paralysis results from the community's questioning the Public Bridge Authority's decision. That's the cause of the paralysis: we challenged, we're bad. Not a moment's thought to any possible defects in their plan, any possible harm their plan might have cause the city, any possible right to partake in the process for anyone other than the people who stand to make money from it. Wow!

WHAT INDECISION?

All this talk about "indecision" is smoke and mirrors. There is no indecision, none at all, and there never was. It was just people trying to force things on other people who decided to ask, "What are you doing to us, and why?"

In the case of the twin span, the PBA decided that they wanted to erect a steel companion bridge. They held public discussions and a design charette – but those events were just for show, they were phoney, the decision had already been made and at the end of the process the PBA arrived where they started. People looked at their decision, said it stank, it served only a few steel builders and maintainers and would cost us far more in the long run than a whole range of rational alternatives, and they began to take action.

The New Millennium Group wasn't indecisive. Its highly educated and highly motivated members knew exactly what was wrong with the PBA's plan, and said so at length and in depth.

Neither did the Episcopal Church Home and the Olmsted Parks Conservancy suffer indecision. They looked at the plan and saw the further harm it would do the community. They tried talking to the PBA and got nowhere, so they went to court, along with the City of Buffalo. That was a specific and decisive act.

Judge Eugene Fahey wasn't indecisive. Indeed, the instrument he delivered is called and is in fact a decision. He said the problem wasn't with the people asking questions. The problem was that the PBA hadn't obeyed the law.

The Commercial Slip? Unless there's some further surprise, it turns out that the Preservation Coalition was right on the money about that one. We've got a real historical site there that is perfectly capable of being made accessible and the Empire State Development Corporation and whoever joined them in wanting to bury it were wrong.

Move Children's to High Street? There are strong arguments on both sides, and both can't be right. Once the proposal to move Children's was made, it would have been irresponsible to just go along with that proposal without considering the questions by the people on the firing line – the doctors who take care of those children – and the effects on the two neighborhoods involved. How could that *not* be worth extensive discussion and consideration?

Build a new convention center in the Electric District? Do we need a new convention center at all? If we do, is that the right place to put it? Little surprise that the developers squalled when newly-elected County Executive Joel Giambra put that on hold until he saw the results of an environmental impact study.

Do the various community groups affected by those decisions have a right to have their voices heard in the decision-making process? If, as *Business First* would have it, none of those constituencies deserves to be heard in any of those key decisions, then who would *really* own Buffalo and what would that mean for the rest of us?

THE REAL PROBLEM

Times have changed and James Fink and other people frustrated by community involvement in project design haven't a clue how or why. People are demanding a different way of doing public business. People are tired of having some committee or corporate executive announce what and how something is going to be done and then the community gets stuck with the cost of the error.

Ordinary people are more connected than they were because communication is easier and faster than anyone could have imagined a few years ago. Anyone can set up a listserv, web site, hit "reply to all" and send a note to everyone on the bulk email that just came in. Concerned citizens are no longer impotent against the corporations' and authorities' teams of press agents, legions of clerks and limitless budgets. Information access is no longer controlled by a powerful few.

People say, "You don't own the city, the county, the region; we ALL do. The days when you announce and we obey are over. The way it works now is, you suggest and then all of us consider alternatives and try to find among them the option that does the most good and the least harm."

Our problem isn't the community's resistance to projects that seem to make no sense or to promise real harm. The problem is the inability or the continuing refusal of the people who want to do those megaprojects to talk with and listen to the rest of us from the very beginning. The problem is them forcing the rest of us to put up all these damned signs, go to court, circulate and sign petitions, set up those web sites, have meetings. We'd rather be doing something else too, we'd rather have nothing on our lawns but grass.

MUGGERS AND MUGGEES

Why didn't you complain about our plan five years ago? the Peace Bridge staff says. Why are those awful people interrupting things now? *Business First* says.

In hockey or football or boxing, if you're not out there when the game starts you default: the other guy wins and you don't get to say later, "But I'm better" or "They should have waited for me" or "I thought it was the next night." What you are or why you weren't there doesn't matter in games if you're not there in the beginning.

In real life, it's when you find out what's going on that determines what you're going to do next. If the "you weren't there at the beginning" rule applied in real life, then every sneaky hidden backstabbing opening gambit would determine everything. Some people might like it that way, but fortunately they don't always get what they want.

Blaming the rest of us for being slow to learn what hard questions we should ask about huge construction projects that will change the character of our community is like saying you shouldn't seek treatment for cancer because it provided no symptoms for a long time or you shouldn't fend off a mugger because you didn't see him lurking in the shadows. You deal with disease and evil in the personal or social sphere when you become aware of it. You can't deal with it before you know it's there; it's irresponsible not to deal with it once you do know it's there.

LEADERSHIP

Leadership isn't doing any old thing; it's doing the right thing. Leadership isn't forcing people to do what some committee that meets and makes deals in secret wants done; it's finding ways to get done what needs to be done, and doing it the right way.

In each of these current imbroglios I find real leadership: the New Millennium Group, the Episcopal Church Home, the Olmsted Parks Conservancy and the City in challenging the PBA about the steel companion bridge; the Physician's Coalition in asking for specificity in the proposed move of Children's Hospital; the Preservation Coalition in asking what the truth was about the viability of those aged rocks at the terminus of the Erie Canal; everybody who asked questions about the proposed convention center that weren't asked last time and resulted in that dog on Franklin street.

The leadership problem here isn't a lack of people willing to get things done; it's an overabundance of people anxious to do things in a hurry, people anxious to do things so they will get their way and make their money without having to consider the needs of everyone else.

Sometimes what needs to be done is stopping someone else from doing something that shouldn't be done. Was it leadership when some morons decided to cut the city off from its waterfront with the Thruway or to destroy Frederick Law Olmsted's Front Park so there could be more space for trucks to idle their engines and pour noxious fumes into Buffalo's West Side? Was it leadership when city officials decided it was okay to plop a huge bank building across the foot of Main Street and a profoundly ugly convention center across one of Joseph Ellicott's radials? Was it leadership when state officials and local real estate developers put UB on an Amherst swamp rather than in downtown Buffalo, or when they ran a four-lane highway through Delaware Park or built Route 33 so it would be easy for people who worked in the city to live in Amherst? How would life on the Niagara Frontier be different now if the kinds of opposition to questionable public works projects that so upsets *Business First* had been possible back then?

At a conference on leadership at USC a few weeks ago, Tom Peters, Peter Drucker, Warren Bennis, and a dozen other theorists on leadership and bigshots from industry and public life talked about leadership in this new age of rapid communication, development, and transportation. The thing that impressed me most was, the smartest people there weren't talking about leadership in terms of herding cattle – getting them where you want them to go, which is what Jim Fink's *Business First* article was about – but rather in finding the right thing to do and helping your firm or community do it or get there.

A few days ago, Bennis said to me, "What leadership is really about is simple: having willing and inspired followers." Which is to say, having people who care about what is going on and who care about being part of it. That doesn't come from ukases delivered from on high; it comes from honest exchange of ideas and needs and options; honest examination of causes, costs, benefits and risks; honest respect for one another's needs and information and legitimacy. And an honest willingness to let ideas endure the glare of uncorrupted examination.

That's what all those groups have been trying to do. Everyone one of them has shared a key line in their list of wants: "Let's talk honestly." Is that obstructionist? Is that outrageous? Is that wrong?

No, of course not. The only wrong is what's always been wrong in these affairs: being silent, advocating silence, endorsing victimage. We have plenty of real leaders in Buffalo these days. They're not the people saying "Shut up and let us do what we want." They're the people who are saying, "Let's talk."

39 THE PBA'S PRIED DOUBLE MACDONALD'S TRUSS

For months there have been rumors of behind-the-scenes negotiations between various officials and the Buffalo and Fort Erie Public Bridge Authority. Supposedly, they were hard at work trying to resolve the stalemate imposed by the Canadian members of the board and endorsed or passively accepted by the American members of the board.

No way: the city, the county, and all the other groups may have been negotiating seriously, but the PBA was just stalling for time. They're still hoping to wear us down.

A month ago, according to former Peace Bridge employee Jay Malone, the Bridge's general manager Steve Mayer announced at a staff meeting that if they couldn't get out of the environmental impact study ordered by Judge Eugene Fahey any other way, they would abandon all plans for a new bridge and would just widen the present bridge. That way, they could avoid any environmental evaluation or supervision by anybody. The most important thing, Malone said Maier said, was to get six lanes of steel bridge out there fast, without interference.

Tuesday's *Buffalo News* revealed that the PBA has a land-based plan too, a cockamamie new design they hope will neutralize all the objections to their old inadequate designs. It's really the same old design, same old plan, with one difference: instead of two Parker trusses, the engineers have planted the ends of two suspension arcs between the bridges on the river side and outside the bridges on the Buffalo side. It's like two MacDonald's arches starting at the same place but pried open to land farther apart. They used the same drawing they used for their earlier twin span design; the only difference is the computer replaced the trapezoidal Parker truss with the Pried MacDonald's Truss. The new design lands right in the middle of Front Park, which suggests they are still trying to get out of the new northern plaza as well.

PBA chairman Victor Martucci told the *News*, "From my perspective it's truly a signature bridge." So it is: but from someone with abominable handwriting.

The PBA and the Peace Bridge senior staff remain desperate to avoid the environmental impact study ordered by Judge Eugene Fahey last spring. Lately, they've been trying to negotiate with the Episcopal Church Home and the Olmsted Parks Conservancy, hoping to get them to abandon the lawsuit. The Pried MacDonald Truss is an attempt to pressure the city to back out of the lawsuit.

Legally, it doesn't matter at this point if the city, the Olmsted Parks Conservancy, and the Episcopal Church Home cut deals with the PBA. The lawsuit belongs to Judge Fahey and his decision stands until he decides to reverse himself or a higher court decides he was in error. Both of those contingencies are unlikely.

The question remains: why is the PBA fighting so hard to avoid New York environmental law? It's not because of Fort Erie's immutable love for the old bridge, that dreary song chanted without cease these past few months by Fort Erie mayor Wayne Redekop. Redekop wasn't singing that song a few years ago. He's singing it now only because Fort Erie is a company town, totally dependent on Peace Bridge largesse. The Bridge is the major employer, and it generates most of the town's other income. The PBA built all of Fort Erie's major public buildings. Fort Erie officials sing whatever song the PBA tells them to sing. If the PBA decided a six-lane concrete bridge was the way to go there would be a new songsheet on the mayor's music stand in seconds.

This isn't about aesthetics, it isn't about tradition, it isn't about sentiment – and it isn't about getting traffic moving quickly. It's about money. Money. It's about who is going to get rich doing what. Far more people, most of them Canadians, will get rich with a steel bridge than with a concrete bridge. There have been direct links between PBA board members and Canadian steel manufacturers. A steel bridge is like a Polaroid camera: the cost of getting it is nothing compared to the cost of operating it. If the old bridge is kept going and a new steel bridge is put next to it, millions of dollars in profits and payoffs will go to people who will get nothing at all if an efficient and economical six-lane concrete bridge replaces the old bridge.

It's hard to believe that the members of the Public Bridge Authority and their two general managers really think that Buffalo citizens and officials are so stupid we will look at this cobbled design and think it's the solution to our problems. They know that this plan addresses none of the issues raised by the Public Consensus Review Panel or by Judge Fahey's decision. This Pried MacDonald's Truss is just their most recent way of saying "Up yours" to us all. This is just a new phase in the stonewalling. It's their way of saying to us all, "We don't care about you, your city, your children's health, the local or the regional economy. What we care about is the money that will be made by building an anachronistic steel bridge and by maintaining this decrepit old bridge. And we're going to stonewall and fight anybody who tries to stand between us and those profits, between our friends and that patronage."

Mayer's comments to the bridge staff about widening the present bridge if everything else fails and this absurd design are proof enough that the PBA and the senior bridge staff have no intention of doing anything decent for this area unless the law forces them to do it. They're hoping that our local government officials are so weary of all this and have so much other work to do they will let them slip on by. Now, more than ever, Tony Masiello, Joel Giambra and Jim Pitts have to remember what this fight has really been about.

July 27, 2000

40 The *Buffalo News* and the Peace Bridge: Prospering and Pimping

KANE: The "Chronicle" has a two-column headline, Mr. Carter. Why haven't we?
CARTER: There is no news big enough.
KANE: If the headline is big enough, it makes the news big enough.
Citizen Kane 1941

All the News that Fits What We Want You to Know

You must have seen it: a three-column two-tier headline piece by Patrick Lakamp in the *Buffalo News* of Wednesday, July 19: : "Peace Bridge Authority's new design: 'Gateway Arches,'" There were three color photos, one of them three columns wide and in full color, all of it above the fold on page 1.The three-column photo was a computer-generated drawing of twin bridges curving south into Front Park, not a truck in sight. Three-column head and photo, all above the fold: that's major stuff.

And, in the *Buffalo News* for Wednesday, July 19, 2000, major stuff was a slight modification of a design found faulty by everyone except the Buffalo and Fort Erie Public Bridge Authority, the Buffalo Niagara Partnership, and the small group of mostly-Canadian steel fabricators who stand to make millions of dollars if that anachronistic design is selected rather than the cheaper and environmentally superior six-lane concrete bridge advocated by the Public Consensus Review Panel, the New Millennium Group, and just about everyone else who doesn't stand to make a buck from this nonsense.

The lead paragraph told us not only what the Buffalo and Fort Erie Public Bridge Authority was up to, but what seems to be the official *Buffalo News* response to the design: " The Peace Bridge Authority has come up with a new, more appealing twin-span design, and community leaders such as Mayor Anthony M. Masiello think it has the potential to break the deadlock over what kind of bridge to build."

Lakamp quoted Public Bridge Authority Chairman Victor Martucci (though he got the name of the organization he chairs wrong): "'From my perspective, it's truly a signature bridge,' said Peace Bridge Authority Chairman Victor Martucci. 'There's nothing like it in the world. And if that's the definition of signature bridge, this certainly fits the bill.'"

I think it takes more than being unique in all the world to be "the definition of signature bridge." The root canal my dentist did the other day was unique in all the world, at least as far as I'm concerned, and I wouldn't want to share that with anybody.

"Authority officials have had the new design in their hands for several months but had declined to release it," Lakamp wrote. They should have declined: there's nothing there. They probably feared that if the public got hold of the design they'd be ridiculed before they got a chance to elicit one-line statements from local politicians trying to be nice. "Martucci agreed to comment on it only after The *Buffalo News* obtained a copy from another source." Well, Authority officials have been peddling the design to every politician in town, and the Buffalo Niagara Partnership sent Natalie Harder down to Washington in the vain hope that she'd be able to convince Senators Moynihan and Schumer that this idea had some substance. If the *News* got it from "another source" it was probably the PBA's partner in bad design and dysfunctional citizenship, the Buffalo Niagara Partnership.

Lakamp's article in praise of the new design consisted of 34 paragraphs of which

> 18 described the PBA's design or the general situation leading up to it;
> 11 described it positively or quoted people saying positive things about it;
> 1 said people questioned whether the steel twin span design was better than a 6-lane concrete design;
> 1 quoted someone who didn't think it was a very good design at all.

A single directly critical paragraph and one oblique critical paragraph out of 34 paragraphs of description and praise! And this is about a drawing – that's all it was, a computer-generated drawing that glued MacDonald's arches on the twin span design the Public Bridge

Authority has been trying to peddle for the past six months, a design already faulted because it would pour heavy truck traffic through Delaware Park for six years, would enormously increase air and noise pollution on the far West Side, would send millions of dollars to Canadian steel companies and construction workers, would consume even more of Front Park, and would saddle all of us with decades of rehabilitative surgery on the geriatric bridge now spanning the mouth of the Niagara River.

THE MEDIUM IS THE MESSAGE

The only report in the *News* of Natalie Harder's failed mission to Washington was in "New design fails to impress senators," an excellent story by Doug Turner, which appeared in a single column back in section C on Saturday, July 22.

"The Peace Bridge Authority's new twin-span design failed to draw support from the state's two US senators, despite efforts of the Buffalo Niagara Partnership to sell the plan in a closed-door meeting, it was learned Friday," Turner wrote. He quoted Senator Moynihan's chief of staff, Tony Bullock, who called the proposal "the same old plan with a new hairpiece.... It doesn't change any of the complex issues that afflict the (Peace Bridge) authority's original plan.... The bridge still takes off from the same place and lands in the same place. It still uses a 75-year-old bridge that is in poor repair."

Marshall McLuhan's famous line – "The medium is the message" – is usually brought up in conversations about television and film. It's equally appropriate to newspapers. Content is not just what is said, but where it is said and how it is said.

Saturday is the least-read day of the week for the *Buffalo News*. The greatest readership is midweek. You can't miss three columns of text topped by a three-column color photo and three-column double deck headline on page one in the middle of the week. Even if you don't buy the paper you see those big letters when you walk by the newsstands. It's easy to miss a single column in section C on Saturday, no matter how well it's written, how important its content. There were far more important questions raised about and cogent criticism of the PBA's MacDonald's Arches plan in Doug Turner's Saturday single-column Section C piece than in the entire three-column above-the-fold (continued on p. A6) Wednesday page-one piece.

EDITORIAL GUSHING

Patrick Lakamp's Wednesday article in praise of the pseudo-design turned out to be only foreplay. The lead editorial in the following day's paper praised the design and suggested that anyone objecting to it should just shut up.

The title told you where it was going: "Making a statement for Buffalo." *That's* what the PBA and the Partnership were about in this design, making a statement for Buffalo?

> Light traditionally appears at the end of the tunnel, not the bridge. Nevertheless, the metaphorical glow emanating from an alternative design for a new Peace Bridge could be a sign that the agonizing, endless debate over the Niagara River crossing is - maybe, finally, may it please the Lord - drawing to an attractive conclusion.
>
> The Peace Bridge Authority this week reluctantly acknowledged the existence of a striking and hitherto secret design that represents a more-than-acceptable compromise on the contentious issue of a twin-span bridge versus a single, six-lane "signature" bridge, a design this newspaper would have preferred.

"More-than-acceptable compromise" to whom? To the New Millennium Group? No. To people concerned about disruption of major traffic patterns in Buffalo for up to a decade? No. To American skilled laborers who would watch almost all the decent construction jobs go out of town if the steel bridge is built? No. To epidemiologists and parents concerned with increasing levels of lung disease among children on the West Side? No. To people who would like to see less of the Peace Bridge's toll money poured into patronage and good-old-boy checking accounts? No. To people who want to see Front Park restored? No.

> Based only on appearance, it looks like a winner. "It's truly a signature bridge," Authority President Victor Martucci said, without exaggeration. Both Buffalo Mayor Anthony M. Masiello and Common Council President James W. Pitts, who had criticized previous plans, praised the new design.

Of *course* Martucci was exaggerating. And Masiello and Pitts were both qualified in their comments on the bridge, they did not simply praise the new design. And even though the *News* editorial writer may like and respect Victor Martucci, there is no reason for the *News* to promote him from chairman of the Authority to its president.

The editorial sounded only one sour note in this ecstatic paean:

> Still, there are critics. Elissa Banas of the New Millennium Group, which has fought for a dramatic single span, dismissed the design as having "the same old problems," even though it is better looking. But the group should reconsider.
>
> It has performed a valuable public service by forcing the authority to retreat from its hideous proposal of mismatched arches, but if this bridge debate is going to be resolved before the next millennium, everyone will have to give up something. On the issue of design, it is time for the New Millennium Group and its allies to declare victory. They have already exerted a profound and positive influence.

Banas never said that the MacDonald's arches are better looking that anything; that's the editorial writer's opinion tacked on to her criticism. More important, this design *does* have "the same old problems." Why should we forget about them because of two MacDonald's Arches? If everyone is "going to have to give up something," what is being given up by the Buffalo and Fort Erie Public Bridge Authority? What is being given up by Fort Erie or any Canadian political or private organization? All the giving up is by the citizens of Buffalo and Erie County. What kind of lousy advice is that for an American newspaper to give its readers? Why should the New Millennium Group, without whose heroic persistence in this sordid affair we'd be seeing twin spans already under construction, roll over and play dead now? That is insulting, patronizing, and contemptuous counsel.

> But with looming business concerns threatening to overtake the issue, and cause this area significant economic harm, further insistence on a single span will soon devolve into obstructionism.

I'll translate: At the beginning of this, long before the public was involved, the PBA decided to build a steel twin span. Once people found out about that and suggested alternatives, the PBA had fake public hearings but never for a moment considered anything else. Once public opinion against the twin span plan really got going, the PBA started stonewalling. It stonewalled on the basis of a theory that went, if you stonewall long enough people will say that opposition to the steel twin span plan "will soon devolve into obstructionism." Stonewall long enough, spread the patronage far and wide enough, and you'll get what you wanted, only a little later than you would have liked.

What's really weird about this is how quickly the *News* editors – both those who decide where and how much space a news story gets and those who control the editorial page – went intergalactic over this pseudo-design from the PBA. After all, the design was nothing new; it was, as Tony Bullock so eloquently put it, "the same old plan with a new hairpiece." The *News* had never taken seriously any of the designs from internationally-known bridge designers like Eugene Figg and T.Y. Lin, yet here it was taking with utter seriousness and unrepressed gush something that was nothing more than the old drawing with two new computer-generated MacDonald's Arches.

I didn't understand the hyperbolic coverage on page 1 of Wednesday's paper until I read the adoring editorial in Thursday's paper. The Wednesday article had taken a non-event and turned it into news. What is news, after all, but what a newspaper publishes? That's what Orson Welles' Charles Foster Kane is talking about in the quotation at the beginning of this article. And once something is news, then there's all the justification in the world for an editorial telling us how important that news is.

THE RECORD

It was, in a word, astonishing – unless you've followed carefully the *News*'s coverage of the bridge affair the past several years. With rare exceptions, that coverage has been as mouthpiece for the Buffalo Niagara Partnership and the Buffalo and Fort Erie Public Bridge Administration. Articles in *the News* present Authority and Partnership staff assertions as if they were fact; *News* reporters rarely question any of those assertions, even in open press conferences. They gloss over or trivialize challenges or opposition to those assertions. The coverage of this story has never reflected the level of journalistic inquiry a project of this magnitude would have received from a major paper in any other American city: this is a $200-million project, it will last most of a decade, it is the largest single construction project in this area since the construction of the UB North Campus.

Columnist Donn Esmonde, editorial cartoonist Tom Toles, and Washington reporter Doug Turner have asked cutting questions and pointed to the real issues, but as far as the front page and editorial columns are concerned those guys might as well have been writing for another paper. Editorials have, with very few exceptions, told Buffalonians to shut up and take it; sometimes they've said that the News would prefer something nicer but since the PBA wasn't going to give it to us we shouldn't interfere with progress.

Lest you think I'm being cranky or engaging in name calling, I'm going to give you a several examples of the kind of distorted and misleading coverage I'm talking about. Keep in mind that a newspaper distorts not only when it says something that isn't true or presents something in a slanted way, but also when it doesn't say things that are true but which are not to its editorial liking, or when it consistently fails to interrogate public officials whose utterances it publishes without question or qualification, or when it consistently distorts or limits statements from legitimate individuals who have another point of view. You present both sides of a complex issue and you're doing journalism; you present one side and you're doing public relations work or pimping.

What's in a name?

At the risk of being repetitive – because I have mentioned this before in these pages – I have to note the single distortion of this story the *Buffalo News* engages in more frequently than any other: the matter of the name.

Buffalo News editorial writers and news writers almost never refer to the Buffalo and Fort Erie Public Bridge Authority by its correct name. Columnist Donn Esmonde gets it right. Cartoonist Tom Toles gets it right, or is playing when he doesn't. Why does the editorial page always call it the "Peace Bridge Authority"? Why does the news department always call it the "Peace Bridge Authority"? Why do the headline writers always call it the "Peace Bridge Authority"?

Because that is the name preferred by the Buffalo and Fort Erie Public Bridge Authority, which has come to hate the attempts of citizens' groups and public agencies to shine some light into its operations. You call the PBA office and the automated voice tells you that you've reached the "Peace Bridge Authority." You get their press releases and you see they're coming to you from the "Peace Bridge Authority" and they're telling you what the "Peace Bridge Authority" is doing for you or to you. I can understand why the PBA would want to try to get the public to forget that they are a public agency; if I were engaged in all the flim-flam going on down there, I wouldn't want the public poking around either. But why should a free and independent newspaper collaborate in that endeavor?

Recent Exempla

Okay, let's look at some examples of the pattern of coverage:

A March 31, 2000, page 1 story by Patrick Lakamp, headlined "Authority Stands by Support for Twin Span":

> "The (authority) is extremely disappointed as to the outcome of a process that held great hope of bringing this debate to a conclusion," Martucci said. "Although the Public Consensus Review Panel could not agree to ratify the engineers' report, a clear and strong consensus did emerge among elected leaders, business, labor and the community at large, on both sides of the border. The consensus was in support of the engineers' report."

Martucci *said* this, but it isn't true. LaFalce is for the twin span but both senators are for the signature span; the Erie County Legislature is for the twin span but the Buffalo Common Council is for the signature span; Fort Erie citizens are for the twin span but Buffalo citizens are for the signature span; the Public Consensus Review Panel voted 20-1 against the companion span; the mayor and county executive endorsed the Panel's vote. There's disagreement over, not support for, the engineers' report. There's nothing wrong with Lakamp including this long quotation from Martucci, but wouldn't it have been responsible and fair to pair it with a quotation from someone who might have had another opinion? Why quote only one side of the argument?

> A unified, full-blown environmental review on a new Peace Bridge and US plaza.

Lakamp used the same phrase in his 19 July piece praising what the PBA calls their 'Gateway Arches': "The authority's opponents said they want the authority to undertake a unified, full-blown environmental review before building a new Peace Bridge." "Full-blown" – why that modifier? What's the alternative – "half-baked"? Why not just say what it is, "a unified environmental impact study"?

A March 31, 2000, editorial, "Time to Move Forward":

An excellent editorial, one of the most well-balanced the *News* has done on the subject. It says that the companion span would be the lesser choice. The graphic accompanying it, however, was of the companion spans without a single truck in sight. Why mix the message?

> In rejecting the recommendations of a bi-national team of

> engineers for twin spans with matching arches and a new northern plaza, the panel decided to fight for what it considered best for the region. It could easily have accepted the recommendations and claimed credit for getting the project off dead center and ending the acrimony of this long-running debate. It could not have been pleasant to reject the recommendations of the engineering team that the review panel itself had put together.

Not true. The Review Panel hired one team, the PBA hired the other. According to all reliable accounts, the two teams cut a deal: the Americans agreed to let the PBA have its steel companion span if the PBA would relocate the current plaza to the north, thereby freeing Front Park.

> The panel believes, as we do, that the single-span concrete bridge is preferable in cost and beauty to the compromise twin-arch proposal. Our concern is that the single-span concept is not worth a protracted delay – as predicted by the authority – that would cost the area economically as we lose business to other crossings. In addition, the deck of the current bridge is deteriorating and may not last until a new bridge is built, causing increased traffic problems.

Where's the logic? If the old bridge is going to need serious reconstruction then we won't be at 6-lane capacity for several years after the new bridge is built anyway. If more open lanes will benefit the area economically, it's the 6-lane concrete bridge that will get there first. They say this later in the editorial; why doesn't it figure here as well?

Donn Esmonde, "Citizens Panel Puts Sense into Bridge Process," April 3, 2000:

> The authority's Steve Mayer admitted Friday what everybody already knew: The authority decided to twin the Peace Bridge six years ago. Suggestions otherwise since then were circular-filed. Alternative-design bridge charettes have been charades.
>
> "The decision was made (in 1994) that the Peace Bridge was in good condition," said Mayer. "We excluded (considering) a six-lane bridge because it didn't make any sense to us to take the Peace Bridge down."

No. Some insiders may have known this and some other people may have suspected it, but Mayer's statement that Friday was the first time a bridge official admitted it in public. This key item appears here in an opinion column three days after Mayer made the admission in response to two questions from a reporter from *Artvoice*. It was not quoted or cited in the news articles about the press conference, even though reporter Patrick Lakamp was there when Mayer made the statement.

A page one story on April 7, 2000, with the headline "Court Order to Delay Bridge Project":

This headline belies the article and the event it reports. Why not "Court Orders PBA to Obey the Law" or "Court Orders Environmental Impact Study"? Why headline the PBA's lament rather than what the judge did or what the public got? And even if this were the substance of the story rather than the PBA's take on it, does the court order delay the start or the finish of the bridge project? Headlines should inform, not mislead. This one misleads.

Patrick Lakamp's page 1 story on April 8, 2000, "Ruling Casts Doubt on Bridge Project," begins:

> A judge's ruling Friday not only delays the building of a new Peace Bridge but raises the question of whether it will be built at all.

Judge Fahey's ruling delayed the *start* of construction, not *completion* of it. More important, the headline and the lead focus on how the PBA's plans are impeded, but not on the decision itself, which said that the PBA hadn't followed the law and now would have to. This is like telling us Fred's plans for a trip to the Riviera were ruined and not telling us that the reason they were ruined was because Fred got caught stealing the money for the trip and is locked up in jail.

Lakamp goes on to quote PBA chairman Victor Martucci rather than any of the plaintiffs in the lawsuit, the guys who won: Andrea Schillaci of the Olmsted Parks Conservancy or Edward Weeks of the Episcopal Church Home or Mayor Masiello. The heart of this story, told much later on in it, are about the benefits of the environmental study the PBA tried to avoid and what was wrong with how they tried to avoid it. But the lead paragraphs and headline focus only on the PBA.

> "I don't think this is a day for celebration," Martucci said after State

> Supreme Court Justice Eugene M. Fahey ruled the authority's previous environmental study for a new bridge was flawed.
>
> "It's a sad day for this region and the other side of the border," he said. "We're not a community that has a surplus of jobs and growth that we can afford to delay the promise of new jobs and economic growth this project would result in."

What jobs are lost because of the judge's decision? Martucci didn't say and Lakamp didn't ask. The PBA and Andrew Rudnick of the Buffalo Niagara Partnership are always going on about the jobs that will be lost if a steel twin span doesn't go up yesterday. They always avoid being specific about those claims.

An editorial titled "Starting Over" on April 8, 2000:

> Now that a court has ordered a full environmental study for a new Niagara River bridge and plaza, the Peace Bridge Authority ought to view this as the chance for a new start, not the start of a new fight.
>
> In a decision that was widely expected, State Supreme Court Justice Eugene M. Fahey ruled that the authority must conduct a comprehensive environmental impact statement for construction of both a new bridge and plaza. He said the separate environmental assessment for the bridge and a more thorough environmental impact statement for the US plaza violated environmental law.
>
> What that means is that we are very nearly back to square one in the planning stage for a new bridge.

This isn't true. The only thing it meant was that the PBA couldn't begin construction in June. A huge amount of planning and research has been done since "square one," much of it applicable to any new design, much of it useful for the EIS.

> We have long felt that the single-span bridge was preferable, but the compromise of a twin-arch – replacing the ugly mismatch of an arch and the Parker Truss – along with a new northern plaza was an acceptable compromise to a bitter feud.

It's not just a matter of bridge aesthetics. The PBA's companion span design would result in huge disruption on the city's West Side for years. Thousands and thousands of huge fume-belching trucks would be routed through city streets and along Route198. Large numbers of jobs and manufacturing orders that would otherwise come here will go out of state and out of the country. Why does the

News say that is "acceptable"?

> The authority has legitimate complaints. Americans got into this debate late,

It was only *after* the decision to twin the bridge was made that Americans were allowed to participate in the process, which is to say, the only process they were allowed to participate in was no process at all.

> and often were not sensitive to Canadian feelings. John A. Lopinski, authority vice chairman, reflected that sentiment with his response to Fahey's ruling: "Canadians are very disappointed that some people in Western New York don't realize that there are two countries involved in this decision."

Yes, but the United States is one of them too. Where has there been an iota of interest in American feelings or needs from the Canadian side? What Lopinski is saying is, "We said what we want and we're disappointed the Americans haven't given it to us." We don't expect more from Lopinski, but why would the *News* quote him on this and not note the hypocrisy?

> We can't undo the past. But we can learn from it. That means a fresh start and an open mind on both sides of the dispute. As long as we have to start over, let's do it right this time.

Other than work and play well with others, what does "do it right this time" mean?

In a huge Sunday Viewpoints column titled "More bridges to cross: There's no way now that a beautiful new Peace Bridge can be built expeditiously, but perhaps we can look at what went wrong in the planning process and learn from our mistakes," on April 9, 2000, Mike Vogel, a member of the editorial board, wrote:

> The Peace Bridge debate was done wrong. In the process, a major opportunity – an opportunity to quickly build a bridge both beautiful and functional - was lost.
>
> It was lost because the Peace Bridge Authority focused so narrowly on its mandate to move traffic across the Niagara River that it failed to see a chance for greatness. It was lost because political leadership faltered during the long community battle over the fate of its bridge. It was lost because the commu-

> nity itself didn't awaken in time to make a real difference.

This isn't true. It wasn't just greatness they missed in their narrow focus: their design would have been the slowest, most costly, and most disruptive of all of them in delivering six lanes; it would have been far longer delivering increased traffic flow than a 6-lane bridge. And more important, the community *did* awaken in time to make a real difference. If it hadn't, we'd be seeing construction trucks in Front Park this summer.

> The Peace Bridge Authority did reach out to the community, after the first concerns surfaced. Late in 1996, a public design competition hosted by the authority and this newspaper drew 479 entries and a symposium brought architects, engineers and others into the process.
>
> It was too little, too late. The authority estimated a gateway bridge could add seven years to a timetable that already included two years of design and 3½ years of construction. To meet the authority's 2002 deadline for opening a new bridge, that symposium should have been held in 1989.

Also not true. The PBA didn't "reach out"; it reacted defensively, and, more important, it reacted deceptively. Why not note that in all of these discussions and competitions the PBA excluded anything but a companion span – as admitted publicly by Bridge operations manager Stephen Mayer at the PBA press conference March 31 attended by *News* reporter Patrick Lakamp?

And why not question the 2002 deadline? Why should that drive everything? Why not ask questions here rather than simply accept at face value everything said by the PBA?

> The authority never wavered. Within a few months, it released its design for a "twin." If the contest and symposium were late, the first serious challenge – businessman Jack Cullen's push for a "SuperSpan" in October 1997 – was even later. So were newspaper crusades and a 1998 public design charette hosted by the Buffalo Niagara Partnership.

Which also, according to Mayer, was a sham, because of that same early decision to exclude anything but a twin span. As noted, Patrick Lakamp was there when Mayer said it; this isn't arcane or suspect information. It had been published in *Artvoice* and in Donn Esmonde's April 3 column.

> While Sen. Daniel Patrick Moynihan championed a bold approach fit for an internationally significant bridge, the authority commissioners responsible for the decision sought security. The two-bridge concept keeps the crossing open if an accident or terrorist attack closes one span; there was comfort in a tried-and-true, conservative approach to everything from building materials (steel, like the Peace Bridge) to aesthetics (a near-mirror of the 1927 span).

Surely he jests. The PBA came up with this paranoid rationale only after opposition to the twin span built up. It's fine to report they claimed it, but hardly responsible to ignore the disingenuousness of it.

Business reporter Chet Bridger, on April 14, 2000, in an article titled "Forum hears ideas for an 'open border,'" wrote:

> State Supreme Court Justice Eugene M. Fahey recently ordered the Peace Bridge Authority to conduct a more comprehensive environmental impact study before constructing a new bridge and plaza. The ruling is expected to delay construction of a new bridge across the Niagara River by at least a year.

How could the judge order them to "conduct a *more* comprehensive environmental impact study" when they never did one and were fighting not to do one? This paragraph gives the impression that Judge Fahey is merely imposing a more difficult EIS on them when in fact he's telling them they can't avoid the law by refusing to do one at all. Furthermore, the delay depends on what the PBA does next. According to the panel of engineers, the total construction time will be reduced, even with the EIS, if they build the 6-lane concrete bridge.

HISTORICAL PERSPECTIVE

All of those examples are from one two-week period last spring. They're recent, the pattern is not. One example: the April 25, 1999, man-in-the-street article by Patrick Lakamp titled "In Fort Erie, twin span support is solid."

The longest quotations were from a man identified only as "Fort Erie lawyer John Teal." Teal got to kvetch without challenge about all the things presumably wrong with the signature bridge proposal and its proponents. "Where were these people seven years ago?" he said. "I see no reason to debate this. It starts to strain the bounds of logic.

This is just politics on the American side. It's a shame that would get in the way of this project."

Lakamp didn't include in his article two key facts he knew about John Teal: John Teal is the brother of Public Bridge Authority board member Dr. Patricia Teal. He was also the mayor of Fort Erie from 1988 to 1997. John Teal was not a random man-in-the-street. He was a spokesman for the Canadian half of the PBA and the *Buffalo News* knew it and didn't tell you.

The *Buffalo News* never saw fit to do a man-in-the-street piece story on the Peace Bridge on this side of the river, even a straight one.

WARREN BUFFETT'S LESSON

So the *Buffalo News* consistently favors statements and positions from the Public Bridge Authority and the Buffalo Niagara Partnership over statements and positions that challenge those two organizations. So it suppresses or misrepresents key facts and gives far greater voice to one side than the other. So it never examines any of these issues in depth and its reporters just parrot what they've been told.

That's deplorable, but it's not what's *really* bad here. What's really bad is that the editors of the only newspaper in a town with only one daily paper have a terrific amount of power and a huge public responsibility. The *News* has utilized its power, and it has abandoned its public responsibility. It is perhaps fulfilling its responsibility to interests and powers it does not care to name in public. It is surely fulfilling its responsibility to its owner, Warren Buffett's Berkshire Hathaway. Buffett bought the paper in 1977 for less than $90 million in current dollars; it has a profit of well over a million dollars every week. (There's an excellent article by former Buffalonian John Henry on Buffett and the *News* in the November/December *Columbia Journalism Review*: "Buffett in Buffalo: His paper prints money. What else does it print?" You can find it online at www.cjr.org/year/98/6/buffett.asp.)

The lesson we should take from this goes beyond the Peace Bridge. We've been lucky with the Peace Bridge because the lawsuits, the work of the New Millennium Group and the work of the Public Consensus Review Panel brought to light huge amounts of information that would otherwise have remained hidden. We've been lucky that the county, the city, the Wendt Foundation and the Community Foundation were willing to underwrite that PCRP inquiry.

We know how badly we were served and how consistently we were misled. But what about those instances where we haven't benefitted from those huge voluntary assignment of resources? How shall we regard the coverage by the *Buffalo News* of other major public issues that will affect our lives and our children's lives – the waterfront, the proposed convention center, political campaigns, corruption or the presumed lack thereof?

With care. With great care. And with reservations. Read the *Buffalo News*, for it is the only daily paper in this one-paper town, but always remember that it prints what it chooses to print, not what you ought to know or what you need to know. It is a source of information, but don't – when the issues at hand matter to you – let it be your only source of information. Keep in mind that the worst sins of bad journalism don't necessarily occur in distorted news stories and editorials, but in stories that aren't covered at all, in important issues that are ignored or downplayed or simplified.

"Freedom of the press," the great journalism critic A.J. Liebling wrote, "belongs to those who own one." And, we might add, is in great danger when there is no real competition. When Warren Buffett drove Buffalo's other daily paper, the *Courier-Express*, out of business in 1982, Buffalo's citizens lost more than a choice in whether they would read the morning paper or the evening paper. We lost the only real check on journalistic laziness and cronyism: competition. "Buffett must be mindful of the potential threat to journalistic quality when he publishes the only newspaper in town," wrote John Henry. He quoted what Buffett said about the paper in his 1984 annual report, two years after the death of the *Courier*: "'Once dominant, the newspaper itself, not the marketplace, determines just how good or bad the paper will be. Good or bad, it will prosper.'"

Prosper it has. What else has it done?

41 THE PBA AS PINOCCHIO: LYING ON THE WEB

You've perhaps heard of Mary McCarthy's famous comment on the Dick Cavett show about Lillian Hellman: "Every word that Hellman wrote was a lie, including 'and' and 'the'." Hellman sued her for the comment but it never came to court because McCarthy died while it was still in process. Too bad because she had solid evidence on her side, the most notable having to do with Hellman's autobiographical *Pentimento,* part of which was made into the film *Julia,* starring Vanessa Redgrave as Julia and Jane Fonda as Hellman. The real Julia was a woman named Muriel Gardner who always insisted she'd never met Hellman. She was about to initiate her own lawsuit, but Hellman died, mooting that out, as they say.

I thought about all of this when I looked through the Buffalo and Fort Erie Public Bridge Authority FAQ page on their website. FAQs are common interfaces in the web age: they give corporations and other institutions a way to post the answers to frequently-asked questions (hence FAQ) so they're not wasting time providing the same information again and again and so the public can have easy access to the questions other people are asking.

Some FAQs – like the PBA's – aren't really FAQs at all. They're composed not so much of questions a lot of people ask, but rather questions the company wishes a lot of people would ask so they could get to say the things they want said without it seeming like what it really is: a public relations screed.

The basic difficulty with all FAQs is a problem endemic to the web: there's no check on erroneous or deliberately misleading information. To trust information from the web you either have to know enough about the source to trust it, or you have to have enough collateral information from other reliable sources to check what you're reading.

Generally, corporations don't put up pages of lies. They may put up fluff, but they avoid flat-out lying because it's too easy for people who know something to annotate those pages and circulate them to other people to show the authors up for the liars they are. But sometimes corporations think the public is so stupid they can say whatever they want and get away with it, or they have such contempt for the public they don't care about being caught out. I don't know which is the case with the Buffalo and Fort Erie Public Bridge Authority – contempt for our intelligence or contempt for our opinion. But it's surely one of the two. Or maybe both.

I'm not saying everything on the PBA web site is a flat-out lie or a serious distortion of the truth. There is no problem with names because the web site does not list the names of any members of the Authority, or, except for incidental mention in four press releases, the names or titles of any members of the staff. The entire web site provides only four telephone numbers: two numbers to help you get trucks and cars through Customs and the office phone numbers of the PBA's Canadian and American press agents. The map showing the location of the bridge seems to be right and so is the description of what goes on at the commercial vehicle processing center in Fort Erie.

I'm sure the toll schedule is correct. After that, things get slippery. By the time you get to the FAQ (www.peacebridge.com/ ecFAQ.-html), it's time to hold on to your wallets and keep your eye on the silverware.

That's all description. Lets look at the half-truths and lies they hope you and the various legislators voting on their proposals and judges voting on their lawsuits read and accept as gospel:

LIE #1:

> Why is there a need to build a new bridge at all?
>
> This is an issue of trade, tourism and traffic between Canada and the United States. The Peace Bridge is the economic link between the two countries.

THE TRUTH:

All studies about traffic patterns show that heavy truck traffic stifles tourist traffic. If they want to encourage tourists coming across the Peace Bridge they should route the trucks elsewhere.

More important, this isn't "the" economic link between the two countries anyway (oh, Mary McCarthy – I finally understand what you mean about lying with the definite article). There are three other motor vehicle bridges within 20 miles of Buffalo, and five more further east of us in New York State. The two countries are linked by about 80 fully functional border truck crossings and an equal number of commercial railroad crossings in Washington, Idaho, Montana, North Dakota, Minnesota, Michigan, New York, Vermont and Maine.

The Department of Transportation statistics bundle crossings in a metropolitan area – Buffalo-Niagara is counted a single crossing, for example, even though there are four vehicle bridges and two international railroad bridges here – so the actual number of truck bridges and commercial goods railroad lines is greater.

(You'll find the USDOT map of those ports and find links to commercial crossings online at: http://www.bts.gov/programs/itt/cross/ports/port.html.)

LIE # 2:

What is Continental One? What is the Peace Bridge link?

The Continental 1 corridor, developed to encourage trade and tourism between Miami and Toronto is relying on the completion of a new bridge so that trade traffic is not rerouted to other areas of the US and other areas of the Province of Ontario. If the bridge is not started soon, the stream of international trade will be negatively affected by increasing the time taken and the cost of carrying goods. Truckers and tourists will be hurt. Jobs in both countries will be negatively affected, but particularly in the Niagara and Western New York area.

THE TRUTH:

It's true that if there isn't easy flow between Canada and US, shipping goods back and forth will cost more. It isn't the least bit true that if a new bridge isn't built in this one place that "truckers and tourists will be hurt" (truckers will truck wherever the bridges are, they'll be paid in any case, and Continental 1 has nothing to do with tourists), or that "jobs in both countries will be negatively affected" (the jobs would be equally well served by a bridge expansion at any of several other locations) or that the effect will be particularly severe in this area (hardly any of the jobs resulting from bridge expansion will come here, except for jobs at the Peace Bridge itself and patronage jobs handed out by Peace Bridge management).

LIE #3:

Why is it important to build a bridge now?

As an economic catalyst for the region, it is vital that Buffalo and Fort Erie move quickly to prosper from this explosion of trade and commerce between the US and Canada. The opportunity for job and tourism increases will be forfeited as well as the opportunity to grow this region into a vibrant gateway for both countries.

THE TRUTH

Tourism, again, is negatively effected by the glut of trucks. Very few jobs will come to this area as a result of the increase in truck traffic, except perhaps pulmonary surgeons to deal with the almost-certain increase in lung cancer, emphysema, asthma, and other respiratory diseases.

LIE #4

> What is the Peace Bridge Authority's proposal?
>
> The Peace Bridge Authority's proposal is to build another 3 lane bridge that would act as a companion to the existing bridge. This design is affordable, can be built immediately and has the ability to ensure traffic continues to flow even under maintenance conditions. The Peace Bridge would have liked to begin construction on the companion span in July 2000. The new bridge would have been completed in 2003.
>
> Upon completion of the new bridge, the existing Peace Bridge will undergo maintenance and redecking. Upon completion there will be an immediate doubling of capacity.

THE TRUTH

It is indeed true that, if they were allowed to build the twin span and then set about rehabilitating the old bridge they would have, at the end, "an immediate doubling of capacity." They fail to mention that such rehabilitation is eight to ten years away, and so, therefore, is the "immediate doubling of capacity."

The companion bridge can be built immediately *only* if the PBA is allowed to build it in total contravention of New York environmental law, which Judge Eugene Fahey told them on April 3 they could not do. A lot of things would be easier or more profitable if you didn't have to obey the law; every crook I know is fully aware of *that*. The new bridge would have been completed in 2003 only if the PBA had been allowed to ignore environmental law and if everyone in Western New York had said, "Okay" to whatever the PBA proposed. Once questions about environment, traffic flow, encroachment and such were introduced, that schedule was out the window.

LIE #5:

> Why are two bridges better than one?
>
> A key reason the Peace Bridge Authority's companion span option makes sense is that the new span will require minimal traffic disruption and construction phasing, thereby allowing for essen-

> tially unaffected and unimpeded use of the existing bridge during construction. Having two bridges ensures that traffic will always be able to flow in both directions.

THE TRUTH:
The twin span project will cause far *more* traffic disruption during its construction phase than a six-lane substitute bridge. A six-lane bridge can be built with no significant rerouting of traffic. The day it is finished, as both Bruno Freschi and Eugene Figg said, you open up the new bridge and you shut down the old bridge. The PBA's plan of building a new bridge, then shutting down the old bridge for years of rehabilitation, will require up to seven years of routing trucks through Buffalo streets and Delaware Park. The other concern here – that "having two bridges ensures that traffic will always be able to flow in both directions" – is no better served by two three-lane bridges than by one six-lane bridge. Traffic has always flowed in both directions during the entire lifetime of the Peace Bridge, even before they converted it to three lanes thirty years ago, so why should that become any more iffy with twice the number of available lanes?

LIE #6:

> Can a single-span bridge can be built in the same amount of time as a companion span?
>
> No. Any new plan or option will require a new Environmental Impact Statement (EIS) and only then can construction begin. This will take eight to 10 years.

THE TRUTH
An EIS Will take 8-10 years only if the PBA continues dragging its feet, as it has been. If it had done an EIS six years ago, that would be an historical issue now.

Other communities get an EIS done in 2 years or less, and given the huge amount of work already done, all experts except the PBA staff says that it would take far less than 2 years here. But only if the PBA stops trying to block the EIS.

More important: the EIS isn't some voluntary detour being suggested by arcane activists. It's not an option; it is required by New York environmental law. The PBA made a conscious decision to avoid the legally-required EIS, which is why they have the judgment against them in state court and the new lawsuit against them in federal court filed last week by two West Side community groups.

LIE #7

> Also, because the Peace Bridge qualifies as a historic structure, any consideration to razing the bridge requires the completion of an additional Environmental Impact Statement, (parenthetically known as a 4F review) which would be extremely time consuming and negate the scheduled completion of the companion span, or any other bridge proposal, for several years.

THE TRUTH:

What "scheduled completion of the companion span"? Always get suspicious when people wanting something drop into the passive voice. The only schedule here is the one they made up, which Judge Fahey has already told them isn't going to happen. Not a single agency in the US or Canada with jurisdiction over historical sites have ever declared the Peace Bridge an historic structure. The PBA and its friends began lobbying to get it declared an historical site in both countries as an attempt to block signature span designs.

LIE #8

> Will tax money and toll increases be needed to pay for a new companion bridge?
>
> Under the Peace Bridge Authority's proposal, no tax money and no additional tolls are required for construction of the new bridge or redecking of the existing bridge.

THE TRUTH

They have no ideas what toll increases will be needed because they have no idea what the new bridge will cost or what it will cost to repair the decrepit underwater structure of the old bridge, should they decide to keep it going. The PBA will pay for repairs and bridge construction out of tolls, but millions of US tax dollars will go into the access roads and the plaza. This is like saying "We're giving you the car for free – but you have to pay us $35,000 for the tires." Moreover, the US government is required to pay millions of tax dollars for offices and processing areas in any new plaza, while the Canadian government gets comparable space for nothing. The Canadian government space is underwritten by tolls, which means Americans are paying for their own office space with tax dollars and for half the Canadian office space with toll dollars.

LIE #9

> Is the Peace Bridge historically significant?

> Yes, the Peace Bridge qualifies as an historic structure in both the United States and Canada.

THE TRUTH

This is like saying your uncle Fred is historically significant. Significant Fred may be, but absent a declaration of historical significance from an agency with some responsibility for such declarations, the utterance is of interest only within the family. And, as I noted above, the only people who have thus far asked that the Peace Bridge be declared a historical monument were doing it as part of the PBA's attempt to avoid consideration of alternative designs and to avoid US environmental laws.

LIE #10

> If the current bridge is designated as a historic site how can the current Parker Truss be replaced with a new arch?
>
> The existing Black Rock Canal arch, known as the Parker Truss, is the least historical aspect of the bridge structure, with the five river arches as the most historic. If it is economically feasible to replace the Parker Truss, the PBA would approach the appropriate historic preservation agencies in Canada and New York State to begin the process to gain approval to replace this part of the bridge. This process would require community and government consensus on both sides of the border.

THE TRUTH

Unless the PBA is successful in its attempt to have the bridge declared an historic site, there is no need to approach anybody because nobody has ever declared it sacrosanct in any regard. If the Parker Truss is not historic, then the rest of the bridge is not historic. They can't have it both ways. The notion that you can cut away one-fifth without changing the whole is like saying the look of your leg won't be altered by the amputation of your foot.

LIE #11

> Is the Peace Bridge structurally sound?
>
> The existing Peace Bridge is in good condition and is carefully inspected every year. According to New York State Department of Transportation (NYSDOT) sufficiency ratings, the Peace Bridge is on par with the Rainbow, Lewiston-Queenston and Whirlpool bridges. Based on a 1999 inspection report, the bridge received a 55% out of 55% rating for structural integrity.

THE TRUTH

How a New York State agency compares the Peace Bridge to other bridges in the area doesn't answer the question of whether or not the bridge is structurally sound. The Federal Highway Administration gave the Peace Bridge a sufficiency rating of 17 (on a scale of 1-100) in 1997. That jumped to 58.9 the following year

How did the bridge improve so radically in a single year? It didn't; the only thing that changed from 1997 to 1998 were the numbers the PBA submitted to the FHA. The FHA ratings are based on self-reports; they don't go out and do any inspections, they trust the bridge managers to tell them the truth and to be consistent in their reporting. The PBA simply doubled its reported numbers and thereby went from a terrible to a passable rating.

More important, there has not been an underwater inspection of the piers since 1981 – almost 20 years ago, though the condition of the piers is key to estimating maintenance costs and life span of the bridge. If those piers are deteriorating or are unstable – as many engineers insist – the maintenance costs would be enormous, the bridge would be out of commission a great deal of the time, and, like a tooth that looks okay on the surface but is rotted on the inside, would not be salvageable for the long term anyway. If you look at the piers, you can see deterioration at the waterline; there is almost certainly far more damage from the swift current beneath that.

LIE #12 (well, this is really just a red herring)

Will piers for a new companion span raise water levels?

Canadian regulators required the Peace Bridge Authority to achieve a "no net change" or zero percent change in pre-and post-construction water levels which resulted in a redesign of the piers to comply. This has been achieved. The International Joint Commission (IJC) in January 1999 stated they agreed with the hydraulic report as submitted by the Authority. In addition US and Canadian agencies conducted their own independent review to confirm the Authority's results.

THE TRUTH

So what? Any number of other designs could result in exactly the same or diminished effects on water flow.

LIE #13

Additionally, removal of the Peace Bridge would be extremely

> expensive and has the potential to greatly impact lake and water flows and levels.

THE TRUTH

Indeed, it would be expensive to remove the Peace Bridge, but not nearly as expensive as keeping it going as it grows ever more decrepit. Removing it is a one-time expense; keeping it going is perpetual, and ever escalating.

It is true that it has the potential to impact water flows and levels – but so what? Should we fret if the old bridge comes down and a new bridge goes up and the net effect on flows and levels is zero? Since the PBA has given serious consideration to no design other than a twin span, they cannot say what design would provide optimum flows and levels. The EIS they have worked so hard to avoid might give us that information. Or they might say to the engineers, "Make sure whatever bridge you build has a zero impact on water flows and levels," to which the engineers would reply, "No problem."

LIE #14

> The hydraulic impacts of one large pier as proposed by alternative plans coupled with the removal of the existing bridge, will require in-depth, time-consuming study.

THE TRUTH

A single pier support system is a characteristic of only one of the several designs suggested. And what's wrong with study when you're undertaking a $200-million project? Why are they so afraid of study? Much better to study before you build than doing corrective surgery afterwards. Corrective surgery takes study too.

LIE #15

> Has the Authority completed the required environmental procedures?
>
> The Peace Bridge Authority has completed all required environmental procedures with respect to construction of the proposed companion span. In accordance with the National Environmental Policy Act (NEPA) of 1969, the Canadian Environmental Assessment Act (CEAA), and the New York State Environmental Quality Review Act (SEQRA). The project has received all necessary environmental permits to commence with construction. The companion span plan has undergone extensive review involving environmental, economic and engineering considerations.

THE TRUTH
This makes sense only if you believe a bridge can be built without landing anywhere. An unambiguous judgment rendered in New York State Supreme Court last April 3 says this is not true. Are they saying here that they are immune from judgments in New York Supreme Court?

LIE #16

> Why was the decision made to build a companion span steel bridge?
>
> When considering design symmetry, climate and weather patterns, maintenance and long-term viability, it was determined a steel structure would add to the design integrity of the new bridge because it would match the existing bridge as well as provide the most feasible option for long-term life and care of the bridge.

THE TRUTH
The decision was made to build a companion steel bridge because some people would make more money that way. There is no other context in which a companion steel bridge makes sense, other than lack of imagination. Lack of imagination would explain why they went that way in the first place, but not why they've held on to what is so obviously an inadequate design so desperately for so long.

LIE #17 (this isn't a lie, but I didn't want to start a new category; it's really a threat)

> If a new companion span bridge is not built, how will maintenance be conducted on the existing bridge?
>
> The Peace Bridge Authority works hard to ensure that maintenance on the bridge is done at times it will have the least impact on traffic. However, if major repairs or decking has to be completed prior to a new bridge being completed, this work may require closures of one or more lanes on the bridge at any given time.

THE TRUTH
What they're saying there is, if you don't let us build the bridge we want to build now, we're going to punish you with lane closedowns, the way we did earlier this summer, just to let you know who's boss.

LIE #18

> If alternative plaza recommendations require the removal of residences and businesses: how will revenue losses to the City of Buffalo be replaced?
>
> The Peace Bridge Authority does not have powers of eminent domain or right of condemnation so the assemblage of property necessary to create new options would be the responsibility of others. For example, the City of Buffalo would have to take the lead with respect to issues of condemning properties and businesses, funding the purchase of existing properties, replacing lost revenue, making decisions regarding moving existing roadways, etc. Moving the US Plaza has many considerations that will have to be studied in order to determine if such a plan is economically feasible and socially desirable.

THE TRUTH
This answer says nothing about Buffalo's revenue losses. What it says is, the PBA has no intention of paying for shifting the plaza north of its present location. If Buffalo wants to have Front Park back, then Buffalo is going to have to find the money to do it.

LIE #19

> The PBA states that a bridge can be built now and still allow for a North or East plaza location. How is this possible?
>
> Bridge construction for the companion span can begin at this time while still preserving the final location of the US Plaza. If an alternative plaza location is selected after the companion span is completed, the option remains open to change the curve of the bridges on the US side to accommodate a new plaza location.

THE TRUTH
Sure, and you can fly to Mexico City by heading north – only it takes you about 22,000 more miles than going the other way and nobody sane would do it that way unless they were getting paid by the hour for making the trip.

LIE #20

> Why is a South plaza location option not being considered by the Peace Bridge Authority?
>
> Considering a location to the South is not feasible with respect to land acquisition, connecting roadways, and most significantly does not provide for an appropriate or safe road grade for a bridge

landing.

THE TRUTH
They can't have a south plaza because the community won't let them destroy any more park land than they already have.

LIE #21

> Which plaza option will allow the City of Buffalo to restore Front Park and Fort Porter?
>
> All plaza location options allow for restoration of Front Park. If it is recommended that the plaza location remain where it is presently located, Baird and Moore Drives will be removed from the park thereby giving back 3.6 acres of land for park use. The Olmsted Parks Conservancy has completed a restoration master plan in the event this option is selected. If the plaza is moved off its present location, additional acreage can be restored for park use and the possible restoration of Fort Porter.

THE TRUTH
How could they add three lanes of traffic and give up 3.6 acres of their current plaza site with no other acquisitions? The real bottleneck now isn't on the bridge but on the plaza; surely they're not going to shrink it. *ONLY* a northern plaza will allow for restoration of Front Park and the space they need for their expanded operation. PBA doublespeak notwithstanding, the laws of physics continue to apply: you can't have two objects occupying the same space at the same time.

LIE #22

> How will the PBA finance any plaza alternative that exceeds their bonding limits?
>
> The Peace Bridge Authority has a limited amount of bond capacity in order to fund all of its long-term capital initiatives without raising tolls or using taxpayer money. Given that, it may be necessary to stretch out the timing of the US Plaza until the necessary resources needed to pay for alternatives are accumulated, or it may require identifying other sources of funding (tax payer money or toll increases) outside the Authority's bond capacity.

THE TRUTH
I'll translate this for you: "We're ready to build the bridge we have wanted to build all along and we have no intention of building a

northern plaza. That was all body lotion we poured out in the hope of softening you up."

LIE #23

Can truck detours be avoided with other plans?

> No matter what plan is ultimately used, detours will have to be utilized during this construction period. The proposed Scajaquada/198 detour route is related to the US Plaza and connecting roadways modernization and construction project and not bridge construction.
>
> The detour route was developed and used successfully by the NYSDOT for reconstruction of the Southbound ramp from the US Plaza to the NYS Thruway in the early 1990s without any problem or fanfare. The plan uses NYS highways designated for commercial traffic. Construction incentives to complete the project early could shorten the time frame needed for the detour.

THE TRUTH

The bridge and plaza work are inseparable, so they can't dismiss this disruption of Buffalo life by saying the 198 detour has nothing to do with the bridge construction. That's what Judge Fahey told them on April 3 and that's what is at issue in the new lawsuit filed in federal court last Friday. Their Alice-in-Wonderland-doublespeak notwithstanding, the PBA's twin span plan will require routing trucks through city streets and Delaware Park along Route 198, and it will vastly increase the number of trucks on route 33. None of the plans for the six-lane bridge require that.

LIE #24

Peace Bridge Authority

That wasn't a question, but it was an incorrect phrase they used 12 times in the FAQ. As I've several times said in these pages, their name is not the *Peace* Bridge Authority, it's the Buffalo and Fort Erie *Public* Bridge Authority, which phrase they never used at all in their FAQ. They'll do anything to keep the public out of their business.

So much misinformation and deception on a single web page! It's exhausting to deal with people or agencies who lie all the time. They force you to test everything and keep safeguards in place every moment. It would be nicer to if we could deal with the PBA the way we deal with the public library, say, or our mailman. But we can't.

That's why those lawsuits in state and federal court are necessary.

What's astonishing about these guys isn't that they work so hard to get their way and deceive the rest of us in pursuit of that goal. People do that all the time. It's not nice, but people do it. What's astonishing is that even though we catch them at it again and again, even though there are hearings like those organized by the Public Consensus Review Panel or court decisions like Judge Fahey's, they keep plodding straight ahead, telling the same lies again and again and again.

The other thing that's astonishing is that not one of the five American representatives to the Public Bridge Authority seems to be concerned about this kind of misrepresentation of fact and gross distortion in their names. The relationship between the Authority and its staff is mostly hidden from public view, so there's no way to tell if this bunch of lies is something the staff slipped by the Authority or if it's something the Authority told the staff to try slipping by the public. They're all political appointees, but many people hoped that when Attorney General Eliot Spitzer replaced a steel executive with Barbra Kavanaugh in February 1999 the public would finally have someone on the board who would represent its interests, someone who would at least let the rest of us know what goes on in those closed meetings. There's no gag order keeping any member of the Authority from telling the truth about what goes on in those meetings; they're all silent by choice.

There might as well be a gag order, because Kavanaugh has been dead silent since she joined the board. The Attorney General's office fought *against* the city of Buffalo, the Episcopal Church Home and the Olmsted Parks Conservancy in the lawsuit against the PBA in New York Supreme Court. We don't know if Kavanaugh has simply absented herself from all PBA activities to avoid a conflict of interest – in which case the public has been deprived of the one potentially independent vote there – or if she is remaining silent by choice or policy. What we do know is that no one on that board cares enough about the public interest to stand up outside of that board room and say: "What they're doing is wrong. They've got to start telling the truth for a change. They've got to stop stonewalling."

Which is why the lawsuit decided by Judge Fahey in April and the lawsuit filed in federal court last week are so important to us all. The halls of Justice – unlike the board room of the Buffalo and Fort Erie Public Bridge Authority – are open to everybody. The justice system provides a forum in which voices can be heard and in which egregious lies can be awarded the punishments they deserve. Neither of those lawsuits is about bridge design. Both are about air and noise pollution, disruption of city traffic patterns, consumption of public land, and other violent incursions into the quality of life in this community.

Which is to say, those two lawsuits are about all those factors of daily life that are of vital concern to all of us and which the Buffalo and Fort Erie Public Bridge Authority would prefer be discussed by none of us. The Buffalo and Fort Erie Public Bridge Authority would like dead silence about anything that would stand between it and the anachronistic, expensive, environmentally degrading steel twin span it so desperately wants to build.

And if it can't shut down conversation, then maybe it can at least contaminate the quality of it by a disinformation campaign, such as the misleading television barrage last spring just before Judge Fahey rendered his decision and this web page of good questions and misleading and mendacious answers.

Want to know more? Call the only two numbers on the PBA's entire web site that will connect you with a person who will talk to you about anything (other than getting your truck through US Customs via the Commercial Vehicle Processing Center in Fort Erie or acquiring a CanPass to help you move through both customs operations more quickly). Those two numbers will put you in touch with the PBA's two press agents: Susan Asquith, Collins and Company, 716.842.2266, or Catherine Clark, Clark and Associates, 905.608.1055.

Call them. They'll tell you everything the PBA thinks you ought to know.

42 Robert E. Knoer: Making a Federal Case Out of It

The lawsuit brought in United States District Court on July 28 by the Buffalo West Side Environmental Defense Fund, the Peace Bridge Columbus Park Association and several individuals was drafted by Robert E. Knoer, a Buffalo attorney who specializes in environmental law. He's taught an undergraduate course in law and the environment for the past decade at UB and has long been active in several professional and community groups engaged in environmental issues.

Knoer is the author of a key document in the current legal discussions about the PBA's attempts to avoid state and federal environmental law, the March 1998 "Report on Environmental Assessment for the Peace Bridge Capacity Expansion Project" The "Knoer Report," as it is usually known, was undertaken at the request of businessman Jack Cullen and architect Clint Brown and their SuperSpan Niagara group. It is tough, specific, and scholarly, a first-rate piece of legal analysis and social commentary. Among other things, the report confronts one of the PBA's key avoidance strategies: insisting that the bridge and plaza construction projects were separate from one another and so did not have to be evaluated by the same environmental impact standards at the same time. The Knoer's Report's preface is as relevant now as it was when it was first released 30 months ago:

> A review of the Buffalo and Fort Erie Public Bridge Authority's Environmental Assessment shows a systematic attempt by the Authority to avoid proper environmental review.
>
> The Public Bridge Authority attempted to, in their own words, "pursue the lowest path allowable." Although such a stated philosophy by a private entity might be understandable, it cannot be condoned from a public Authority. The Buffalo and Fort Erie Public Bridge Authority is a *public* authority. It is not a private developer who has the luxury of complying with the bare minimum letter of the law. As a public agency, the Authority's responsibility to the public is to comply with the letter *and the spirit* of the environmental review process.
>
> This systematic avoidance of the legal requirements under Federal and State law began with the Authority's decision to eliminate certain reasonable alternatives from review altogether.

> ...Exploration of alternatives and their environmental impact is at the very heart of the purpose behind environmental review. From the beginning, the Authority made a determination that it would do an assessment which resulted in a "Finding Of No Significant Impact." It is improper, and in contradiction to Federal and State environmental review laws, to make a decision on whether or not there will be environmental impacts prior to conducting the investigation.
>
> The PBA at every turn attempted to evade its duty under State and Federal laws. The PBA attempted to ram through a project based on a faulty environmental review. The Authority does not even attempt to hide this disrespect for its legitimate responsibility. In meetings as early as August 1995, the Authority declared that its *policy* will be to "follow the lowest path legally allowable" in conducting its environmental review.
>
> The United States Coast Guard has been instructed by the Secretary of Transportation, who has authority over the United States Coast Guard, to be an 'environmental leader.' Instead, the Coast Guard appears to have agreed here with the Buffalo and Fort Erie Public Bridge Authority to look the other way while the Authority attempted to force through a flawed environmental review.
>
> The Public Bridge Authority made a determination early on in the process to improperly segment the Plaza and Bridge projects. Even though this issue was repeatedly brought to their attention by numerous public and private voices, the Authority continues forward on its improperly segmented environmental review.
>
> The attempt by the PBA to avoid legitimate environmental review, to improperly segment the Plaza and Bridge projects, and to improperly dismiss alternatives to the project without the requisite public input leaves the Authority's Environmental Assessment, and thus the entire Bridge/Plaza Project, open to significant costly and time consuming challenge under the National Environmental Policy Act, State Environmental Quality Review Act and Department of Transportation Act.

Much of Jack Cullen's cover letter to the report (sent to Robert W. Bloom Jr, chief of the Bridge Branch in the Coast Guard's 9th District) now reads as prescient, for example this warning about liability:

> Who would want to be involved in this project when six or seven years from now a court lambastes the entire project as an example of narrow-minded and arrogant bureaucrats purposely avoiding the environmental responsibilities with which they have been charged?

> This would not just be an embarrassment to those involved but it will cost hundreds of thousands of dollars and years of litigation to get to that point.
>
> Wouldn't that time and money be better spent doing the review right the first time and building the bridge and plaza which is acceptable to an entire community, in a place acceptable to the entire community, such that it will serve the entire community?

Cullen was wrong about only two things: it took two, not six or seven, years for a court to find against the PBA and, by the time this second case is done, the legal costs will be in the millions, not hundreds of thousands of dollars.

The Coast Guard, like the PBA, ignored Cullen's letter and the report. But now the cat comes back: much of the substance of Knoer's July 28 brief in Federal court comes from that 1998 report for SuperSpan.

Since Bob Knoer identified and analyzed these key issues so early, we asked him to talk about the current Federal case and the PBA's continuing reluctance to take seriously any of the community's concerns.

Could you outline the main issues in the current federal lawsuit, and how it's different from the state lawsuit?

I think the main difference, and the main issue that this suit raises – it hasn't really been raised by anybody in the past – is the basic assumption: do we want the trucks? The prior suits don't deal with that. The prior suits deal with, "Give us a choice between infrastructures to bring all these trucks here. Give us a nice-looking bridge – depending how you define 'nice-looking' – give us a plaza a little bit this way, give us a plaza a little bit that way." But they never really ask, "Do we want the trucks at all?" This suit raises that issue very squarely by saying that the environmental impact of the traffic has not been looked at in enough depth. Forget about *what* is crossing, forget about exactly geographically *where* it's coming across. What is the impact of pulling that much traffic into a concentrated area? I think that's the biggest difference between these suits.

It basically attacks the Authority and what the problem has been all along, which is not the individuals at the Authority. I don't attribute nefarious motives to these people. It's the corporate culture. I use that word "corporate" specifically because they act like a corporation. Although they are a public authority, they act like they have profit motive and they have a need to get bigger. They forget what the product is – the product is benefit, and that's economic, and it's environmental and it's social.

The PBA is in competition with bridges down the river, which creates a gotta-get-better, gotta-do-more attitude internally. They just took without question the assumption, "There's trucks out there; bring 'em here." That is something that I think needs to be examined. Nothing there allows anybody to step in and say, "Well, you know what, what you're doing doesn't benefit the region in general, or New York State, or the United States." I think it's a product of the way the Authority is now designed, which is "You are in charge of the bridge, you have responsibilities from one plaza to another. Have at it." The Authority believes that themselves. They've rebuffed our requests for FOIA [Freedom of Information Act] information. They say, "We're above it."

Because they're bi-national?

Because they're bi-national. That's what they believe. I don't think they're right. That is actually part of the suit, whether or not they are subject to the same New York statutes that any other public benefit corporation is.

That's how they've also avoided all the sunshine information laws.

The Public Bridge Authority was created to operate this bridge, but it was created as a New York State municipal corporate body – that's the language that's used. And then there was a companion body created on the other side of the river, and the two joined to form a board which runs this bridge. But that doesn't detract from the fact that there is a New York body that has all this information, that goes to all these meetings. That body, I think, is subject to FOIA, and hopefully we'll find that to be the case.

And to other New York State laws.

And to *all* New York State laws. I think if it's not, it should be. That will help alleviate not only this immediate problem but help us into the future, because it really has to be addressed. The bridge is supposed to be a vehicle to benefit the region. It is not a money-making mint, that's just not the purpose of government, that shouldn't be the purpose of government. It should be to employ our money, our taxes, our resources that the government holds in trust for us, to employ them to benefit our lives.

We all have different opinions as to what that means. To some people it means making more money, to some people it means "give me more free time," to some people it means – whatever. The government should be, however, making that their point and not handing over a very expensive piece of infrastructure to a group whose mission is defined as "Make as much money as you can."

That money doesn't come back to the community anyway. It's totally internalized.

Right, not a bit. And that's a big problem, that's a major problem, because it is a *lot* of money, by any estimates. I can't say I have any inside information on how much money it is, but from the outside analysis that I've seen and the people I've spoken to who have looked at it, it's a serious amount of money, an amount of money that could do a lot of good in a lot of ways. And it's being internalized, and it doesn't make sense. A lot of the money is from trucks crossing, which do not in and of themselves benefit the immediate area. There's no economic benefit that I've seen or that anybody else has been able to point out to me – and I've asked the people who I think have the most at stake in telling me why I'm wrong, and I haven't heard the answer yet.

Andrew Rudnick of the Buffalo Niagara Partnership and the people at the bridge keep yelling, "This brings us jobs, jobs, jobs." I haven't found them.

I haven't located those jobs myself. There are legitimate, good, blue collar jobs that are associated with maintenance of the bridge. I don't think those jobs justify the idea that we should increase the bridge capacity just to make jobs to maintain the bridge. I mean, it's kind of like the lady that ate the fly. It just keeps going. At some point you have to say, "Well, what's the premise and why do we believe it?"

Can you talk to me about the segmentation issue and how that applies to your lawsuit?

"Segmentation" is a dirty word in environmental law.

We have to step back to why we have a National Environmental Policy Act. Prior to the National Environmental Policy Act, the government agencies said, "We can't look at the environment because we looked at the legislation that created us, and the legislation that created us says nothing about our obligation or responsibility for environmental issues. We have to do what we were told to do because we are a government of limited, delegated authority and we only have the authority given to us."

So the National Environmental Policy Act came around and said, "Okay, we are going to bless all of these government agencies with not only the authority but the *responsibility* to at least consider what their actions are going to do to the physical and social environment. The idea in these statutes is, create a process. It's a fifth-grade math, "Show your work" kind of a thing: "Don't just give me the answer, that's not going to get you full credit. You've got to show your work."

That's really what this process is: "I want to see that you have the scoping and that you looked at what this proposal was, I want to see that you held public hearings, I want to see you produce a document. All these steps." And then at the end it doesn't matter what your answer is, it truly doesn't under these statutes. If you do this process right, no one is going to overturn the answer, even if everybody else around you thinks the answer is wrong. The process is not meant to produce a particular answer elevating the environment above the economy or anything else, that's not the idea, The idea is to integrate thinking about the environment into the process.

"Segmentation" is a dirty word because segmentation says, "Well, here's my project, but I'm going to look at just a little bit of it right now because I think I can swallow a little bit of it, and justify it to everybody because it's not going to be that bad. Then I'm going to go back and take a little bit more of it, and then I'll take a little bit more of it." Well, that perverts the process because you're not looking at the whole process.

When I teach my students about it, the example I give them is a roadway. The classic example is a roadway. You want to go from X to Y, but there's a wetlands or other preserve area in the middle. So instead of going from X to Y you say, "Okay, I'm going to go from X to X1," which ends just before this little piece of preserve land, and I'm going to build myself a nice, big road. Then I'm going to go from Y1 to Y on the other side of this thing and build myself a nice little road. And then I'm going to go and say, "Look, we've got these two beautiful pieces of infrastructure. We just have to cross this little bit." And then it's much easier to justify. Instead of saying, "Okay, this is my project," and people say, "Well, you could go up and around" or "Do you really need it?" or whatever.

In real life you see segmentation all the time in sprawl situations, because in real life you see someone come to a town and say, "I want to build a golf course." Golf courses are not objectionable to people, in fact they're included in the greenspace acreage counts in a lot of towns. When they say, "We have this much greenspace," they're including golf courses. You got to a community and say, "I want to build a golf course in your backyard," and people say, "Gee, that's better than some great big box store or something, so sure, that's fine." But what they really have somewhere in some back room is a drawing of a golf course with high-priced homes around it, and then strip malls around that. They really have this whole community concept, but they're going to get in with this golf course. Then they're going to say, "Well, you know, we'd like to build low-density homes," and for the most part the town thinks, "That's a nice tax base." Then "We need roads to service these homes, and, gee, these people need dry-cleaning and drugstores and things." Whereas if they had gone to the town and said originally, "We want to increase your population, we want to increase retail acreage, and so forth."

So segmentation is the idea that, you have to look at the whole project, you can't just look at part of it. Unless you look at the whole project, you don't get the whole environmental impact. So it's a dirty word. You can't segment.

Divide and conquer.

Divide and conquer, that's a good way to put it.

There's a sister concept to segmentation which comes into play here, and that's the idea of cumulative impact. Cumulative impact says, "Okay, you have a single project, but other people also have other projects going on," and in order for the government agency to assess the environmental impact you have to look at all of them cumulatively.

Both of those apply to the bridge. Even if they got by this – what I believe to be – tortured legal argument that the bridge has separate utility from the plaza – and that's where they claim they can segment because they claim the law provides that if you have separate utility, you can do them separately – -even if they get by that, they can't get by the fact that they have cumulative impact. Those are two different concepts but related.

They've been insisting on segmentation, saying the bridge and plaza construction projects were independent of one another, for a long time now. They had to have known that as soon as someone filed a lawsuit they'd have to really deal with it. Do you have any sense of why they thought they could just breeze through with this?

Number one, I think that traditionally we don't have very good record in Buffalo of getting enough public opinion rolling to stop an agency as powerful as the Public Bridge Authority. I think they felt going into this process, and in their own notes they readily admit – and I have to give them that much, they're very up-front about what they're doing – they readily admit, "We are going to make as a corporate policy, an Authority policy, to follow the lowest path legally allowable." The shocking nature of *that* coming from a public benefit corporation! That goes to the reason why they thought they could segment. They sat down and they said, "Okay, where's the bar? What do we have to do to get over? We're going to just barely scrape the top of it."

Unfortunately they underestimated where the bar was, and the bar was up a little higher. I think that going into this they truly thought that they could do what they wanted, and they thought that they could segment because they didn't believe that there would be sufficient opposition, and they felt that they had at least a legal argument – they did get an opinion that said that, provided the bridge and the plaza have separate utility, you can segment. There are cases where that is true in the law, but in this particular project, it was a misguided legal position for them to take. I think they're paying for it now.

You have to remember why they segment, why they felt it was important. If you think about it, they're doing an EIS for the plaza now, so it's not like they thought they could minimize their paperwork – if anything, they doubled their paperwork, because they're doing a whole EIS process for the plaza anyway. Why not just go ahead and do it for the bridge? What happens is, if they get the bridge through, and they get the permit, the bridge has to be there. That takes away any alternative of moving the plaza. You're locked in. The plaza has to be where the bridge lands. So by segmenting it, they narrowed the choices, the alternatives, which is a key issue in environmental review. They narrowed the alternatives automatically. That was the only benefit segmentation, that I can see, would have for them. Because it certainly didn't speed up the process. From a practical standpoint, what they're trying to do is get trucks across the complex. It doesn't do them any good to get them across the bridge if they're stuck at the plaza, or vice versa; the idea is to get them from Fort Erie to Buffalo and beyond.

You have to get them from the QEW to the Thruway.

That's an even better way to look at it. So to say, "Look, we gotta move the bridge, we gotta move the bridge, we gotta move the bridge along" doesn't make sense, because that doesn't address what their own studies going back to the 70s say. It's not the bridge, it's not the lane capacities; it's the plaza. So why do they focus on the bridge, and why do they push the bridge first? I think the reason was, they figured there was less environmental impact, because if you look at just the physical bridge, it's in the water. It's really not going to change much, and as long as they could deal with the navigational issues, they figured the bridge was easy. It's hard to say with a straight face that the plaza is not going to have an environmental impact, so they didn't even try it there. Plus the plaza is in parkland, or impacts parkland directly and is contiguous to parkland, so the Department of Transportation act comes into play, and they can't get around it. But if they force the bridge and focus on the bridge and make the bridge a rush that needed to be done, once that was in place, the plaza is a lock. Once you have all that infrastructure going into the bridge, who can argue "Gee, maybe we ought to send these trucks elsewhere."

This is like your metaphor of the small bit of wetland.

Yeah, it is, and it's a classic example of segmentation.

So you think they entered this without expecting any real opposition. They just figured they'd get through on the bridge and the rest would naturally follow.

They expected opposition, because they refer to the opposition, and they refer to why they had to separate these projects, because if they looked beyond the bridge to the plaza, they expected significant neighborhood opposition. I mean, they refer to it in their papers. So they expected opposition, but they felt that they could overcome it. I can't speak for them, these are my assumptions, and I don't want to attribute thoughts and motives to them that I can't necessarily prove. But in looking back on it, that's my assumption, that they were moving forward, that they felt that, yeah, there's going to be opposition, there is to pretty much any project anywhere, that's just the nature of the beast. People *should* speak up when they have opposition. I think that was it, I think they felt they could overcome the opposition. They didn't expect people like the New Millennium Group, people like the Buffalo West Side Environmental Defense Fund, people like SuperSpan – which was a group of businesspeople, to a great extent – they didn't expect them to rise up in opposition.

You mention SuperSpan. Can you tell me how and when your report for them came about?

That was submitted to the Coast Guard prior to the end of their hearing process in March 1998. What happened was, I got involved in SuperSpan and went to some of their meetings, and was intrigued by the fact that the Authority was moving forward in the light of a significant amount of what I think is respectable and important opposition, opposition you wouldn't expect. It wasn't a bunch of people running around and complaining about small issues. It was people who were really trying to make a point, people who were trying to work within the process.

It got to the point where the Authority clearly was not listening and clearly was going to move forward with its project the way they envisioned it, and I was asked, "What can we do? From a legal standpoint, what can we do?" So I said, "Let's review what they've done to date and see whether they've complied."

I acquired as many of the records as I could, including the draft environmental assessment, and the FONSI and I reviewed all the minutes that I could of the Authority, and came up with this report. The purpose of the report was to submit it to the Authority to tell them "Here's why we think you really need to step back and do it right. Because we think that if you don't, you're putting yourself in line for a lawsuit which is probably going to set you back." We wanted to make that clear to them. We did make that argument back then. Then, in order to protect the ability to go to court in the future – SuperSpan was never interested in actually litigating, in actually going to court – but in order to preserve the public record, this was submitted to the Coast Guard, so that the Coast Guard had this whole document when they made their final determination that there wasn't going to be an environmental impact, that their segmentation was not a problem, and that the Authority could move forward.

That's important to us now, because when I go to Judge Arcara in federal court and say, "Judge Arcara, I think we're entitled to relief," one of the judge's standards, one of the things he's going to look at, is "What did this agency or these agencies know at the time they made the decision?" And "Did you participate in the public process to let them know this?" Because it's not fair to the agency or the project sponsor to come in later with new ideas and new concepts that they weren't even asked to address. It was very important that this be filed and be made part of the record. That's the genesis of that report.

This document, in essence, put you into the process. It legitimizes your entry now.

Legally? Yes. In environmental law a lot of times what you have is an administrative agency taking some action, or failing to take some action, and you're going to the judicial branch as the check on that agency. But when you do that, the judicial branch does not want to step into the shoes of that agency and say, "Well, I think this is right or that's right." They're looking at whether the agency fulfilled its duties under the law and under the Constitution. The court does not want to hear new arguments when you're looking at the review of an agency. The court wants to hear "This is what the agency saw, this is what the agency heard." The court is looking at the record of decision, the record that was before the agency. This document is part of that record. This document lets us say to the court, "Your Honor, they did not do it right even though they were advised what the problem was."

Did you get any response from them at all?

No.

How could they read this and not know there was a huge chance of finding themselves in court based on exactly these issues? Your report isn't just opinion, you keep referring to specific law, you show how the PBA at every turn attempted to evade its duties under state and federal laws, how the PBA attempted to ram through a project based on a faulty environmental review.

I attempted through most of that document to use their words, not mine. I think that was important, because it was not a matter of a rant about why they're wrong, it was more a matter of asking them to reflect on what *they've* done, what *they've* said, and the information *they've* had. It is my opinion that they didn't do what the law requires. I don't think they could have read it and not known they were facing the challenge.

The first day that they met with the Coast Guard and Nick Mpras said, "Can you separate the bridge and plaza?" The Coast Guard had just done the Blue Water Bridge, a US-Canada bridge and plaza. It's not like this was a brand new concept they were inventing. And his first statement to them was, "Can you separate the bridge and plaza?" So I think it was pretty clear from the beginning. But they decided as a calculated decision that they were going to follow the lowest path legally allowable, and if they could get away with segmentation, and if they felt it was going to advance their interests, that's what they were going to do.

That gets back to the first thing I said when we sat down here: it is a culture of benefit the authority, not benefit the public. That's unfortunate, because it's a major piece of infrastructure that we need to advance, if it's used right. I don't think they could have read that document. I certainly gave it to them. Whether or not they took the time to read it, I don't know, but I never did get a direct response to it.

Could you just briefly say what this report addresses and what it finds?

The report first of all reviews what's happened. It goes back and asks, Why are we here? What are we trying to accomplish? What is the Authority trying to accomplish? What is the problem that was identified? How have they handled it? Then the report looks at what the law says needs to be done in this scenario. Then finally the report says why what they did didn't meet with what the law requires

I reviewed the traffic studies that were done by New York State Thruway, Department of Transportation, Canadian Ministries, the capacity expansion studies the Authority did, the minutes of their meetings, and so forth. I then identified what they did in this process. One of the first things they missed, in my opinion, was a failure to provide all alternatives to the people. The idea of environmental review, the concept is the same as our free speech concept: lay it out there. The best way to combat bad speech is to allow more speech that's better. But what you *have* to do is lay it out there. In the environmental review arena what you do is you lay out the alternatives.

You and I drive across the bridge perhaps but we certainly don't spend our time and don't have people experienced in what traffic is around us. We needed *them* to lay out the alternatives, including the alternative of doing nothing. What if we don't do anything? They didn't lay out the alternatives because they determined that their world is limited to that short corridor between Fort Erie and Buffalo, and in their world, in the universe of alternatives, you could tweak the bridge to the left, you could tweak the bridge to the right, you could put more lanes on that bridge or you could build two bridges. Their alternative never included the real problem, which is getting traffic from Canada to the US, and maybe we can do it downstream, and maybe we can do it *way* downstream. That alternative was never addressed in their process.

I'm sure what they're going to respond is "We don't have the authority. We couldn't put a bridge downstream. We couldn't put a bridge there." But that doesn't cut it. They're a *public* benefit corporation, and even though they might not have been able to do it, they had to lay out to the public, "You ought to think about this."

Even if they didn't have the authority to actually build these alternatives, they had the obligation to at least advise people and put it out on the public kitchen table, so to speak, and let the public address it, let the public say, "That's a good idea" or "That's a bad idea" in an open way.

That was the first mistake they made, not giving the alternatives.

Secondly, they then decided "We're going to cut this in half and we're going to do the bridge separate from the plaza." That created a false picture of what the project was. The project is not to expand the bridge itself. I've been somewhat – I won't say amused because it's a serious issue, but it's struck me that as we went through the process in the beginning, when opposition began to fester, the Authority said, "We've got to do this! We've got to do this now! The bridge is falling down, it's got to be done now!" And as we get through this process and the idea of a twin, they're trying to say, "You can't take down this beautiful bridge! Why waste this beautiful bridge?" When the study came out that supported their original idea that the bridge was falling down, they said, "No, no, no, those studies don't know what they're talking about, they're statistical anomalies."

So, if it truly is not the bridge – and I think that's the position they're taking now, that the bridge isn't falling down – then why are they pushing the bridge instead of doing the plaza first? Do the plaza first. It's still segmentation, but I think it gets you more to the heart of the impact.

In the report, I identified those issues, I identified where in the process various people in legitimate government agencies pointed out to them, "This doesn't seem right. Can you do this?" They did it enough times, it was mentioned enough times, that it's difficult for me to believe that the Authority did not expect that they would have opposition. I think that their hope was that it wouldn't gel enough to rise to the level of legal opposition and somebody actually pursuing them.

They're very powerful. The Authority is very powerful, the people who support the Authority are very powerful. Some less powerful than they believe, but still very powerful. I really think that they believe, "We can do it." And they moved forward.

I conclude with the fact that they did it wrong, and I point out where I think the flaws are in the process. Not to berate them, and not in any way to denigrate their jobs and their efforts, but to point out to them that they can do better. And I hope the litigation is going to result in something better.

When this report was delivered two years ago, there was still time to fix it.

There's still time now.

They knew about the problem in 1995; that's the first time somebody said, "Light bulb go on; you can't separate bridge and plaza." That's five years ago. Five years from now, if we're sitting here, or you're sitting here interviewing my son at that point, we're going to be saying the same thing. So there's always time to fix it. Certainly when that report came out I would have hoped they stepped back and said, "Okay, let's open up the process, let's combine the EIS for the bridge and the plaza since we're doing an EIS anyways. Let's go back and utilize the information that we have, the air studies that we've done to date, and the other things that we've done to date, and expand on it, and take care of it and do it right.

A lot of people were drawn into this by the aesthetics question: which bridge shall we have? But that's not where the most important part of the argument seems to be now.

Not for my clients certainly, and I think not for me. It's not aesthetics. I think the bridge became a symbol for where we are. A hundred years ago, the Pan-American Exposition was here, and we were the fifth largest city in the country. We were at a great point in our history, and now we're at a very low point in our history. We're losing population, the region in general has not been seeing economic benefits. I think the bridge and what we do with the bridge became a symbol to young people here asking, "Do I want to stay here? Is this a place I want to live? Does this place exhibit to me values and things that I want to experience, that I want my kids to experience?"

It became a gauntlet: is this region going to be satisfied and just take care of a mechanical problem, or is this region going to soar above and say, "We're better. We can do better"? The bridge became a symbol of that. We don't want to just drive trucks across, we want to make a statement to the rest of the region and the world: we are moving forward, we are progressive, we are trying to get back to our heights. I think to a lot of people that is what it became: a question of whether we wanted to live with mediocrity, or say to our young people, "Look, we are progressive and we're moving forward."

People then became more interested in and more knowledgeable about the parks. I think Olmsted's name has been used, if you ran some sort of database over the past few years – people who'd never heard of him, people who never understood the great work that he did, not only locally but everywhere – are now aware of him. So if nothing else, that came out. As people began to see these drawings, and as people started to identify the geography around these drawings – Fort Porter and Front Park and LaSalle Park – they all of a sudden started to realize, "There's more here," and they start to look at it in greater detail.

The suit that the West Side Environmental Defense Fund has brought in Federal court deals with an issue which has become more and more prevalent nationally, and that is the issue of these diesel fumes concentrating and their impact on neighborhoods and their impact on citizens. The information on it, I think, wasn't as readily available as the project was going along. Nonetheless it is an important issue, important to the people who are experiencing it.

So in a way it's been a wonderful issue, in the sense that it's drawn a lot of people into a lot of questions and involvement in the community.

Wonderful in the sense of after a hurricane you feel good about the work the citizens are doing.

It has energized people, and I think one of the things that's wrong in the PBA's and Buffalo Niagara Partnership's opposition to this is that they keep trying to say that everybody's coming to this from the wrong point of view. But there's a real spectrum of interests now.

When I first started getting involved in this thing, when it was signature bridge versus twin span, the Partnership came out – and people think of the Partnership as this unified voice of business – and it said in the paper that "the Partnership supports...," "the Partnership does this," "the Partnership does that." Well, there's a lot of people who were members of the Partnership that I had been talking to that I would ask, "Did anybody ever ask you which bridge you support?" No. So when you say "the Partnership," you're talking about the people that *run* the Partnership and their positions and opinions. I'm not a member of the Partnership and I've never attended any of their meetings and I can't speak to the democratic process that they have or don't have, but it seem to me that there was a large number of people who were at least nominal Partnership members, who paid their dues, whose opinion was never solicited. So when you say that there is a unified voice of any group, you have to look beyond that, you have to look at where that comes from: Is it legitimate, and does it really express what the members of that group believe?

Tell me how to characterize this. Is it narrowly fitting within what they think are the boundaries of the law or is it is deliberate avoidance of legal requirements? Is it like a taxpayer disagreeing with the IRS about the legitimacy of a claimed deduction – which is perfectly legal – or a taxpayer lying about deductions he didn't really have – which isn't at all legal.

I believe in my heart that they felt they had a legally justifiable position. Going back to the quote that I referred to earlier – "We're going to follow the lowest path legally allowable" – you've got to remember the "legally allowable" part of that quote. I think they did try to comply with the law as they read it. The problem is they were looking at the words and not the spirit of the law, and they were forgetting that they are not a private entity. But I don't attribute any criminal or similar motives of any type.

These PBA board members are, to a great extent, businessmen, whose job is to maximize revenues. I think that these people have come together, and they came to this bridge, and they're volunteering their time, to a great extent – I don't think they're making a lot of money off of being there – and they said, "Okay, what's the project? We're going to roll up or sleeves and we're going to do it." And they did. They did it in a business-like way, trying to minimize costs, maximize revenues and do it in the shortest time possible.

The element that was missing was the public nature of what they were doing. That, I think, is what happened. I don't think anyone set out to break the law, I think they set out to comply, as they had to, *minimally* with the law, but they forgot that they were not acting as a private entity. I guess, to finish off the corporate analogy, they should have looked to their shareholders and asked what they wanted: that's you and me and everybody else.

The PBA was created by men and women that you and I and our forefathers put into office, and they were given the right to do that by us. So members of the PBA do have a responsibility to us as the shareholders to open up the process to the public and truly get public input on the whole issue, to they let the public know *all* of the alternatives.

43 The Great Autumnal Peace Bridge Q&A, Part I

This article and its companion in the September 28 *Artvoice* consist entirely of responses to questions submitted by *Artvoice* readers about the operations of the Buffalo and Fort Erie Public Bridge Authority.

The questions are in italics. Some of them are very brief; some contain comments. None of these questions is me pitching slow balls to myself. Every one is authentic and the texts of nearly all are exactly as they came in. A few were asked by two or more people, so I combined the phrasing. Some recent questions had been asked and answered in one of the two previous Peace Bridge Q&A articles ("The Great Summer Peace Bridge Q&A," May 3, 1999 and "The Vernal Peace Bridge Q&A," March 30, 2000). I included some of those questions, with updated answers, in this round.

My thanks to all the *Artvoice* readers who took the trouble to voice their concerns and all the public officials and private citizens who helped me find the answers.

I wish I could thank equally the Peace Bridge officials for their help but I cannot. Peace Bridge general manager Earl Rowe neither accepts nor returns my telephone calls. The PBA's web site invites questions from the public and promises answers. I've asked questions there and have gotten no responses at all. I don't know if this is how they deal with everyone who isn't asking a question about truck cargo processing or buying or selling something, or if it's just how they deal with members of the press who are not in their pocket or on their payroll. As I've pointed out here previously, the informational pages on their web site are full of misinformation, and they won't even list the names of the current membership of their board of directors.

If PBA board members were forced to live on Buffalo's West Side, how many of them would do an about face and demand an environmental impact study?

Not one of them lives anywhere near Buffalo's West Side, so not one of them has to wrestle with that sticky question.

The policy choice of the PBA board to follow what they called "the lowest possible path" has resulted in all these delays and huge expenses. What they did was maybe legal, but it's surely irresponsible. If they were in a private corporation they would be tossed out and maybe sued for wasting the stockholders' money. But since they are a public benefit corporation, shouldn't they be charged with gross negligence? They're all political appointees, so the voter has no direct access to them. Isn't there anything that can be done about people who are negligent, incompetent handlers of the public trust? If they were members of an ordinary corporation stockholders could hold them responsible for costing the company all this money.

Not if the corporation had the total indemnification clause the PBA tucked into its by-laws in 1995:

> The Authority shall indemnify and hold harmless every member of the Board and officer of the Authority and their respective heirs, estates successors and assigns who is or was a party, or is or was threatened to be made a party, to any action, suit, proceeding or investigation, whether civil, criminal or administrative, arising under or in connection with the activities of such person as a member of the Board or an officer of the Authority (any such person is hereinafter referred to as an "Indemnitee") from and against expenses (including attorneys' fees, judgments, fines and mounts paid in defense, settlement or judgment actually and necessarily incurred by the Indemnitee) in connection with such action, suit, proceeding or investigation, or any appeal therein or thereto, to the fullest extent permitted by applicable law, but excluding any liability resulting from the willful misconduct of such Indemnitee.

The import of the last phrase is this: unless you get one of them standing up and saying, "I did what I did to screw everybody because I'm malicious" or "I did what I did to make a lot of money," they are liable for nothing. Peace Bridge toll money will pay for their attorneys, their taxis to and from court, presumably even their meals while trial is going on or while they're meeting with their attorneys. That total coverage continues through appeals, and payment of any judgment against them should they lose at appeal. So long as the harm they do is done out of ignorance, stupidity, naivete, or blockheadedness, or seems so, the individual members of the Buffalo and Fort Erie Public Bridge Authority are responsible for nothing.

(A curious collateral note: that indemnification appears in the minutes of the same meeting at which the term "twin span" is recorded for the first time. I wonder what might they have been thinking, to deal with both those issues the same morning.)

I'd argue one of the premises undergirding this question: the five *Canadian* members of the Buffalo and Fort Erie Public Bridge Authority are *very* competent handlers of the Canadian public trust. They've gotten the PBA to pour millions of dollars taken from toll revenue (half of which is paid by Americans) into Fort Erie construction projects: a new Fort Erie city hall, a new Fort Erie double ice rink, a new Fort Erie courthouse, a huge Fort Erie duty-free shop that employs dozens of locals. The PBA contributes to just about every Fort Erie civic activity and organization. Just last week the Public Bridge Authority sponsored "The Mayor's Millennium Dinner," an event honoring Fort Erie's last *nine* mayors. Do you remember the last time the PBA poured money into an event honoring any of Buffalo's mayors?

The PBA provides jobs – directly or indirectly – for a least one person in nearly every family in and around Fort Erie. The PBA is Fort Erie's WPA. The PBA is *good* to Fort Erie. Fort Erie will do anything the PBA wants or needs. It's a beautiful symbiotic relationship, over there, on that side of the river, on that side of the border.

The Canadian members of the PBA have also delivered on a regional level. They've held fast to insistence on anachronistic steel construction, which will not only cost more to put up but which will cost a good deal more to maintain. Individuals in profit-making industries are always interested in keeping costs to a minimum. But some government agencies often think exactly the other way: the more things cost, the more money there is to spread around, the more jobs there are to be handed out. The Canadians want to see money spent: money for steel fabricated in Canada and money for people to scrape and paint and derust and repave two steel bridges for decades. The more maintenance the bridge requires, the more patronage there is to hand out. The Canadians want to see money spent building an entirely new bridge in twenty or thirty years from now when the decrepit old bridge gets too expensive to justify maintaining. If the absurd twin span were built now, then how could they put up anything but a second twin span to match that absurd twin span then? They couldn't, so that would be more money going to the Canadian steel fabricators, more money in patronage maintenance jobs.

What about all the harm done to the environment and infrastructure in Buffalo by all those trucks passing through here on the way to and from cities to the south and Mexico? That's an American problem, so why *should* Canadian politicians care about it? How excited would Washington get or would Buffalonians get if a construction project that would send a lot of money our way would poison the air in a residential area in Hamilton, Ontario, or clogged an otherwise adequate highway in Toronto?

It's the American members of the Buffalo and Fort Erie Public Bridge Authority we have to wonder about. Why would they accept a plan that has a hugely negative impact on the city of Buffalo, that pours enormous amounts of noxious fumes into residential areas, that sends jobs to Canada and states other than New York, that is willing to permit years of trucks routed through Buffalo's West Side?

I can give a partial answer to how they let it happen. For years the Americans came to the PBA board meetings unprepared. They were more concerned about what was for lunch than what was on the agenda, one person who was there told me. The Canadians came in and regularly voted as a bloc, obeying the marching orders sent down from Ottawa. For the past year, we've been told, the Americans have been one vote short: the representative of the New York Attorney General hasn't been present for key votes because the Attorney General was representing the PBA in court fights *against* the citizens of Buffalo. To avoid a conflict of interest, she didn't vote, with the result being that there was always one more Canadian vote than American. Since the death of American board member Louis Billitier, the balance is even more skewed: it's now 5 to 3, in favor of the Canadians, a lousy score for Buffalo in hockey and in bridge politics both.

Not that the American board members would vote to support regional interests anyway. Governor Pataki has said he supports a signature span and environmental issues, but not one of his PBA appointments has publicly reflected either concern. His appointment to fill the late Louis Billitier's seat will be indicative of where he really stands: will he give us another captain of industry and major Republican fund-raiser like Victor Martucci, who was immediately elected chairman and who immediately began endorsing the Canadian twin span plan, or someone who understands the issues and really has the public interest at heart, like Jeff Belt?

Why! oh why! would a project sponsor (the PBA or any developer for that matter) EXPOSE itself to such a simple legal challenge and on such fundamental legal grounds (as SEQRA [State Environmental Quality Review] and NEPA [National Environmental Policy Act]) after spending $MILLIONS on studies, plans, permit apps, modeling, etc...$MILLIONS!!! What kind of lawyer would advise this? They had to know that the path they took was *wide open to a legal challenge that could render the ENTIRE project dead in its tracks. They had to know that some environmental body or at least one of their neighbors would and could sue!. (If not, they have a fantastic lawyer malpractice case.)*

This writer went on to answer his own question:

> A: (in my opinion) if they knew -- and they did --then they continued that path because they already know the answers to a full study. And those answers probably prohibit this project from going forward at all!! The (dirty little secret) studies they did in the pre-approval days, probably seriously jeopardize the expansion in the shape and form they are financially viable to undertake. Their secret studies probably revealed thresholds well above SEQRA/NEPA standards, meaning costly mitigative measures and therefore arrived at their current "scheme" in order to skirt the overall (comprehensive EIS) results. It is the only way I can fathom why they proceeded down such a legally risky and expensive path.

Is the author of that letter right on the money or speculating wildly? The PBA's board members and attorneys will never tell, so we can only speculate. Had the PBA done the legally-required EIS in the first place, there wouldn't be cause to ask such questions or need to speculate on the answers.

What is the PBA's rationale for appealing Judge Fahey's ruling that they have to submit to a full environmental impact study for the bridge and plaza, that they can't segment them into two parts? Is their argument procedural or substantive? If the latter, they almost certainly lose, but are they perhaps arguing that a state judge lacks jurisdiction over an international body?

Nobody other than the PBA and its lawyers knows what grounds they'll argue, or even if they'll actually file an appeal. All the legal people I've talked to (the ones who don't work for the PBA) say they would lose any appeal that tried to attack the issues: if they want to put a new bridge here, they've got to do the legally-required environmental impact study.

It's equally unlikely that they'll win on a jurisdictional issue. They've frequently taken the position that they're above various state and even federal laws because they were authorized by two countries, but that's never been tested in court. In point of fact, two separate entities were created by two separate countries and the two separate laws said that both of those entities would work together. (New York, 1933, chap. 824; Statutes of Canada, 1934, chap. 63). Nothing in either law said that the combined entity would be exempt from the laws that apply to everyone else. One side of the bridge isn't even subject to the law controlling the other side of the bridge: the American law exempts the bridge from property taxes, the Canadian law does not; the American law says the books and records shall be open for close inspection by the New York treasurer and comptroller whenever they wish, the Canadian law says nothing about checking the books.

Because it exists on the basis of two separate laws in two separate jurisdictions, the Buffalo and Erie County Public Bridge Authority is subject to more, not fewer, laws. It's understandable that they'd prefer things were otherwise but, happily for us, they're not.

I wrote a letter to the editor of the Buffalo News *in which I referred to the "Public Bridge Authority." They published my letter, but they changed "Public Bridge Authority" to "Peace Bridge Authority. I had the name right. Why would they change it like that?*

Two other people asked similar questions, one about a letter, another about a "My Turn" column. It's one thing for the *Buffalo News* to use the PBA's preferred name for itself in news articles, something else to alter a reader's letter so it says something the reader never wrote or intended. I asked *Buffalo News* editor Margaret Sullivan to explain that policy. She responded immediately, saying she'd pass my inquiry along to Copy Editing Coordinator Scott Thomas. This is Thomas's response in its entirety:

> Editor in Chief Margaret Sullivan has asked me to respond to your e-mail regarding our style on the Peace Bridge Authority.
>
> We're aware that the legal name of the entity is Public Bridge Authority. We adopted "Peace Bridge Authority" as the newspaper's style partly because that's how most Buffalonians commonly refer to the authority. (We call many entities by names that differ somewhat from what they call themselves; the University at Buffalo and Buffalo State college are two examples.) We also recognize that the divergent names for the authority carry some weight on different sides of the Peace Bridge debate, and we wanted to avoid any appearance of favoritism.
>
> We may revisit this issue the future. Thanks for your

inquiry.

That's really loopy.

First, the parenthesis. "University at Buffalo" *is* the name of the University at Buffalo; the *Buffalo News* is only calling it what it calls itself. What's the big deal about calling a major public institution by its correct name? The *Buffalo News* cites this as a policy decision? Astonishing. The official name of Buff State is State University of New York College at Buffalo, which is not only ungainly, but might confuse some people who don't know that SUNY contains scores of colleges but only four universities, which is why SUNY lists it on its web site as "Buffalo State." Calling UB by its correct name and Buffalo State by the name Albany and everyone in this region uses for it doesn't hide anything; rather, it makes things clearer.

But the *Buffalo News* has joined the Public Bridge Authority in using a name that hides what the PBA is and does. Clarity with the schools and misdirection with the PBA – not the same thing at all.

It's true that "Peace Bridge Authority" is the name commonly used in Buffalo. The reason that is so is the Public Bridge Authority and the *Buffalo News* call it nothing else. How are ordinary citizens, who trust the *Buffalo News* to call things by their right names, to know that the *Buffalo News* is working with the Public Bridge Authority to mislead?

It's like the Mafia calling itself the Greater Buffalo Neighborhood Improvement Society and the *Buffalo News* telling us about the activities of the Greater Buffalo Neighborhood Improvement Society. Anybody or anything can call itself whatever it wants, but calling yourself something doesn't mean you *are* that thing. A newspaper should be opting for the side of clarity, not misdirection. This isn't at all a matter of "divergent names." This is a matter of a legal name clearly there in every legal document and a made-up name used to keep the public as far away as possible.

The Buffalo and Fort Erie Public Bridge Authority is a public agency that works very hard to act as if it were a private construction and real estate firm. It is neither. I've long thought it was lousy journalism for the *Buffalo News* to print that misleading name in anything but direct quotations of Public Bridge Authority officials and documents. I thought it was editorial laziness, but after reading Scott Thomas's letter I know it's not the least bit lazy: it's a considered and deliberate decision, and I think it's reprehensible.

What is the limit of the PBA's authority? Is it the only entity empowered to construct and manage a bridge between Buffalo and Fort Erie? What if the bridge is two miles downstream? (In this regard, maybe we oughtta let them call themselves the "Peace Bridge Authority" and then give them no authority over a separate bridge.)

There's no legal reason they or another company – public or private – couldn't build and operate a bridge at the International Bridge (sometimes called 'The Railroad Bridge') a mile to the south, or any other place between here and Niagara Falls. But there are massive political reasons. Any such company would have to get US and Canadian federal and state/provincial governments to provide customs and immigration facilities for the new bridge and get those same governments permit the new bridge to at the new locations and hook up with the Thruway on this side and the QEW on the other. Without those federal control services and highway hookups, it would be a bridge from and to nowhere. And that's where the PBA gets its nose back in: the only way another crossing would get through either government at present would be if the PBA defined it as an expansion of the current operation. Could that change? Sure, but making changes like that is hugely difficult, especially since the master manipulator Robert Moses went to the Great Construction Project in the sky.

I would like to know why they do not join up with the rest of the region to come up with a plan to manage the commercial traffic.

Because then they might have to consider the possibility that this isn't the best place to put the new truck bridge, and they'd lose all that potential patronage and power.

There is some interest from a federal supporter of ours to try to get a full audit of the PBA. Any suggestions on how that might be brought about?

Ask New York comptroller Carl McCall to do one. Write your state legislators and the governor and ask them to ask McCall to do one. It's a month until elections: they'll promise you anything now and some of them might even deliver on those promises later. McCall has the authority. The only other time he did one, I've been told, the PBA managed to steer him into asking only benign questions. A lot more is known about PBA operations now and there are far more tough questions a serious audit would have to ask.

My question to you and to the Public Bridge Authority is this: What benefit will this region derive from the increased truck traffic as a result of building the twin span or any other bridge. Although I am strongly in favor of the Signature Bridge and creating a northern plaza (thereby returning to us the complete Front Park), I am anxious to find out where are the economic benefits that the PBA is so anxious to bring to this region.

The PBA and Andrew Rudnick of the Buffalo Niagara Partnership have consistently said that an expanded bridge will bring huge numbers of jobs to this area and any delay in construction costs us jobs. Though asked repeatedly for specific data and projections, neither the PBA nor the Partnership has ever produced anything other than vague claims.

Surely there will be some jobs for customs brokers and attorneys and transfer agents. There will be more bridge jobs because two steel bridges will require far more maintenance than the present single steel bridge or a brand-new concrete bridge. There will be more maintenance work on I-190 because of the increased truck traffic (the Canadian government projects an annual passage of four million trucks through Buffalo every year a decade from now). There will be more work for health care providers dealing with childhood lung disorders occasioned by the huge increase in noxious fumes from those trucks being routed alongside residential areas.

Beyond that: nothing. There is no indication that any of the thousands of manufacturing jobs created by NAFTA will come anywhere near Buffalo. Nearly all those heavy trucks will use Buffalo as a border crossing and highway connector, nothing more. They won't even stop for lunch or fuel here.

Rudnick and the Buffalo Niagara Partnership may have reasons for supporting distant manufacturers and purchases, but so far as we can tell, the primary beneficiary of increasing the truck traffic through this port will be the Buffalo and Fort Erie Public Bridge Authority itself: as it has more lanes and more trucks and more maintenance to oversee, its own staff and bureaucracy will grow. The basic law of most species is the law of Self-Preservation; the basic law of most bureaucracies is the law of Self-Expansion.

What is done with Buffalo/Western New York's share of Peace Bridge profits? Improvements to the Canadian side of the Peace Bridge are highly visible: an attractive plaza, a new duty-free shop, a new truck processing facility, and an ever-expanding Town Hall and sports complex on Highway 3. There is nothing comparable on the American side of the bridge. Where they have showpieces, we have shabbiness. If we assume that an equal amount of money was made available to the American OWNERS over the years, what was done with our share? What continues to be done with it?

It doesn't work like that, alas. Those construction projects on the Canadian side were parts of deals for land and to curry political favor. The money never went directly to Fort Erie to do with as it wished; the PBA just picked up the tab for each of those expensive construction projects.

The profits of the bridge go into public coffers only when the PBA retires all of its debt, and the PBA has no intention of ever doing that. If the PBA retires its debt, it dissolves on the spot and all its holdings revert to New York and Canada. Every time the PBA has gotten close to being able to retire its debt, it refinanced or issued new bonds creating new debt.

The current legislation specifies that New York and Canada each get $200,000 per year from PBA profits. Canada has told the PBA it can keep its $200,000; the American $200,000 goes to the NFTA.

If the Authority paid real estate taxes, the way you and I do and the way any profit-making corporation does, it would have paid about $6 million last year instead of the $783,000 it divided equally between Buffalo and Fort Erie in lieu of taxes.

Assemblyman Sam Hoyt has sponsored a bill (A08811) that would have the PBA pay Buffalo and Fort Erie $2 million for improvements to the parks adjacent to the bridge and $1 million a year thereafter. He notes that the $200,000 Buffalo gets as a result of the 1970 legislation is worth $857,304 in current dollars. The PBA has raised its tolls considerably since then, but it hasn't raised its payments to the communities where it resides. We'd all love to have our real estate taxes set at the 1970 level, but there's no reason the PBA should be the only guy in town getting that sweet deal.

Who are and will be the firms or people to benefit from the "kickbacks" or outright bribes to the PBA? By this I would like to know specific names and dollar amounts involved if available.

So would I, but they aren't telling, and never will. An in-depth full-steam audit by New York Controller Carl McCall might provide a start.

The PBA still insists that we will be suffering economic harm if we do not build a new bridge now. In addition to the obvious arguments against this claim that the bridge is not the bottleneck and the Commercial Vehicle Processing Center in Fort Erie remains grossly underutilized, I challenge the PBA to back up this claim. The PBA has said in the past that GM, Southwest Airlines, and UPS would not come here if we didn't commit to the Twin Span. Last I checked, GM committed to the investment at Tonawanda,

Southwest will be flying this fall, and UPS has expanded in the area through local acquisition. The PBA*'s own study emphasizes the minuscule economic impact of the small number of trucks that do not simply pass right on through. Is there so much positive economic impact that we should ignore everything else, including our community's health and well-being (not to mention all of the other issues)? I would argue that no amount of economic impact is worth the lives of people.*

You answered that question as well as I could.

The engineers recommended a new plaza to the north of the current plaza and a companion span. Why did the Public Consensus Review Panel accept the first recommendation and not the second? Weren't they bound to accept whatever the engineers said?

If they were bound to accept whatever the engineers said then there would have been no need for them to vote on it. That was just a scam floated by PBA chairman Victor Martucci in his statement the day after the Panel vote. I can't speak for all members of the Panel (only one of the 20 voted for the engineers' report – Natalie Harder, Buffalo Niagara Partnership president Andrew Rudnick's representative) but what I heard was that the engineers cut a deal. The American engineers hired by the panel gave in to the Canadian engineers hired by the Public Bridge Authority. The Canadians wouldn't go for a new plaza unless the Americans traded the bridge. The American engineers voted as they did for political, not engineering or ecological reasons, and they said so in their report. The PCRP rejected that recommendation because they had hired the engineers to make technical, not political recommendations. The engineers hired by the PBA were of course under no such restriction, which is probably why they held out for that political deal.

Do you have an understanding of how the PBA *was created? Who initiated the legislation that created them? Was it Canada and the US that created the Authority or just NY? If the Authority's business practices were to be changed, would it have to go through Canadian parliament?*

There were conversations about and attempts at building a bridge between Fort Erie and Buffalo going back at least to 1851, but nothing happened until 1919 when a group of twenty-five Canadians and Americans set up the Buffalo and Fort Erie Bridge Corporation. They wanted a bridge that would get let them move between the two communities faster and more flexibly than did the ferries then available. (Did you ever wonder why West Ferry Street has that name?) They put up $50,000 of their own money to get the corporation going, and then set out to raise $4,500,000 in bonds for the actual construction. The bonds were mostly sold locally, and the offering was oversubscribed before the first offering day was out. People on both sides of the river really wanted the bridge, they liked the idea of the bridge, they were willing to invest their own money to have the bridge.

The Bridge opened to the public on June 1 of that year. At the August 7, 1927, opening ceremony, which was attended by the Prince of Wales and the Vice President of the United States, John W. Van Allen, one of the original incorporators of the Peace Bridge said:

> Hereafter, this bridge belongs to the public. Our sole remaining function is to collect the tolls and pass the money back to those who advanced it. The construction problems are over; [there] remains now only its dedication to service, and we wish to all, great joy and the convenience in the use of it.

Would that it had been so. The public has yet to get ownership of the bridge, the people who run the bridge have and are doing far more than collecting tolls and giving money back to investors, and the construction problems have never been worse.

Because of the Great Depression and the end of Prohibition, the numbers of vehicles making the crossing plunged and there was a real danger that the company would default on its bonds. For a time, Frank B. Baird, president and prime mover of the project, put his own money into the struggling company but not even his great private wealth could keep it going in the face of declining income.

In 1933, the Bridge Company sought governmental salvation. Over the next year, three separate pieces of legislation in Ottawa, Albany, and Washington, D.C., created the Buffalo and Fort Erie Public Bridge Commission, a public benefit corporation. Public benefit corporations exist in a land of deliberate legal ambiguity: they aren't government agencies and neither are they private corporations. Their profits in theory belong to the governments that created them, but they behave more like ordinary corporations than an arm of government. They can partake of some of the benefits of government status – their property and bonds are tax exempt, for example – but they have separation from primary agencies of government not enjoyed by organizations that really belong to the public, such as SUNY or the New York Thruway. A public benefit corporation controls its own resources.

The Authority at first had nine members, six from the US and three from Canada. It acquired all the assets and debts of the bridge company. With the debts restructured and almost no taxes to pay, the Authority was on firm financial footing.

In 1957, New York State created the Niagara Frontier Port Authority (now the Niagara Frontier Transportation Authority) and tried to tuck the now-profitable Peace Bridge into it. The Bridge Authority balked. It asked New York Attorney General Jacob Javits for a ruling: could the New York Legislature take over an organization created by the government of New York, the government of Canada and the US Congress? Javits said no, the power play wouldn't hold up.

So the ownership papers were redrawn another time. There were four key changes:

> Total board membership would increase to ten, with five members from each country.
>
> The bridge would be directly tied to no other agency so it could remain fully independent.
>
> The two countries would divide excess revenues equally.
>
> And the sunset, the date everything would be turned over to the two governments, was extended from whenever the outstanding bonds were paid off to 1992.

By 1970, it was clear that bridge capacity would have to be expanded, which meant the Authority would have to issue new construction bonds. The governments extended the life of the Authority to 2020 and raised the debt limit. The Canadian government and the State of New York, instead of splitting whatever cash was left over after payments were made, would each get a flat $200,000 each year, and the Public Bridge Authority would keep the excess for development.

Does the Peace Bridge make money?

Yes. Huge amounts of it.

The 1998 annual report, issued November 5, 1999, shows total assets of $120,013,088, up more than $10 million from the previous year. It shows an excess of revenues over expenses of $9,180,378. It shows a fund balance (assets less liabilities) of $64,668,291.

They're bloody rich.

They received $13,952,982 from truck tolls and $4,996,828 from passenger vehicle tolls in 1998, a total of $19,949,810. The previous year's numbers were $11,614,065 for trucks, $5,326,049 for passenger vehicles, a total of $16,940,114. That's a jump of more than $3 million – almost 18% in one year. Who but a dot-com wouldn't be ecstatic about that kind of profit increase? The trend for auto traffic is fairly flat, but the truck revenues are going nowhere but up. With what it is making on truck traffic, the Authority could dispense entirely with the income from automobiles and still turn a huge profit.

One interesting entry in their statement is "Rentals: $4,866,212." The two duty-free shops paid $3,660,000. In case you wondered why the PBA moved the American duty-free shop from Porter Avenue to the bridge plaza, where it creates constant traffic jams, that's why: they make more money having it on the bridge than having it off, and they're not the least bit bothered by the traffic jams.

The report doesn't give the source of the other $1,206,212 in rentals, but since the Canadian Customs Act prohibits any toll-collecting border crossing entity from charging the Canadian government any rental or any other fees for the space it uses for inspections, that is probably all American rental money. That is, the Americans have to pay rent for space the Canadian government gets for free, which means American tax dollars pay that rental money and American toll dollars pay half of the support for the Canadian facilities. Who ever told you Americans were canny businessmen?

Is there any way to make the PBA responsive to community concerns?

The Canadian representatives are all dancing to Ottawa's drum, so there's nothing anyone this side of the border might do to influence them unless our own Federal government decides Buffalo's interests matter as much as stroking Ottawa's feelings matter. Thus far, that hasn't happened. Madeleine Albright will talk to anyone in the Balkans, but no one on the Niagara Frontier has ever been able to get her attention. Our two US senators have been strong advocates of a rational aesthetic and environmental policy here, but neither of the two current candidates for Moynihan's seat has said anything but mush. The problem is, the bridge patronage is above partisan politics. The PBA is about big money, not Republicans or Democrats. As of this writing, neither Fazio nor Clinton is willing to take them on.

Sam Hoyt has sponsored a bill (A09910) that would increase the board to 12 by giving Buffalo and Fort Erie each a seat on the board. The same bill would require the two cities to approve all major capital projects undertaken by the PBA. The one vote wouldn't make much of a difference in voting – there would still be 6 Canadians voting as a block – but at least we'd have a channel through which we might receive information about what is going on in the closed meeting rooms. The approval provision is bigger. It would give Buffalo direct involvement in expansion matters instead of having to go to court to stop further mutilation of the City, the City could just say no.

I'm sure the PBA will pour buckets of money into blocking Hoyt's very reasonable bill. As far as the PBA board is concerned, what they do is none of our business. PBA board member and Fort Erie resident Deanna DiMartile told *Time Magazine* correspondent Stephen Handelman, "It's only when outside influences step in that things break down." This, I think, is the heart of the current problem with the PBA. It regards nearly everyone outside its own boardroom as "outside influences." Senators Moynihan and Schumer, Governor Pataki, the Buffalo Common Council, the region's delegation to Albany, you and me – we're "outside influences."

Sure we are.

I would like to know how we turn the Public Bridge Authority into a responsible group with the vision and leadership necessary to make this project the best it can be.

And we'd all like to know how to make silk out of a sow's ear. You want silk, go to silkworms, not sows.

44 Where Things Are

An Encounter at City Hall

The question I hear more than any other about the Peace Bridge affair is not "What is going on?" (which presumes people are doing something) but rather, "Is anything happening?" (which asks if anyone is doing anything at all).

"Only the PBA can build the bridge," Buffalo Common Council President James Pitts is fond of pointing out, which is to say, all these lawsuits and court orders and public posturings wind up with a single question: Is the Buffalo and Fort Erie Public Bridge Authority doing anything to resolve the legal, political and economic mess that it created and only it can end or is it going to sulk and wallow in accusations and endless court procedures?

There's a slight chance that something did happen this week. If it did happen, it was small. A ten-minute encounter embedded in a two-hour flurry of rhetoric and performance. It may, as Macbeth memorably put it in his eulogy for his suicide wife, have been only "sound and fury signifying nothing." Then again, it may very well, in its relative minority, have been the first indication of the sweet presence of reason in this wearisome affair. The encounter took place in the course of the first meeting in 15 months of the Buffalo Common Council's SuperSpan Signature Bridge Task Force.

Seven or eight of the two-dozen people in the room spoke at length during the meeting. Some shared useful information. Some argued for one position or another. Some asked pointed questions. One (a real estate salesman) tried to ensure future business.

The encounter that mattered, however, directly involved only two men: Tony Bullock (Senator Moynihan's chief of staff) and Victor Martucci (chairman of the Buffalo and Fort Erie Public Bridge Authority).

Current Conditions

It's been more than three months since New York Supreme Court

Judge Eugene Fahey issued his preliminary order telling the New York Department of Environmental Conservation and the Buffalo and Fort Erie Public Bridge Authority that the PBA could not ignore federal and state environmental law, that it would have to do a full environmental impact study on its bridge and plaza expansion project before it sliced the earth. The day after Judge Fahey's preliminary order, PBA chairman Martucci told a press conference that if the PBA had to obey state and federal law it might not build any new bridge or plaza here at all. Or, he said, the PBA might appeal.

Shortly thereafter (primarily because of relentless pressure from the five Canadian members, all of whom are appointed by and follow the orders of the Canadian federal government), the PBA filed a notice of intent to appeal Judge Fahey's ruling. That doesn't mean they're going to do it, only that they're keeping their options open. They have until March 2001 to file the appeal and provide the court and opposing counsel all the supporting documents. If they do file, everything will be on hold for years while this thing wends its way through the courts, and at the end of the process they'll almost certainly lose. Judge Fahey's decision can force them to obey environmental law, but only if they want to build the bridge badly enough to let some daylight into their heretofore sheltered operations.

It all got racheted up a jurisprudential level on July 28 when a lawsuit against the PBA and several federal agencies that had lazily gone along with the PBA's plans was filed in Federal court. The PBA and the federal agencies have requested an extension until the end of November to answer the federal lawsuit. That process too could take years before it is ever resolved and, like the state case, even if the federal court finds for the people and against the PBA, the decision matters only if the PBA decides to do something.

Senator Chuck Schumer, who has been a great advocate for Buffalo interests in this matter, wearied of the PBA's inaction and set up a second visit from the operators of Detroit's Ambassador Bridge. The Detroit gang visited a year ago at the invitation of Senator Moynihan and proposed building an entirely new bridge a mile north of the present bridge, adjacent to the International/Railroad Bridge. On May 26, at the Adam's Mark, they presented a revised version of their proposal, then they went away, and that was the end of that.

There have been a few flutters out in the countryside, but none of them seems to be going anywhere either. Alan Gandell, head of the Niagara Falls Bridge Commission, said that putting a trucks-only crossing on the Whirlpool Bridge could relieve the current Peace

Bridge congestion quickly, and maybe there were even ways to divert nearly all the increase in truck traffic occasioned by NAFTA to other area bridges. Then, on August 15, without any public explanation, Gandell resigned. Some people say it was because once the recent development of the Rainbow Bridge plaza was completed his board wasn't ready for any new projects, and after 10 wearing years on the job he saw no reason to hang on just to do maintenance work. Whatever his reason for stepping down, nobody's doing any serious talking now about putting a truck deck on the Whirlpool Bridge or any place else up in Niagara County.

From all I've been able to find out, things down at the PBA are a mess. The five Canadian board members perform, as always, as a monolith, demanding everything, considering no compromise, in perfect lockstep to orders from Ottawa. The three voting American members of the PBA are reeling under greater and greater pressure. I know of no one this side of the border who is not in the debt or control of Canadian steel interests who any longer supports the anachronistic and hugely expensive steel twin span design the Canadians so desperately want. The Americans on the PBA can't outvote the Canadian bloc, and they don't want to be mere flunkies for the Canadian steel interests either. I've heard that one member of the board has been going around town saying that the current mess is the fault of the staff (the term "staff" at the PBA always means the two bridge managers, Earl Rowe and Stephen Mayer, not the scores of other people who work down there). David Crane, economics editor of the *Toronto Star* keeps writing articles full of misinformation about why no bridge expansion is going on. I'm pretty sure he's being spoonfed by the Canadian Minister of Transportation's office, because what he writes is all partyline stuff. "The Canadians," one Buffalo businessman who has closely watched the bridge expansion project get stuck in glue said to me, "don't have a clue that their plan is dead in the water."

Our local pols either waffle on the Peace Bridge question or are waiting for someone else to do something interesting. When was the last time you heard anything invigorating on this key issue from the mayor's office, the county executive's office, from anyone on the Common Council other than Jim Pitts or Dominic Bonifacio? Likewise the candidates for US senator: neither Hillary Clinton nor Rick Fazio will go near the Peace Bridge issue: their big money providers have told both of them to stay out of it and they are.

THE ROOM WHERE EVERYBODY GETS TO TALK

Three exceptions to the political hiding and waffling pattern are Common Council President James Pitts, Assemblyman Sam Hoyt, and Senator Daniel Patrick Moynihan.

Moynihan was instrumental in making Buffalo area residents aware that an expanded Peace Bridge could provide an architectural work that would fire the imagination rather than a dull structure that would just occupy space and squander money. Hoyt has submitted three bills to the New York legislature that would alter the composition and accountability of the PBA and of late has been having conversations about further restructuring the Authority and ways of ending its penchant for operating in near-total secrecy. And Pitts, who established the SuperSpan Signature Bridge Task Force, has maintained what turns out to be the only public forum in which the various players can sit around a table and talk about where things are and how things might be moved along.

Pitts called a meeting of the Task Force last Thursday, the first since June 3, 1999. PBA chairman Victor Martucci came and sat across the table from Senator Moynihan's chief of staff, Tony Bullock. Bullock came up from Washington especially for that meeting. It was, so far as I know, the first time the chairman of the PBA has participated in any public conversation about the Peace Bridge since the first meeting of the Task Force on March 19, 1998.

Since then, PBA chairs have appeared at press conferences or have stood up and made speeches, but not one has taken part in any public discussion about the issues. Not one has listened to what anyone else has had to say. PBA chairman John Lopinski said at a Buffalo City Hall press conference last October 13 that the PBA was going to participate in the Public Consensus Review Panel and there was a lot of glad-handing immediately thereafter, but that participation turned out to be nothing more than sending in a team of engineers who hyped the PBA's position in technical meetings. No member of the PBA ever took part in any of the Consensus Panel's deliberations. Victor Martucci made a statement at a televised meeting of the PCRP, but that's all it was: a flat statement of what the PBA was doing and intended to do. There was no pretense of listening or considering other opinions or other data. The PBA's two general managers – Earl Rowe and Steven Mayer – and its attorneys and press agents came to the meetings and took notes on what people were saying, but with one exception not one of them ever participated in anything. The exception was on December 16, 1999, when Rowe and Mayer did a dog-and-pony show for the panel. It was a lot of noise, giving nothing, asking for everything, slipping and sliding and lying about this and that and blowing smoke every direction but up.

Pitts put the Task Force on hold while the PCRP did its work. He was at the City Hall press conference where Lopinski announced that the PBA would cooperate with the PCRP, but he didn't speak. He was fairly sure, he told me, that the PCRP process would not result in any change of behavior by the PBA, but, "I thought it best to let that process unwind." And now that the PCRP is over and the PBA's involvement in it turns out to have been just a delaying action, and now that the courts have enjoined the PBA from doing anything without an EIS and the PBA has made no steps to begin one, he decided to start the conversations on the 18th floor of City Hall going again.

MOYNIHAN'S MAN

There was a lot of talk in the meeting by various people about work that had been done that could be folded into an EIS but Tony Bullock would have none of it: as far as he's concerned, nothing has been done about the EIS. Some bits and pieces of past studies may be useful, but not until the EIS starts will anyone know what information will be necessary and valid, what will be acceptable, what will have to be redone, what will have to be done for the first time. "The process," Bullock said, "has not started." Every day the PBA does nothing, he said, is another day that nothing happens. The PBA can complain and sulk and fight in court, he said, but there was no way it could avoid the legally-required EIS if it is ever to put a spade in the ground. The choice is entirely theirs: if they want to build a bridge, obey the law. They knew beforehand that an EIS was required, they went through huge machinations to avoid it. A New York State Supreme Court judge has told them they have to do it and a federal judge will probably tell them the same thing.

"Not unless the PBA changes the way it thinks about these issues," he said, "is anything going to change." They can talk about delay, but the sole reason for the delay is their continuing refusal to obey environmental law. If they really think bridge expansion matters, then it's time to stop evading and deflecting, time to start being honest and open. If they really believe there is economic value to expanding bridge and plaza capacity, it's time for them to do the EIS and build a bridge. They can't blame everyone but themselves for the delay because they are the cause of the delay, no one else.

Bullock was lucid, direct, tough, and unambiguous.

Bullock sat directly across the table from Martucci, but he never (as far as I remember; my tape recorder malfunctioned and because I thought it was running I didn't take as detailed notes as I should have) addressed Martucci directly. Bullock referred to positions the PBA had taken and actions the PBA wanted to take and things it would have to do. He never said, "You're the chair. Why don't you stop jerking everybody around and get them to do the right thing?"

He said, as did James Pitts several times in the course of the meeting, that no one could build a bridge at that site except the PBA. If there was going to be expanded bridge and plaza capacity only the PBA could do it. He said that Senator Moynihan could not understand why, if increasing capacity was so important, the PBA had done nothing in the months since Judge Fahey's order.

When he was done, Martucci said he wouldn't respond, other than to say that his function that day was to listen and to bring back to his board his impressions of what he had learned. Ordinarily, that kind of statement would be totally without meaning, but Martucci is a man who is fully capable of being lucid and tough. He has in the past argued the PBA's position as if it had been ordered from On High. He argued nothing at last week's meeting of the Task Force.

After a good deal of information-sharing and speechifying and questioning by various people, Pitts said he'd hear one or two more comments and then the session would adjourn. Buffalo architect Clinton Brown – who had made a long speech when the meeting began – got up and made a long closing speech. He thanked Martucci for having attended and pontificated on the meaning of the "dialog." Several other people before and one or two after Brown said nice things about the "dialog."

Nonsense. There had been no dialog. There was only a presence and the paying of attention, and a few small questions answered.

But, as things go around here, that's not bad, that may even be a step forward. To my knowledge, this is the first time since things went sour that the chairman of the Public Bridge Authority has taken part in any community conversation that wasn't a performance, fix, or a lecture. Yes, Martucci spent almost the entire time listening. But at least he did that, and he did none of the preaching and lecturing and hectoring and stonewalling and smoke blowing that have been hallmarks of PBA chairmen's public utterances these past several years.

SO?

I've several times said in these pages that I thought things had reached a turning point, that the PBA was finally starting to participate for real, that the current move wasn't just another ploy. Every time I've said something like that I've been wrong. The PBA has never used up its capacity to astonish me with its disingenuousness and dishonesty and greed.

Are things different this time? Who knows who was sincere and who was posing and who was there for the record and who was there to start a conversation and who was there to perform and who was there to learn? All I know is, Victor Martucci came and listened, the conversation was for the most part serious and focused and rational, Tony Bullock read out Pat Moynihan's riot act on the need for a decent job here and the need to stop stalling, and it was, as they say, a start. A real start or one more jive start – only time will tell. I'll let you know when and what I find out.

45 THE GREAT AUTUMNAL PEACE BRIDGE Q&A, PART II

This article and its companion in the September 21 *Artvoice* consist entirely of responses to questions submitted by Artvoice readers about the operations of the Buffalo and Fort Erie Public Bridge Authority. The questions are in italics. Some of them are very brief; some contain comments. None of these questions is me pitching slow balls to myself. Every one is authentic and the texts of nearly all are exactly as they came in. A few were asked by two or more people, so I combined the phrasing. My thanks to all the *Artvoice* readers who took the trouble to voice their concerns and all the public officials and private citizens who helped me find the answers.

I have long understood concrete to be a poor choice of building material, especially in extreme climate conditions. I have also been told the Tampa Bridge is not holding up well, and its concrete construction is to blame. Do you have any information about this ? Also, ironworkers in Buffalo support the use of steel; if the reality is a twin span will use non-local materials and labor, why is this so? And why would concrete be more advantageous to the local economy?

Staff members at the PBA have floated this canard about steel being a much better construction material for our severe climate for years, but it's just not true. Concrete is the construction material of choice for just about all recent major bridges in Scandinavia, the North Sea, and Canada, all places with weather far more severe than ours. The Confederation Bridge, for example, which connects Borden-Carleton on Prince Edward Island and Cape Jourimain in New Brunswick, the longest bridge over ice-covered water in the world (12.9 kilometers), is precast concrete.(Visit www.confederationbridge.com/en/accueil/-index.htm for more information.)

The reason a concrete bridge uses 80% local construction workers is because the bridge segments are cast locally and are then floated or trucked to the construction site. A steel bridge uses at most 20% local workers and almost no locally manufactured goods, unless you consider the steel firms up in Ontario local. There are more jobs in maintaining a steel bridge, but they're make-work jobs, jobs that are created only because steel is a far less efficient material than concrete for building bridges.

I asked Shri Bhide, program manager of the Bridges division of the Portland Cement Association, for help on this.

> According to National Bridge Inventory data, concrete bridges outperform steel bridges. This trend is true in all types of climates, cold, dry, hot, humid. In fact, colder climate is better, since concrete's diffusivity, ability to let harmful chloride ions get to reinforcing steel, is lower at lower temperatures. Properly designed and constructed concrete bridges are virtually maintenance free.
>
> Steel bridges, on the other hand, require regular cleaning, and painting. Fatigue critical steel bridges need to be inspected much more frequently than the mandated minimum two year interval. Concrete bridges are not susceptible to fatigue. Also, inspection of steel of bridges is more time consuming. Concrete bridge decks supported by steel girders don't last as long as the same decks atop concrete girders.
>
> The life cycle cost, which is the sum of initial cost of construction and the recurring cost of maintenance is thus lower for concrete bridges.

I suppose one might say that Bhide is biased, given that he works for the cement industry, but he backs up his summary comments with primary data gathered by public agencies with no connection to the cement industry. That's the opposite of the people who spread rumors about the defects of concrete bridges but don't come up with any data at all to back it up.

I haven't heard of any problems in the Tampa bridge caused by the concrete.

How much does it cost to maintain the Peace Bridge and who's getting the maintenance money?

The PBA's annual report says that maintenance of the bridge system costs about $4 million each year, but it doesn't say how much of that is for the plazas and how much for the bridge itself. The total has been increasing year-by-year. The prime engineering contractor is DeLeuw-Cather, which has not only done maintenance on the bridge for decades but was also the primary consultant on the steel twin span design. There are many subcontractors involved, some of them local, some of them not; I haven't been able to get a list of who they are and what they get, nor have I been able to learn how they are selected. Some of the PBA's most expert critics have long said that one reason maintenance is so high is that there are so many sweetheart deals passed out among political friends every year. We'll never know the facts about that until there are full fiscal and management audits of the PBA's operations, which they've never permitted.

To put the PBA's maintenance expense into perspective, consider this: the Peace Bridge, made of steel, approximately 2500' in length, costs at least $2 million per year to maintain (and probably a good deal more); the Sunshine Skyway in St. Petersburg, Florida, made of concrete, 22,000' in length, has, over the past 10 years, cost an average of $215,000 to maintain. Simple arithmetic shows that the concrete Sunshine Skyway bridge costs $9.70 per foot per year in maintenance while the steel Peace Bridge costs at least $800 per foot per year in maintenance.

Maintenance on the steel Peace Bridge costs about 82 times maintenance on the concrete Sunshine Skyway Bridge!

Some of that huge difference is a factor of age: the Sunshine Skyway Bridge is new and the Peace Bridge is geriatric. Some of it results from the great difference in maintenance costs for steel and concrete bridges: as Shri Bhide explained in the response to the previous question, steel requires constant maintenance and concrete requires almost none. And some of it has to do with management: how lucrative contracts are given out and who is able to get them and whether or not there is any real competition among possible contractors.

I've heard that the reason we can't move the Customs facilities to the Canadian side is that the US Customs agents won't go unless they can carry their guns and the Canadians won't let them bring their guns because if an American got trigger-happy on Canadian soil it would be a huge mess. Is there any way around this? If the answer is yes, why aren't they talking about moving the plaza operations to the Canadian side?

People *are* talking about it. This is one of those options that might become clearer once a full environmental impact study is undertaken.

The answer to the first part of the question is yes, there are ways around it. One point of discussion is whether or not US Customs agents need firearms to do this kind of work on Canadian soil in the first place. When was the last time you heard of a shootout at a bridge Customs booth? The US already has Customs and INS stations on Canadian soil – if you've ever flown to Buffalo from Toronto or Montreal you've gone through them, and they work fine. Clearing motor vehicles is more complicated, but where's a malefactor to go if he or she displeases any of those agents? You can't back up (all those other vehicles are in line behind you), you can't go left or right (barricades to the left of you, barricades to the right of you), and you sure can't go full throttle ahead (the US troops will have time to be ready for you before you reach the three flags at the midpoint of the river). What if some drug-crazed malefactor does something crazy? Call a Canadian cop to help out. This is manageable.

And if it's absolutely necessary for political or manhood reasons for some of those agents to be armed, there's plenty of precedent for that too. Every embassy in the world is defined as a plot of land belonging to the home – not the host – country. US Marines, in the most unfriendly places, are armed to the teeth. They are equally armed in friendly places – the US embassy in Ottawa for example. In the coldest season of the Cold War commie soldiers at the commie embassies in Washington, D.C., were all armed to the teeth. There's nothing new here, nothing that hasn't been worked out before. This has got to be one of those issues that will be solved in fifteen minutes whenever the senior players involved want to solve it.

Whatever happened to the Public Consensus Review Panel? Could it, should it have continued to work as an organized group? Or was its role subsumed by the old and new litigation? Or does it continue in whole or in part? If it came to a stop before the matter achieved a resolution is that a failure or a great push in the right direction? Did the momentum stall when the formal process reported out? Surely a report was not a goal in itself.

The Public Consensus Review Panel was never meant to be a continuing organization. It had a single function: to find out what bridge and plaza combinations would best serve the interests and needs of all the various individuals and groups ("stakeholders" in the current jargon) affected by the bridge and the two plazas, a job the PBA itself should have taken on but did not. Indeed, in segmenting the bridge and plaza construction projects and refusing to do the legally-required environmental impact study, the PBA deliberately attempted to avoid any input from any groups other than those with whom it already had political and financial relationships. It tried to keep the public at bay while it dealt with cronies. That's the root of the current mess.

The PCRP invited the town of Fort Erie and the PBA to take part in its work; both refused. The PCRP held public hearings, nearly all of which were broadcast by WNED, and engaged an engineering firm to provide technical advice. On October 13, 1999, the PBA, in a failed attempt to get Judge Eugene Fahey to take no action in the lawsuits filed against it by the Episcopal Church Home, the Olmsted Parks Conservancy and the city of Buffalo, agreed to take part in the PCRP process and to abide by its recommendation. Judge Fahey was, happily, unswayed: he threw out the PBA's lawsuit demanding that the city allow it to begin construction and he refused the PBA's request that he throw out the lawsuits from the city, the Episcopal Church Home and the Olmsted Parks Conservancy demanding that the PBA do an environmental impact study.

Then the PBA put the process on hold for two months while it cast about for engineers who would serve its interests. (Some members of the Review Panel later said the real reason for the long delay was tactical, designed to take the steam out of the engineering group that had already begun work. If that was the design it was at least partly successful.) Those engineers joined the engineers already hired by the PCRP in examining bridge and plaza options, but the PBA still refused to sit at the table. At the end, the two engineering teams came up with a compromise: they recommended moving the plaza to the north of its present location, thereby freeing up Front Park for restoration, and building the steel twin span Fort Erie and the PBA were demanding. The engineers said that they had achieved a political compromise by giving the Americans some of what they wanted and the Canadians all of what they wanted.

The problem was, the Public Consensus Review Panel hadn't asked the engineers for a political compromise; it had asked only for technical advice. So the Panel voted to endorse the recommendation for a northern plaza and to reject the compromise recommendation for a twin span. Immediately the PBA dropped into word-play: PBA chairman Victor Martucci said the Panel had promised to endorse whatever recommendation the engineers came up with (which was not true) and since they hadn't done that, then the PBA didn't have to deliver on its promise to follow the Panel's final recommendations. Everything degenerated in press conference he-said/we-said disingenuousness.

Judge Fahey's subsequent ruling – that the PBA couldn't go forward with anything unless it engaged in the required EIS – doesn't itself address any of the issues that the Public Consensus Review Panel focused on. Judge Fahey said nothing about what kind of bridge had to be built, where the plaza should be, or whether there should be an expanded bridge or new plaza at all. He simply said that if the PBA wanted to do any of these things, the PBA would have to obey the law. That was almost five months ago. Since then, the PBA has done nothing.

So, yes, a report was the goal of the Public Consensus Review Panel. The idea was that once all reasonable positions had been given voice and serious attention had been paid to technical matters, then the PBA would take a less closeted view of these matters and would start behaving like a public benefit corporation, which in law it is. I don't think any of the foundation and public officials who funded the Panel had any idea of the extent of the PBA's resistance to that idea. They all really thought that if the information were made available, then the PBA would perform in the public interest, they thought the Americans on the board would take an active interest in protecting Buffalo and that the Canadians on the board wouldn't be so relentlessly self-serving. They were wrong on both counts.

But the long effort accomplished the one thing the PBA most hoped to avoid: thousands and thousands of people on this side of the river had an opportunity to learn what was really going on, and many of them found a forum in which they could let one another know that the choices made by the PBA mattered to them, whether or not they ever crossed that bridge.

How long can the plaintiffs in the lawsuits hold out? (I recently received a fund raising appeal from the West Side Environmental Defense Fund.)

If the writer means the case in New York State Supreme Court: for a long time. The Episcopal Church Home and the Olmsted Parks Conservancy have long-term interests and they're not about to fold their tents just because the PBA's highly-paid lawyers make noises about an appeal they're almost certain to lose (if they're not just blowing smoke about filing it, which they probably are). The City of Buffalo will stay in the case as long as the mayor remains convinced that it's in his political interest to do so.

If the writer is wondering about the case recently filed in Federal court, the answer is: for a while. The attorneys in that federal case have contributed a huge amount of their own time to it. The PBA has deep pockets – whenever they get short of money they just jack up the tolls on the bridge a bit and they're no longer short on money. Which is to say, they're taking money from the public's pockets to fight the public's attempt to get them to behave responsibly. The plaintiffs have asked for attorney's fees under the Equal Access to Justice Act. The PBA is great at stonewalling. It's your money they use to do it and, other than care and feeding Fort Erie, there is no tradition of public service there. They'll do whatever they can to starve out their opponents. So my response to the sender of that question is: I don't know how long the goodguys can hold out. You got an appeal for help from the WSEDG, so send them a check.

Can you identify exactly what the Parker Truss is and why it is there and why people want it gone? Is it true that the minimum clearance over the Black Rock Canal at the Peace Bridge is 100 feet? Who makes these rules? Does this explain the Public Bridge Authority's constancy in seeking to land the bridge(s) at the high Bluff at Fort Porter/Front Park? What is the tallest ship/mast to travel up the canal in the last x years? Can an environmental impact statement change an anachronistic requirement? If you built a bridge at the railroad bridge corridor, how could it clear the canal without help from a bluff? (The existing railroad bridge swings to permit boat traffic.) Is this one of those constraints that requires a bad bridge or is it being used to force a twin bridge that many don't want? Is anyone working on this?

Wow.

The clearance issue has nothing to do with one six-lane span versus two three-lane spans. This is perhaps the one part of this mess for which the Public Bridge Authority cannot be blamed.

Almost 75 years ago the US Coast Guard decided that the bridge had to clear the Black Rock Canal by 100 feet, and it has refused to revisit that decision ever since. When the matter of a Peace Bridge expansion came up a few years ago, the Coast Guard said whatever replacement went in that slot had to maintain the same clearance. (No matter that no boat or ship requiring anything near that clearance had passed through that waterway in more than 60 years.) The Coast Guard said that sometime in the future a tall ship might once again come up that canal and head out into Lake Erie so everything here had to be geared to that possibility. (No matter that tall masts on modern sailboats plying inland waterways are now hinged so they don't need that huge clearance anyway.) Bureaucrats make their living thinking about every possible contingency, not about getting from here to there.

An environmental impact study has nothing to do with this because an EIS cannot get the Coast Guard to think about anything. How can an EIS deal with a Coast Guard official who says, "What if they start making tall ships without hinged masts again in 50 years?" How can anybody deal with questions like that?

The first effect of this Coast Guard ukase 75 years ago was that the lovely repeating design of the original bridge was destroyed. The last segment was supposed to be just one more inverted U reaching up to the bank at the Buffalo terminus, but the Coast Guard said that crossing had to be 100' not just in the middle, but all the way across the Black Rock Canal. No way one of those graceful arches could do that. So they had to build that ungainly steel box – the Parker Truss – on top, holding the bridge from above rather than supporting it from below, as do the arches going the rest of the back to Canada. The engineers who built the bridge thought their design had been destroyed even then.

It's not as ugly from the Canadian side because it's much further away from them than it is from us. They get to look at the curved arches; we have to look at the iron box.

If a truck bridge were built adjacent to the International Bridge, it would either have to pivot, as the International Bridge does now and as will the new Woodrow Wilson Bridge in Washington, D.C., or it would have to have the same 100' clearance the Peace Bridge has. The pivot option is workable: the International Bridge has to swing to let a ship pass only a few times a year and the whole operation takes only 20 minutes or so, nothing like the delays truckers experience every time they cross the Peace Bridge now.

I was wondering how much money the PBA takes in on the Duty Free shops. Last Friday the trucks on the US side were lined up so tight outside the barricade to get their booze that there was only room for one lane of traffic to manoeuver around them. No one polices this perpetual mess. So too, none of the fancy computer-generated pictures of the new twin-spans they put out ever show that shoddy building on the American side, hemmed in as it is by broken shards of concrete and littered with filth.

The PBA doesn't run the shops; it rents them out and collects rent plus a percentage of the gross. It's difficult to tell from their annual financial statement exactly what their net on the operation is, which is one more reason a real audit by New York comptroller H. Carl McCall is in order. But it looks like they're grossing $4-$5 million per year.

The computer-generated pictures don't show any trucks either.

What would it take to dismantle the unresponsive Buffalo and Fort Erie Public Bridge Authority and get some responsible adults to play?

Passage of Sam Hoyt's recent bills in the legislature modifying the way the PBA does business, adding a representative of the city of Buffalo to the Board, and requiring the PBA to get approval of the city before it engages in any major building jobs would be a good start. It's a game effort by Sam but political insiders tell me the chance of any major change in the PBA structure coming out of Albany is slim: the Peace Bridge is a money machine, a lot of people in business and politics on both sides of the border and both sides of the aisle reap rich rewards from it as it is right now. Albany has a poor history of interfering with nonpartisan cash cows.

How come neither Hillary Clinton nor Rick Lazio has come out for Buffalo's interests in this affair?

Because the PBA brilliantly kept this a bipartisan issue. Both Clinton and Lazio are getting money and political support from people supporting the twin span project. One of Hillary's oldest political friends and supporters is Congressman John LaFalce, who has been in the Canadians' pocket on this from the beginning. Rick Lazio has his LaFalce equivalents. Since neither of them is going near the Peace Bridge issue, neither of them is going to lose or gain votes by avoiding it. From their point of view, it's a wash on the votes and in exchange they both get campaign contributions. From our point of view it stinks. Senator Moynihan's office has tried to interest Hillary in the Peace Bridge but they've gotten nowhere. They're worried that if she's elected she'll go wherever her old friend John LaFalce points her on this issue and that if Lazio wins he'll just go with the big steel interests. Either way, the election seems likely to leave opponents of the twin span weakened politically.

Why would the PBA disclose its intentions prior to the senate vote and swearing in of a new senator? Won't the absence of Moynihan, Bullock and Kane ease their path? While it won't change the law requirements, it will change the political environment and thus have an impact on how the federal agencies play through the next round. Imagine how much senatorial scrutiny the scoping process would have if that were happening now.

Asked, and answered.

You keep writing about the Canadian interests in the steel twin span and how Fort Erie has a great interest in the status quo because it's bought and sold and how the Canadian Federal government has no reason to interfere with all that money going to one of its small towns and some of its steel companies. I can buy all that. But why would the American members of the PBA go along with an operation that serves people on this side of the border so badly? It just doesn't make any sense.

I can only speculate about this, because none of those PBA board members talks about such matters to people like me. I passed this question along to someone who has long been involved in both politics and business in this region and this is what he wrote in reply:

> The businessmen on the PBA probably get red carpet treatment from the Department of Environmental Conservation and other state regulators in their development projects. That can be huge. I am sure they milk this a lot. It's nearly impossible to determine. As you already know, we will never know what really goes on as the board is protected from FOIA. We were hoping Barbra Kavanaugh was going to start to pry that veil, but she has been a miserable failure on that score. We really need to pressure the atty general to: TEAR DOWN THAT WALL MR. SPITZER!!! (Reagan style.). Sam Hoyt is trying to do this legislatively, but it won't pass. Spitzer, as an atty general, and as a board member, can sue for open meetings and financial disclosure by staff and board members. But he won't do it because he is not being pressured to do so. That is up to us.

When are they planning to do an underwater inspection of the existing structure?

Not until they're forced to as part of an EIS or they get started constructing a new bridge, whether 6-lane concrete or 3-lane steel. As long as they can avoid an underwater inspection of the piers they can keep claiming that the old bridge will last forever. Once they have data about the real condition of those piers they can make that claim only under risk of perjury.

Why have they not accepted the northern plaza option? Even the Buffalo News pushed that. Is it because they are still thinking they can frame a negotiation so that they look like they gave a few things and settle?

On December 16, Stephen F. Mayer, general manager for operations, presented to the Public Consensus Review Panel a plan for a northern plaza. It was a very nice plan. He said that the PBA would like to build their twin span now without doing an EIS, then at some unspecified point in the future, if someone gave them the money to build a new plaza, they would swing the end of the bridge to meet the northern plaza. That is, they offered nothing other than building the steel twin span they've been committed to all along. He didn't explain how you swing a steel bridge weighing a gazillion tons and not the least bit flexible.

The fact is, the PBA has *never* said it would use any of its massive financial resources to move its operations out of Front Park. There is no incentive for it to do so. Such a move serves the community's needs and interests, but it is irrelevant to them. At some point the PBA may offer a new plaza as a trade or it will be forced to do it as a condition of expansion but, given its behavior to date, it will never do that merely because it's the right thing to do.

I've heard a story which, if true, would explain the intransigence of the Public Bridge Authority on the "twin span:" The majority, if not all, of the steel has been ordered AND MANUFACTURED and is sitting somewhere on a Stelco site waiting to be delivered. This report comes to me so many times removed from any authoritative source that I only pass it on as a possible explanation for their attitude ... but could even the PBA be so arrogant??

The PBA surely could be so arrogant but I don't think they've done it. You can't manufacture the steel until you have the final design and there are still too many things that might change – even if they get their steel twin span by us – for the component parts to be locked down. I have heard that there are letters of intent – promises to take delivery of certain quantities of steel by certain dates. Those are necessary because the steel makers and fabricators have to make space for the order in their pipeline and on their lots. The dates on those letters of intent have no doubt been revised several times over the past year as the City of Buffalo and Judge Eugene Fahey blocked the PBA's construction plans, but surely whatever political deals were made in the allocation of those promises to buy are still out there and those steel companies still want to make their profits on this project, whenever it gets underway.

I've also heard that the PBA has paid a good deal of money to lock in the steel fabrication bid prices they got back whenever they thought this was all a lock for them.

Is it true that the PBA takes no tax dollars for its operations?

This question was asked each time we've announced a Peace Bridge Q&A. ("The Great Summer Peace Bridge Q&A," May 3, 1999 and "The Vernal Peace Bridge Q&A," March 30, 2000). The next few paragraphs are pretty much what I wrote in response to the question last time around. Nothing has changed: they're still making the claim, and it's still untrue.

There's a pretty little packet of six prints on heavy 8 ½" x 11 stock the PBA sometimes gives away titled "Construction Paintings of the Peace Bridge by H. H. Green." Green was one of the original incorporators. A note on the inside cover of that packet says, "No public funds have ever been granted or used for construction, operation, maintenance or for capital expenditure. All financing has been done from private and institutional funds." John A. Lopinski's Chairman's Report in the 1997 Annual Report begins, "No public funds have ever been granted or used for construction, operation, maintenance or for capital expenditure. All financing has been done from private and institutional funds." They say that a lot. It is perhaps the lie they have told more than any other.

The PBA exists entirely on public funds. Entirely. If the public wants to cross the river from the Buffalo into Fort Erie, the public has to pay the Buffalo and Fort Erie Public Bridge Authority $2.50 (up 25% from last year) for the right to do it – unless the public is driving a truck, in which case it pays considerably more. Last year the PBA received almost $5 million in rental income, the largest portion of it for space rented at above market rates for use by Customs and Immigration – your tax dollars pay that rent.

The Authority has been tax exempt since 1934. Tax exemption means the government has decided that what an organization does accrues to the public good, so it doesn't take from its profits (profits, not income) the share everyone else pays. Symphony orchestras, social service organizations, museums, churches, schools – all such organizations receive tax exemptions. Taxes are monies that belong to the public. If an organization is declared tax exempt, the government is subsidizing that organization to the extent that the organization is permitted to keep and use for its own purposes the funds that would otherwise have been shared by all of us. All organizations that engage in commerce and are tax exempt are recipients of public money to exactly the extent of the taxes they would have paid had they been treated like everyone else. That includes not only the taxes on profit, but also sales tax and real estate tax from which they are exempt. Bonds of tax exempt organizations are themselves tax exempt, which means those bonds enter the market with a competitive edge over bonds from profit-making organizations. That enables the sellers of nonprofit organizations' bonds to offer them for a lower interest rate, which means the nonprofit gets to rent money more cheaply than you and I do.

Last year, they accepted $2.76 million in grants from the US government to develop a high-speed frequent-traveler lane and an electronic document transmission system. All US government grants come from tax dollars.

Powerful people in those big houses on Nottingham and Meadow and Middlesex and other streets in that area recently forced Nichols and Medaille to back off their plan to turn the old Nichols Middle School campus at Nottingham and Meadow into a branch of the college. They did it because they worried about all those students using new parking lots in their neighborhood at night. How come those same people have done nothing about the PBA's construction plan, which would send heavy trucks roaring along the 198 for most of a decade?

Beats me.

According to the most recent traffic plans, for as few as four and as many as eight years, trucks would be rerouted along Niagara street, through Delaware Park on the Scajaquada Expressway , and out to the Thruway on the Kensington Expressway. That will not only clog traffic on those routes – neither of which was designed to be a major truck thoroughfare – but will also have detrimental effects on Delaware Park, on housing values in the top-dollar locations bordering the Park, on access to the Albright-Knox, Burchfield Gallery, Historical Society, and Buffalo State College, and more. In general, the quality of life in that part of the West Side and the safety of road use on those two highways will be severely degraded.

Who is replacing the late Buffalo businessman Louis Billitier on the PBA?

I heard two names: Victor Rice, retired head of LucasVerity, and Buffalo businessman Paul Koessler. I also heard that Rice was vetoed because of his negative remarks about the Buffalo Niagara-Partnership in his *Artvoice* interview a few months back. It would be nice to think that the people who made those choices read and were influenced by *Artvoice*, but if Rice were in fact a serious candidate and if his candidacy was flushed out, it didn't need Jamie Moses' tape recorder to make it happen. His low opinion of the community role played by the Partnership is well known, which is why he was forced out of the Buffalo Niagara Enterprise chairmanship last year. Koessler is a big Republican fundraiser (last week he hosted an event at his house for Rick Lazio that produced $50,000) and a board member of the Buffalo Niagara Partnership.

Is anything really happening?

Overtly, no: the PBA can't start construction because the city of Buffalo won't give the necessary permits and Judge Fahey won't let them segment the bridge and plaza projects and says they have to do a full EIS. The PBA threatened to appeal, but it hasn't done it yet. The PBA doesn't want to do the EIS because it fears careful evaluation of the truck traffic in a populated area might kill the whole deal. So that's all on hold.

Covertly, things seem to be percolating wildly. There are rumors that the easy collaboration of Canadian and American interests that for so long characterized meetings in the PBA's board room is fraying at the edges. Peace Bridge general manager for operations Steve Mayer, I've heard, is so frustrated he's been muttering about finding more fulfilling employment. His corporate services counterpart, Earl Rowe, is reported to have snapped at board chairman Victor Martucci, "I don't work for *you*!" (So whom does he work for if not the chairman of the board that hired him?) The five Canadian members of the board are fighting desperately to preserve not only Canada's steel interests, but Fort Erie's lucrative franchise. The three active American members (Billitier hasn't been replaced yet; Kavanaugh doesn't vote on anything connected with this) are more and more frustrated by the stasis and the Canadian intransigence.

And the center may not be holding so well on the Canadian side either. The current situation developed six or seven years ago when the Canadian economy was weak; that economy is now booming and some Ottawa politicians are uncomfortable about this bridge expansion being held up because of political promises made to steel company owners in another financial time, and they're even more uncomfortable about having their hugely lucrative trade relationship with the United States whipsawed by the parochial interests of the town of Fort Erie.

That's all rumor – just things I've heard. We'll have to wait to see how much of it is true and if these really are the first cracks in that big stone wall down at Peace Bridge Plaza.

46 What Andrew Rudnick Didn't Say

Editor's note:
One of our readers faxed us the October issue of "Buffalo Bylines," the newsletter of the National Association of Purchasing Management-Buffalo, Inc. The newsletter focused on the Annual International Night Meeting held jointly with the PMA of Canada-Niagara District, Thursday October 19, at the Minolta Tower in Niagara Falls, Ontario.

After a reception beginning at 6:00 p.m. and a dinner beginning at 7:00 p.m., the primary event of the evening would be a presentation entitled "Peace Bridge Twinning Project." The two speakers would be Andrew J. Rudnick, president and chief executive officer of the Buffalo Niagara Partnership, and Steven J. Mayer, of the Peace Bridge Authority. (That's what it said: "Peace Bridge Authority." One would expect international purchasing agents to know there is no such thing as the "Peace Bridge Authority," but there you go.) The newsletter had a two-paragraph bio for Rudnick and a long paragraph describing the Partnership, the Partnership logo and a small photo of Rudnick. It gave no information at all about Steve Mayer other than that single phrase, "Peace Bridge Authority."

Given Rudnick's relentless advocacy for a steel twin span even though that would drive a lot of jobs from this area and would disrupt traffic on Buffalo's west side for years, and Mayer's close involvement with the twin span proposal for several years, we thought this event was important enough to ask our chief international affairs correspondent, Bruce Jackson, to drive up there to cover this story.

Jackson did not, we're sorry to report, get the story we sent him across the border to get. It's the first time since he started covering the Peace Bridge War for Artvoice two years ago that he came back empty handed. All he delivered was this:

Chief –

Sorry to say there's nothing to report about Andy Rudnick, Steve Mayer and the International Association of Purchasing Management-Buffalo and PMA of Canada-Niagara District. For some reason, both men changed topics at the last minute. Indeed, neither Rudnick nor Mayer once in their presentations used the phrases "twinning project" or "twin span."

This all took place – or didn't take place, rather – in the Minolta Tower in Niagara Falls, Ontario. Traffic had been slow crossing the Rainbow Bridge so Diane and I didn't get there until a little after 7:00 p.m. The Minolta Tower is a 28 storey building with nothing between the ground floor and the 25th floor except the stairway and elevator shaft. The ground floor is gloomy and tacky: maybe a dozen huge electronic games that mostly focus on driving fast or shooting people or both at once, a closed popcorn counter, a gift shop full of the kind of dreadful souvenirs that people bring back from trips and wonder why they did. The event was in a restaurant on the 26th floor.

There was a huge mob of Japanese tourists trying to get on the elevator we were trying to get out of. That was pretty scary, let me tell you. They were in kind of a wedge, so they couldn't move back or to the side and they had packages of souvenirs from downstairs, some of which were pointed and sharp.,

By the time we got inside, the purchasing managers were sitting down and getting their soup. Some waiters cleaned off one of the tables the Japanese tourists had just vacated and gave us soup, too.

I managed to catch Steve Mayer's eye after a while and we waved hello. I set up a tape recorder near where they'd be speaking so I'd be sure to get every word they said about the Peace Bridge twinning project. Andy Rudnick came over to Steve's table during the turkey stuffed with apple dressing and they conferred and looked over at where Diane and I were sitting but Andy didn't wave hello to me. Andy never waves hello to me.

During his 20-minute talk, Rudnick mentioned the bridge only once, in passing. His talk was hard to summarize because it was so spread out. I think he said that there are three primary modes of transportation for moving goods – trucks, trains, and air – and we should use all of them. He also said the people around here, on both sides of the border, were wonderful and hard-working. When he finished, he said Steve would talk now and he ran out of the room. At first I thought he was heading for the men's room, then I realized he was heading for the elevator. I thought he maybe had another pressing presentation, but he only stood by the elevator until Mayer started talking and then he came back into the room and sat down in a corner so far away from where I was sitting I couldn't see him.

Mayer had a computer and screen and PowerPoint projector set up. He talked about how important the several bridges in the area were and how manufacturers don't use warehouses so much any more and that's why some of them consider the Thruway an extension of their manufacturing operation because that's where their goods are a lot of the time. He said a lot of nice things about the area and told us things about the importance of transportation. He never did show any of those PowerPoint slides. After a while he said that he'd pretty much covered everything in the slides anyway, and were there any questions?

A purchasing manager from Grand Island said that as far as he could tell all the congestion on the Peace Bridge came from traffic backed up at the toll and customs and immigration booths, so he wondered what congestion there would be, if any, if there were no tolls and no customs and no immigrations booths. Mayer said that was hard to say because it's a three-lane bridge and sometimes you have two lanes in one direction and sometimes you have two lanes in the other direction. The man from Grand Island had trouble grasping how that answered his question, so he asked it again another way. Mayer went into a lot of interesting detail about how much help the Commercial Vehicle Processing Center in Fort Erie has been and how they really could use another truck inspection lane and how if the US and Canada ever get around to declaring the bridge an international facility in which the laws of both countries apply all of the operations that cause congestion on the American side could immediately be moved to the Canadian side and none of this mess in Buffalo would be necessary at all. He never did say what congestion there would be, if any, if there were no tolls and customs and immigration booths.

Someone else asked what was the status of the Peace Bridge expansion. Meyer said there had been two lawsuits filed against the Bridge and the Bridge had filed a lawsuit against someone else and they'd lost that one. They were, he said, trying to decide what to do about those lawsuits filed against them and the one they'd lost and they were "working very hard to break the logjam." He said that an environmental impact study would take three to five years and, if they did have to do one, the whole project would take ten years or more.

Several times in the course of the Q&A Mayer referred to Rudnick and gestured back to the corner where I think he was sitting, but Rudnick never said anything and never took any questions himself. I got to wondering if maybe the reason Rudnick had fled the room immediately on finishing his comments about three primary modes of transportation being trucks, trains, and air and about how wonderful the people around here are, was because he maybe didn't want to field questions from the American and Canadian purchasing managers or other people in the audience.

At the elevator I heard one purchasing agent say to another, "I thought they were supposed to be talking about twinning the Peace Bridge."

"Yeah," the guy he was talking to said. "What we got was Chamber of Commerce."

"Maybe they forgot which audience they were talking to," the first guy said.

"I guess," the second guy said.

As Diane and I were getting into our car, I saw Steve Mayer's assistant carrying the PowerPoint equipment across the street to a parking lot.

So, Geoff, I won't have that piece I promised you for this week's paper. Neither Rudnick nor Mayer talked about what they program said they were going to talk about. You'd have thought that once they saw that Diane and I were there, taping it all and taking photographs, that they'd be especially concerned to stick to the program because both surely know how much the subject interests the readers of *Artvoice*. Maybe we'll catch them doing it some other time.

Scoop

P.S.: I assume I can put all this on my expense account even though I didn't get the story: $2.50 for the bridge, $14 for the turkey stuffed with apple dressing, $6 for the glass of not bad cabernet, and $32.50 for the ashtray, t-shirt and panoramic poster I picked up in the first floor gift shop in the way out. All US $. I left the receipts in the office in the usual place.

The editors wrote Jackson back, reminding him that he doesn't HAVE an expense account, and even if he did it wouldn't cover an ashtray, t-shirt and panoramic poster. As for the rest, we said, his case would be helped if he filed a story about something Andrew Rudnick did do rather than something he didn't do. This arrived Tuesday morning:

You got it, chief.

I saw Rudnick a few days later, at the Monday, October 23, meeting of the Common Council's bridge Task Force. The main subject of that meeting was a draft proposal the Public Bridge Authority sent to the city on Friday offering to do the environmental impact study ordered by Judge Eugene Fahey if the city agrees to certain conditions. Neither Buffalo Corporation Council Michael Risman nor the engineer representing the PBA at the Task Force meeting would say what those conditions were. They said that they only people who were in on the current negotiations were new attorneys from New York City hired especially by the PBA to handle this operation and Buffalo city attorneys.

The curious thing was, just about every time anyone asked a question about the draft and the negotiations, Rudnick's head bobbed up and down or nodded back and forth. Sometimes he answered questions he seemed to think warranted a more detailed response than Risman had provided or questions he had avoided entirely. Rudnick said he was attending the meeting as a representative of the Buffalo Niagara Partnership but he didn't explain why he was providing information on behalf of the PBA or how he knew so much about matters the city attorney said had transpired in secret attorney-only sessions. (Rudnick's degree is in economics.)

Everyone there who had a part in the secret meetings said they expected some kind of resolution to the negotiations within a month, but don't count on it. There's not much trust on either side these days. The last time the city signed on to a draft agreement proposed by the PBA it turned out to be a time-bomb that nearly destroyed the Public Consensus Review Panel. The Canadian members of the Public Bridge Authority are disgusted at the way the Common Council, the Olmsted Parks Conservancy and the Episcopal Church Home have gotten talking rights in what they think should be a deal designed and carried out by the Big Boys in the Back Room, the way it's always been done.

Too bad for them, I think: Common Council president James Pitts and Olmsted Parks Conservancy president Andrea Schillaci made it clear to the Corporation Counsel, the PBA engineer, and Andrew Rudnick that they very much intended to give a very careful look at and have a real voice in any new deal between the PBA and the city of Buffalo.

Stay tuned.

47 The Bridge Authority has a Movement

The Buffalo and Fort Erie Public Bridge Authority announced at a Wednesday morning press conference that it will obey New York environmental law and undertake the legally-required nonsegmented environmental impact study it has, with great effort and at huge expense, tried to avoid for the past six years.

After nearly seven months of stasis, the Authority decided that it cannot undermine, undercut, or otherwise evade Judge Eugene Fahey's order that it must treat Peace Bridge and plaza expansion as a single project.

PBA chairman Victor Martucci said the Authority would give up all permits it has received from all American and Canadian government agencies and abandon all current design plans. He said the Authority would work with the City of Buffalo, the town of Fort Erie and community groups on both sides of the border to evaluate environmental aspects of bridge expansion and to consider all reasonable design options. He said that this time the process would be open to public scrutiny. His remarks were endorsed and expanded upon by PBA vice-chairman John Lopinski, Buffalo Mayor Anthony Masiello, Buffalo Common Council President James Pitts and other officials.

This decision comes after extensive negotiations with Erie County and the City of Buffalo, extensive conferring with twin-span supporters in the Buffalo Niagara Partnership, and payment of huge legal fees to two Buffalo law firms and to the New York law firm the PBA hired to supplement those two firms in the present negotiation.

The PBA needs easements from the City of Buffalo to begin any kind of construction at Peace Bridge Plaza. The City has refused to issue those easements unless the PBA first conducted the legally-required environmental impact study of the entire project. Recently, the PBA attempted to get the City to promise to deliver those easements as soon as the PBA completed an EIS. The City refused, pointing out that it could not agree beforehand to an agreement that might force it to issue easements for a plan the still-undone EIS might show was harmful to the community or otherwise defective.

Important questions remain unanswered:

> Did the PBA decide to obey the law because it decided it had no other choice or were there deals cut with the city and the county that were not announced at Wednesday's press conference?
>
> Did the city, the county and the PBA include the Episcopal Church Home and the Olmsted Parks Conservancy – the two community organizations whose lawsuits stopped the PBA from erecting its costly steel twin span – in all of the negotiations that led to this decision? If not, why not, and what are the legal implications of that?
>
> Is this anything more than the usual PBA smoke and mirrors, like their pretend-participation in the work of the Public Consensus Review Panel a year ago that turned out to be astute sabotage?
>
> What, if anything, does this sudden conversion have to do with the elections now going on in Canada and the senatorial election just completed in New York?
>
> What, if any, deals were made with the American and Canadian steel manufacturers who have for so long counted on huge profits from the manufacture and maintenance of an anachronistic steel bridge here?
>
> The PBA is presently an organization composed of five Canadians (representing Fort Erie development interests and Ottawa trade interests) who consistently vote as a bloc and three Americans who seem incapable of influencing that Canadian bloc on any vote about anything. Can we trust an organization with that kind of lopsided partisan voting power?

It will be some time before the meaning of Wednesday's press conference is clear. By now the *Buffalo News* and the electronic press have told you what the various speakers said at Wednesday's press conference. In the next few weeks, we'll do what we can to find out what those speakers really meant.

48 ROBERT MOSES IS FINALLY DEAD IN BUFFALO: PEACE COMES TO PEACE BRIDGE PLAZA

THE MOSES MODEL

Buffalo and Erie County have long operated on the Robert Moses model of public works. That is a model in which a powerful person or group of persons decides, for whatever reason, that a certain piece of work should be done. That person or group of persons then amasses the financial, legal and political resources to do it. Anyone who gets in the way is ignored and, if that doesn't work, crushed. When the project is completed and doesn't deliver as promised, it is expanded and amended: you pour more concrete, more money. Moses, for example, built parkways and expressways in New York City, which he said were needed to reduce congestion, and every one of them increased congestion, whereupon he built more parkways and expressways, destroying viable neighborhoods and green space in the process. (The best source of information about the way Robert Moses spent $27 billion and reconfigured New York's landscape is Robert A. Caro's magisterial Pulitzer Prize-winning *The Power Broker: Robert Moses and the Fall of New York,* Knopf, NY, 1974.)

Anyone who pays even marginal attention to public affairs in this region can recite our Local Litany of Moses-class Disasters, all those public works and private construction project that promised much but (save for a few people who got rich on them) only subtracted from the quality of our lives:

> the Thruway that cuts the city off from its waterfront and routes high pollution truck traffic through populated areas;
> the Scajaquada Expressway that bisects a middle-class neighborhood and an Olmsted park;
> the suburban UB campus that yanked a huge middle class population group and its associated services out of the city and deprived the university of the great resources a city can provide;

> the dark and ugly and inadequate convention center that breaks one of Joseph Ellicott's functional radii;
> the huge HSBC office building that squats across the foot of Main Street, literally blocking out downtown's light;
> the mass transit system designed to increase access to downtown that all but destroyed downtown retail business and then stopped halfway to its suburban destination because some people felt it would make access to the wealthy suburbs too easy to the nonwhite poor.

There are others, but you get the point and this is already depressing enough.

Buffalo and Erie County are not special in this kind of stupidity: other cities at other phases in their history made the same kinds of blunders. But we managed to do it a lot over a fairly brief period of time, and that time coincided with a decline in the region's economy. Most of the reasons for the decline were beyond they city's or county's control: it wasn't local decisions that sent big steel elsewhere. Who's responsible for that decline is far less important than the fact of it. And the consequence: we can't afford to keep repeating that kind of error.

The decision by the Buffalo and Fort Erie Public Bridge Authority to build a steel companion span alongside the decaying present bridge was worthy of that list of blunders, and it was made in the classic Robert Moses manner. Public hearings were held, but they were sham hearings because the decision about what would be built had been taken before a single public hearing ever took place. Information was given to the public, but it was deceptive and distorted. The PBA was appointed to serve the public but in fact it took marching orders from undisclosed private interests. And the city's only daily newspaper covered the story by interviewing only people the Bridge Authority wanted interviewed and, for a long time, by confining much of its coverage to rewrites of Bridge Authority press releases.

Everything else being equal, the Peace Bridge expansion project would have proceeded just like all those other Robert Moses-type projects: the steel twin span would have gone up, it would have been a disaster, the people who foisted it on us would be dead and gone, and the survivors would be stuck with it forever.

But this time something anomalous happened on the way to the future: an informed, passionate, persistent, and consistently growing public opposition to the bridge project grew and wouldn't go away. With only a few exceptions (Congressman John LaFalce and West Side county legislator Al DeBenedetti are the only two who come to mind), all the local politicians lined up on the side of the people. Lawsuits were filed and the very deep pockets of the PBA couldn't make them go away.

And finally, on November 15, the Buffalo and Fort Erie Public Bridge Authority threw in the towel (or seemed to). Victor Martucci, the Authority's chairman, told a hastily-called press conference that the PBA was abandoning all its current construction plans, it had given back all construction permits it had received from all American and Canadian permitting agencies, it would not appeal Judge Eugene Fahey's order that it perform a full environmental impact study on its proposed bridge and plaza construction projects, and it would immediately begin the EIS it had fought so long and assiduously to avoid.

Everything, Martucci said, was now on the table, nothing was locked in, and the process henceforth would involve the public and would be fully open.

WEIRDNESS

His comments were immediately followed by happy and largely-platitudinous speechifying by John Lopinski (vice chair of the PBA), Buffalo Mayor Anthony Masiello, Common Council President James Pitts (who announced that he was changing the name of the Council's Superspan/Signature Bridge Task Force to the Binational Bridge Task Force), Debbie Zimmerman (chair of Ontario's Niagara Regional Council), and Erie County Executive Joel Giambra.

It was one of those perfectly weird moments politicians and lawyers seem to find less bizarre than the rest of us: people who until moments before had been ripping and clawing at one another are suddenly smiling and purring and making kiss-kiss, people who had vowed never to give up the ship are suddenly asking sailors from the other side to come aboard and help them sail it to a mutually-agreeable lagoon. Golly gee-whiz, it was just swell!

If you didn't know about the huge amount of work put in and great amount of money expended over the past three years by the New Millennium Group, the Episcopal Church Home, the Olmsted Parks Conservancy, a score of other groups and organizations, Bruno Freschi, T.Y. Lin, Gene Figg, and countless other individuals, nothing said by Martucci or any other podium-speaker at that press conference would have let you know that a war had been fought on this territory.

And if you looked at the smiling faces of the PBA's American and Canadian press agents as they listened to Martucci read his prepared statement and field questions from the invited and uninvited press, nothing would have told you that over the past year the PBA had paid and those press agents had spent at least a million dollars of public funds on a huge disinformation campaign that consisted of scores of misleading spot ads on all Buffalo commercial television stations and a series of orchestrated letters published in *The Buffalo News*. That campaign was designed to put pressure on Mayor Masiello and the Common Council so they would issue the permits the PBA needed to start construction, it was designed to embarrass the Olmsted Parks Conservancy and Episcopal Church Home so they would abandon their lawsuits, and it was designed to pressure Judge Eugene Fahey so he would rule in the PBA's favor if they didn't back off.

If you walked in off the street with no knowledge of the bitter history, you would have thought you'd come across a public agency operating in a perfectly fine way: open, fair, collaborative, friendly.

And, if everything Victor Martucci said was true and everything he implied comes to pass, it will be all of that. Right now, we've got to assume that's where things are and will be. In practical politics you get nowhere brooding about the past; you've always got to be looking to the future. The past is for historians – and for those who want to have some sense of what traps might be in the apparently smooth roadway ahead.

WHAT VICTOR MARTUCCI SAID

Martucci opened with this:

> We're here to announce that the Authority has begun a binational integrated environmental process that combines the proposed actions involving bridge capacity expansion, US plaza modernization and connecting roads. We are confident that this binational integrated process will lead to the final binational solution to relieving existing congestion.
>
> I want to make this point absolutely clear at the outset: the Authority is committed to a process that is open, fair and inclusive, whereby every practical option will be given full and thorough review. The Authority has been working over the past several months to determine how best to proceed in response to Judge Fahey's decision. Many meetings have

> been held with Canadian and US officials, permitting agencies and special interest groups to assess possible paths to a resolution. As a board with binational responsibilities we have actively investigated every scenario that would lead to a positive resolution of this issue.
>
> The authority will develop this binational integrated environmental process in compliance with all applicable United States, New York State, City of Buffalo, Town of Fort Erie, and Canadian Federal and Provincial laws and regulations, including, but not limited to the Canadian Environmental Assessment Act, the National Environmental Policy Acts, The New York State Environmental Quality Review Act, and the City of Buffalo Environmental Review Ordinance. We see this as the quickest and most productive path to expand capacity, relieve congestion, and comply with current legal rulings.

This, as just about everything else Martucci said at that press conference, is perfectly reasonable and responsible. But a bizarre fact undergirds it: this is the chairman of an international public agency telling a press conference that the agency he directs will act legally. Normally, we *assume* public agencies will act legally. Public agencies are *supposed* to act legally. Public agencies aren't supposed to be looking for ways to avoid obeying the law.

"Supposed to" and "does" aren't the same. For years, the PBA has been fighting tooth and nail to avoid New York environmental law.

In the rest of his statement and in the question period following, Martucci again and again referred to public involvement in the process. He must have used the phrase "binational integrated environmental process" twenty-five times. He spoke of the need for consideration of neighborhood needs and concerns, for consensus and understanding, for everything – in other words – for which everybody on the other side has been asking, suing, and pleading since this started.

For years, Fort Erie has been the PBA's pawn and Buffalo has been its nemesis. According to Martucci, that's all over: the three are now full partners.

> We strongly believe that the proposed bi-national integrated environmental process clearly defines respective roles and responsibilities of the Authority , the city of Buffalo, the town of Fort Erie, and the public, so that consensus can be reached regarding a preferred alternative. We have a long

> way to go, but we are committed to a process that is open, and inclusive, and whereby every practical option will be given a full and thorough review. Each step will take as long as necessary to get it right and we'll work with our partners expeditiously to ensure the optimum binational plan is reached.

He was, more than anything else, a proponent of clarity and openness:

> I think as long as the process is open and it's fair and everybody has a fair hearing, every option gets its thorough review, I think in the end, although you're going to have some people happy and some people unhappy with the final result, that everybody will come away from the process confident and satisfied that the process was done right. And that's our commitment.

GETTING HERE

In the Q&A following the formal statement, Andrew Siff of WKBW asked why it took more than seven months since Judge Fahey's ruling to get here. "Frankly," Martucci said,

> it's a very complicated process. We don't have, as a binational authority, the very narrow interests that several of the interested parties that have expressed their opinions on this have. We've been having several meetings over the course of the last few months with community leaders, with government leaders on both sides of the border, with special interest groups. And as a binational organization we have a fiduciary responsibility to examine every option that was available to us. And after a very extensive and very comprehensive review, after reaching out to the community on both sides of the border, we reached the conclusion that the best path for us to follow is this integrated bi-national environmental review process. It gets us to expanded capacity the quickest and it addresses the concerns and the needs of the community on both sides of the border.

I'm sure that's all true, but he probably could have said the same thing more economically: "Our lawyers told us there was no way we were going to get out from under Fahey's ruling." That's basically what he said to me when I asked him the same question the following day: "The judge's ruling changed the landscape," he said.

Fahey's ruling didn't happen in a vacuum. Important judicial decisions rarely do.

One of the key factors in keeping this alive was the Public Consensus Review Panel, initiated by then-county executive Dennis Gorski's office and funded by Erie County, the Community Foundation, the Margaret L. Wendt Foundation, and the City of Buffalo.

The PCRP did three hugely important things. It provided a fair forum where a very wide range of community, commercial, and political groups could participate in the process on an equal footing. It provided occasions where the public could listen to and comment on what various groups and individuals had to say about various aspects of this. Perhaps most important, it kept the bridge issue alive while the court cases were being developed and decided. The PCRP was a true educational forum. There was nothing like it anywhere else in the process. The Common Council's Bridge Task Force had some of the same aims, but it never had that kind of visibility or that measure of resources.

Rich Tobe, now at the Community Foundation but previously Gorski's director of development, is largely credited with being the person who got the various elements of the PCRP together. He told me a year ago that one reason he thought the PCRP was necessary was there had to be a forum in which the issues raised by the New Millennium Group could be given fair hearing. All those young professionals, he said, were coming up with important data and raising important questions that were being ignored by the PBA. Another reason, he told me recently, was all four funders of the PCRP thought the courts were the least preferable place to resolve a major public policy issue.

As it turned out, the PCRP didn't only give voice to people the PBA wanted kept silent. It also kept the whole enterprise alive long enough for legal opposition to take shape and become effective.

The Common Council's Bridge Task Force, organized and chaired by Council President James Pitts, bracketed the PCRP: it was there before the PCRP was set up, and it took up the work again when the PCRP was dismantled.

Senators Moynihan and Schumer never wavered in their insistence that Buffalo and Fort Erie should have a bridge that made as much aesthetic and ecological sense as commercial sense. Mayor Masiello, after some initial fuzziness and waffling was key: had he folded on the easement the PBA needed to begin construction much of this would have been mooted out long ago.

Robert Moses knew that once he ripped open the ground it was all but impossible to stop a project. The city's refusal to issue the easements and Judge Fahey's order kept the PBA from breaking ground. The judge's final order meant they could do nothing without going through the very people who had been gathered around the table for the PCRP. Victor Martucci said, "The judge's order changed the landscape," but the details of that landscape had already been elegantly etched by the people who took part in the Public Consensus Review Panel, as was the model of public participation the Panel developed.

With one difference: the Town of Fort Erie and the Public Bridge Authority both refused to join the panel. Now, according to PBA chair Martucci, they're both happy to take part in the continuing conversation.

WHAT ABOUT BUFFALO?

For Buffalonians, perhaps the most important phrase in the first part of Martucci's statement was "the City of Buffalo Environmental Review Ordinance." The PBA never paid any attention to Buffalo ordinances before. It bought Fort Erie and it ignored Buffalo. Judge Fahey's order last spring telling the PBA they would have to do an unsegmented environmental impact study (which meant they could not pretend the bridge and plaza were separate projects) meant that the PBA could no longer ignore Buffalo.

The last time around, the two lead agencies overseeing all of this on the American side – the New York State Department of Transportation and the United States Coast Guard – – couldn't care less about Buffalo's needs, interests or condition. This time Buffalo has one of the primary seats at the table. The City of Buffalo – the mayor's office and the Common Council – will be one of the four lead agencies in the environmental review.

That single fact perhaps makes this announcement from the PBA different from all other announcements by the PBA. This time we will have people answerable to us assigned to watch what those people are doing. The PBA's staff will be legally-bound to answer all questions our representatives think have to be asked. If they fail to provide those answers or if the answers aren't good enough, they do not get to turn one shovelful of Buffalo earth.

ATTENDANCE

Notably absent from the November 15 press conference were three long-term steel twin-span advocates: Fort Erie Mayor Wayne Redekop, US Congressman John LaFalce, and Buffalo Niagara Partnership president and CEO Andrew Rudnick.

Martucci said that Redekop and LaFalce would have joined the proceedings but had prior commitments that took them out of town. Redekop sent a representative;. LaFalce did not. No one said anything about Rudnick or the Partnership. Rudnick didn't show up at the meeting of Pitts' Bridge Task Force the next day, either.

REACTIONS

Thus far, reactions of people who have been deeply involved in the battle have been cautiously optimistic.

Jeff Belt, president of the New Millennium Group, said:

> I am very happy about the Public Bridge Authority's decision to scrap their plan to twin the Peace Bridge and expand the Front Park plaza – at least until a fair, open and non-segmented Environmental Impact Statement can be completed. After the Public Consensus Review Panel voted to recommend an all new bridge and plaza system and the PBA responded by closing a lane of the bridge, I feared that we were in for a long period of punishment." Belt also said that in the course of the long debate he "learned about the importance of parks and communities. If we approach the Peace Bridge project with creativity and boldness, we could not only restore the Front Park Presidio, we could actually implement Frederick Law Olmsted's original vision for that park. We could extend the Terrace to the water's edge by decking over part of Interstate 190. A restored and expanded Front Park would be an enormous asset for Buffalo. It could catalyze renaissance in the West Side.

Andrea Schillaci, president of the Olmsted Parks Conservancy, one of the litigants in the lawsuit that got the PBA to reverse its position, said,

> The way we see it is that we got what we came for. We wanted the PBA to do a non-segmented EIS. That's what Judge Fahey said they have to do and that is what they have said they will do. We are in the process of putting together a team to work on this issue going forward. We are very excited to have an opportunity to participate in this process

> in a meaningful way and intend to devote as much of our time, energy and resources as it needs to be done well. We don't see it as the end but really just the beginning. We fought hard for a chance to participate and that is what we plan to do.

Edward Cosgrove, former Erie County DA, member of the Public Consensus Review Panel and the Common Council's Bridge Task Force, was both hopeful and cautious: "We'll know what this is worth when we see who they name to run it. We'll know the first time you go down to where they're doing the work and say, 'I'd like to see what's in that file drawer.' In the interim, we have to be optimistic."

WHAT'S NEW AT *THE NEWS*

Not much.

The Buffalo News maintains its formal policy of never referring to the Public Bridge Authority by its legal name. Instead it uses "Peace Bridge Authority," the name the PBA adopted as part of its strategy to convince people it was a private corporation rather than a public agency.

And the *News* continues acting as flack for the Authority. There was, for example, "Peace Bridge: Deck fixes delayed; timeline uncertain," a non-story by Patrick Lakamp on page 1 of the second section of Monday's paper. Lakamp's first paragraph was: "The Peace Bridge Authority will try to put off redecking the existing Peace Bridge until a new crossing – whether a twin span or a signature bridge – is built and opened to traffic, officials said Monday."

What's news about *that*? The plan has *always* been to put off redecking until a new bridge was up.

Something worrisome does turn up later in the article: "The new capital plan shows bridge construction starting in 2007, after a four-year environmental review. 'We're all hoping we can pull that back and shorten the amount of time for that (study),' said Stephen F. Mayer, general manager for operations at the Peace Bridge."

Mayer has long rattled the saber of the serious consequences that would follow a legal EIS and he has often doubled the time other experts said it would take to do one. Is he being serious now or is he just setting us up for more high-pressure techniques a year or two from now, when the PBA hopes the organized opposition that got us to this point will have moved on to other community issues?

It's hard to forget how the operators of the bridge tried to whipsaw the community last summer. They shut down one of the bridge lanes for repair work during the heaviest daytime traffic hours and did no repair work in the late night hours when there was little traffic – just the opposite of the way such elective repairs are handled everywhere else. All of Martucci's fine talk of community and openness is swell, but where will we be if last summer's traffic-jam blackmail is repeated a year or so from now?

WHOSE PETARD DID THE HOISTING?

The great irony, which cannot be lost on all ten members of the PBA, is that it was the Canadians who complained for the past two years that the Americans were stalling their truck traffic over mere aesthetic considerations, but in the end it turned out to be Canadian recalcitrance about aesthetics ("a twin span or no span at all") that helped stop this steel twin abomination dead in its tracks.

Had the PBA, the *Buffalo News*, and the Buffalo Niagara Partnership played it straight with the public, had not they mounted and touted those bogus design charettes, had they not treated the American public as if they were morons (who other than a moron would honestly believe for a second that a bridge and where the bridge lands are separate construction projects?), then this entire project almost certainly would have sailed through a far simpler and less suspicious and less informed EIS than the one that will take place now. Recent data about the far higher rates of lung disorders around the bridge plaza and the correlation between diesel emissions and cancer might very well have entered the process anyway, but there would not have been nearly so much time for public knowledge, suspicion and fear to develop about that data.

Most people I've talked to are pretty much convinced that if the PBA had gone for an elegant design back when Pat Moynihan, Bruno Freschi, and the SuperSpan group suggested it, back when the Americans were thinking more of design than environmental impact, they'd be well into construction by now. It wouldn't be the steel bridge the PBA wanted – which would have meant some steel companies that were looking forward to huge construction and maintenance profits would have been disappointed – but it would be a six-lane bridge that could have handled those trucks full of Canadian goods heading south. Their own intransigence and hypocrisy were, finally, their undoing. That and their contempt for the legitimacy of United States law: they saw US law as just another inconvenience; they didn't understand that it is a fact.

The NMG, Episcopal Church Home, Olmsted Parks Conservancy

and Judge Fahey were instruments in what happened, but the lawsuits would have been unnecessary and the NMG would have been deprived of this as an issue had the Canadians and the PBA been willing to settle for a win with a small 'w' rather than going for broke.

QUESTIONS THAT REMAIN

Every environmental impact study begins with "scoping," a process in which the broad boundaries of the project to be evaluated are set out. Everything that follows is predicated on how things are defined in the scoping stage.

The scoping process that will soon start will give us an idea of how serious and honest the PBA is about all of this. Will the PBA and Fort Erie agree, for example, that the EIS should consider not just the best way to expand capacity at the current site, but whether the current site is the best place to expand capacity or whether expansion will fix problems that improving customs operations will not? Will it define the project in purely local terms or as a regional issue?

The answer to these questions will also influence what happens to the lawsuit currently in Federal court. That suit, filed by West Side community groups and residents, has several parts, only some of which were answered at the PBA's press conference. There still remain questions about whether or not the PBA will admit it is subject to the Freedom of Information Act, if it will obey all environmental laws, if the Coast Guard has formally taken back its finding of no significant impact, if it will compensate the litigants for the attorney's fees that got us to this apparently harmonious point."If they don't consider alternate locations," said attorney Robert Knoer, "there will be a lawsuit challenging the EIS at the end."

Someone at the press conference asked Martucci if the crossing could be elsewhere and he responded: "I think the process itself is going to define that. We have to set objectives and criteria for the process and we'll be working with our partners, the City of Buffalo and the Town of Fort Erie, as well as the public, to define what those objectives are going to be." Well, the question was "could." Since Martucci had said that everything was open he could have said "Sure." But he didn't, instead he came up with a carefully-worded answer that was either vague or evasive. Or perhaps he didn't fully understand the importance of the question. We won't know which it was until the scoping process is well under way. Clearly, the scoping process demands careful watching.

SOMETHING HAPPENED

There is a reason the power brokers prefer the Robert Moses model of public works projects: it is fast and it is efficient. You see a need, you decide what needs doing, and you do it. It's all so manly, so godly.

The other way, the consensus way, seems slower and much more cluttered. You have all these meetings and you have to deal with the questions raised in them. You have to listen to all these individuals and groups, some of them thorny, some of them angry, some of them screwball, some of them with interests directly contrary to yours. You have to deal with people who can't be bought and sold. Instead of merely satisfying some agency's technical requirements to get a permit you have to first satisfy a score of groups asking questions about why you should be asking for that permit in the first place. It's such a bother. And it is so damned slow.

It is indeed slower, but that doesn't mean it's less efficient. Getting someplace quickly means only that you got there quickly; it doesn't mean that you were heading in the right direction in the first place. If everything is as it seems, as Victor Martucci and the PBA's chief engineering consultant Jake Lamb say, then there has been a real change in direction of the whole Peace Bridge expansion project and we will all have a voice in where we go and how we get there.

I think something real *has* happened. There has been a major change in the way things are done around here. I think what just happened at Peace Bridge Plaza and what happened when the state was forced to uncover rather than bury the Erie Canal's commercial slip makes it unlikely that any industrial group or public officials will, in the foreseeable future, act in terms of the Robert Moses model in Buffalo. There's a generation of young-to-middle-aged professionals who cut their political teeth on these two projects and they're not likely to go away when the convention center proposal resurfaces or when the preliminary study of Kaleida's plan to move Children's Hospital comes back into play, or whenever the next project down the line gets to the point where it looks at all real.

The Peace Bridge War and the Commercial Slip Affair gave people around here the idea that they don't have to watch and accept while their neighborhoods and conditions of life are altered irrevocably because of decisions made in closed rooms by people who gave no thought at all to their rights and needs. Something good may really have happened here.

Some stories *do* have happy endings. But don't turn out the lights yet: we're still a long way from the end of this one.

49 THE MAYOR, THE GOVERNOR, & DETROIT'S FIRST CASUALTY

THE MAYOR'S PLAN

Buffalo Mayor Anthony Masiello has resurrected and expanded his satisfy-everybody-bridge-plan of two years ago. It's an enhanced version of the plan that produced embarrassed silence and *sotto voce* wisecracks when city attorney Peter Cutler first proposed it on the mayor's behalf at the Common Council's Peace Bridge Task Force on March 29, 1999.

At that time, the mayor proposed ending the Peace Bridge design deadlock by rehabilitating the old bridge and building a three-lane signature bridge adjacent to it. That, Peter Cutler said the mayor said, would let Fort Erie keep the bridge its mayor Wayne Redekop says his constituents dearly love, and it would give Buffalo a decent border entrance. Nobody at the time took the mayor's proposal seriously because it seemed more absurd than what the Public Bridge Authority was proposing.

The February 1, 2001, reincarnation of that side-by-side idea has three important modifications:

> it locates the new bridge south rather than north of the current bridge, which would mean that bridge plaza wouldn't require condemnation of nearly so many current houses and apartments as the northern plaza everyone except the mayor currently advocates, and it could take advantage of alignment engineering and research for bridge footings already done by the PBA;
>
> it doubles the size of the new bridge from three to six lanes, bringing to nine the total number of lanes feeding into Front Park;
>
> it endorses the recommendation many people and groups have been making for some time to create an international

zone on the Canadian side and shift all truck processing functions over there.

There are problems with the mayor's plan.

His nine-lane recommendation is predicated on unquestioning acceptance of the PBA's claim that ever-increasing truck traffic will continue to need more and more lanes across the Niagara River. (Indeed, nothing in his nine-page statement questions any PBA assertion or assumption.) That may have been a reasonable assumption in March 1999, but not now. Truck traffic is no longer increasing here; it went down last year. Is that decline merely a function of the weak Canadian dollar and peaking-out of the NAFTA benefit, as *Buffalo News* business reporter Fred O. Williams suggested earlier this month? Is it a result of a weakened American economy? Is it a result of inadequate staffing at the US Customs facility here? Will that downward trend continue? There are only two lanes of the QEW delivering trucks to the border, so why do we need four or five lanes to get them across less than one mile of river? The Canadians are talking about building a new highway to peel some of those trucks off the QEW, but that's years from happening and, if the economic decline continues, it won't happen at all.

The mayor's document indicates no awareness of what has become one of the key questions in this whole enterprise: Is that area on Buffalo's West Side the optimum place for a huge amount of cross-border truck traffic? The whole idea of necessary expansion must be given a hard look before we increase lanes at the present site or anywhere else, and that is, presumably, exactly what the environmental impact study now underway will do.

The writers of the mayor's report claim that once the old bridge is repaved it will be fully operational for decades to come. The PBA's chief consulting engineer and head of its EIS, Vincent "Jake" Lamb (former executive vice president of Parsons Transportation, the parent company of the PBA's longtime engineering firm, DeLeuw Cather), says that as far as he knows that's pretty much true. Many technical people don't buy it. There hasn't been an underwater inspection for decades, so nobody knows if or how seriously those piers

> have deteriorated. If the bridge's infrastructure has deteriorated as much as has been suggested, we're looking at two long-term construction projects that cannot go on at the same time – it's the city planning equivalent of opening you up to take out your appendix, waiting until you can move without major pain, then opening you up again to take out your gall bladder. Jake Lamb says that core samples from the piers show no deterioration. He said there's nothing else they can do, short of building a cofferdam (a watertight enclosure built to make inspections below the waterline) and going down there for a look. Well, why not do that? This is a $200-million project, so what's a few bucks for a cofferdam to do a bit of essential reality testing?
>
> It's ugly. You paste something beautiful onto something ugly and you wind up with something ugly, only bigger. Gresham's Law works in visuals. You remember the time you looked really good but had a huge zit on the tip of your nose and it was the zit that caught everybody's attention? It works the same when you put an beautiful bridge right next to an ugly bridge.

WEIRDNESS

I could go on scab-picking at the mayor's plan, but it would only be specifics and you've gotten the general point. The real question to ask about it is this: why did they do it? Why come out with a specific plan now, when the EIS is just starting to take shape and even the PBA has taken its steel twin span design off the table?

One city official said at last week's meeting of the Common Council's Bi-National Bridge Task Force that they had these plans around and thought they should introduce them to the conversation because they might help the thinking of the various people working on the EIS. It's just one more idea in the pot, he said, nothing special. In no way was it meant to interfere with or influence the thinking of the EIS. It was just an idea they had, some drawings they had, something the mayor thought we should think about, and he was offering it, just like any citizen might offer a suggestion.

Sure. Just like the guy with the lightbar on the roof of his patrol car who pulls you over on the highway is just one more guy who wants to chat with you, nobody special, and you can ignore him if you feel like it. Sure you can.

This proposal came out of the mayor's office at a formative

moment in the EIS process. Jake Lamb's group is right now considering bids from firms wanting to take on various major roles in the EIS. They are right now formulating the composition of the various groups that will be involved in that process and figuring out how their information will funnel into the final recommendations. And here's the mayor's office with a lot of hoopla giving the press a very specific design, one that takes as givens factors everyone else now agrees are questions. That's not a mere neutral submission of one equal idea among others. It's one of the key players, a major public officer, delivering a plan with two very large and pretty polychrome architect's drawings before anything else is out there.

I've heard three explanations and justifications for this very heavy-handed maneuver, any or all of which are possible:

> The officials of Fort Erie are furious and frustrated at what's going on over here, they don't like having their plans held up by US environmental law, and Wayne Redekop is rolling his eyes to the ceiling and looking dazed during meetings of the tripartite steering group set up by the PBA (public officials from Fort Erie, Buffalo, and representatives of the PBA). This public plan is Mayor Masiello's way of saying to Redekop and his associates, "We feel your pain." That would be neighborly, but if true it's a patronizing model of international relations. Why not just say to them: we don't want an ugly bridge on our property and you're not the only guys who get to say what links your town and our city? Why not say to them: stop sulking because the PBA has been forced to obey US environmental law? Why not say: grow up? Why not say to them: the Bridge supports your whole town, it's made you rich, why can't you at least let us have something nice to look at? Why not say to them: Go... I'll stop before I get into trouble.

> The mayor's staff people working on this problem convinced him that this design really is the best and he wanted to get it to the public before the EIS confused everybody with too many facts or alternative options. Too bad if that's true, because there really are far better designs out there, most notably the one proposed in "The Past and Future Front Park," an August 2000 Cornell University Landscape Architecture and City and Regional Planning MA thesis by Peter William Hedlund. Hedlund came up with a fabulous design that extends the restored Front Park over the Thruway, so

> the Park has direct access to the lake and anybody can reach Front Park and LaSalle Park on foot without being squashed by a semi.
>
> Mayor Masiello isn't the least bit serious about this, but he's been stung by all the comments in the press and by other politicians about his fence-straddling, waffling, and lack of initiative and leadership on this major public works project. He's got an election coming up next fall. Even though he's already got a million bucks in his war chest and there are no realistic contenders on the horizon, there's no point letting this high visibility issue sit out there unattended.

THE EIS

The most important recent public moment in the Peace Bridge War was last November 15, when PBA chairman Victor Martucci called a press conference to announce that the PBA would end its appeal of Judge Eugene Fahey's order that it conduct a full environmental impact study. Not only would the PBA do the EIS, Martucci said, but it would give up all permits it had gotten from all government agencies on both sides of the border. It was back to square one: the PBA would examine the question anew, it would heed community concerns and needs, it would make the process open and accessible.

If true, that would be a radical change from the way things had run thus far. The process Martucci outlined didn't guarantee a signature bridge, but it at least meant that for the first time the various interests would have to announce themselves publicly, and whatever choice the PBA made might be subject to some measure of public scrutiny.

That process is now underway. The PBA has invited bids from consultants to deal with various major functions. The process is being directed by Jake Lamb, which leaves some critics nervous: Lamb, as I noted above, is a senior executive of Parsons, and it was Parsons that came up with and rationalized the steel twin span. Lamb says he and everyone down at Peace Bridge Plaza are now fully behind the EIS process and the process will be so open that no one will have any worry about further chicanery or foolishness. We can only hope. And watch.

The EIS is to be overseen by a tripartite group composed of officials from the Buffalo, Fort Erie, and the PBA. Fort Erie officials, as I said, remain unenthusiastic about the process: it has the possibility of interfering with the output of their beloved cash cow. Buffalo officials, as the mayor's recent effort demonstrates, can be erratic. The Common Council's representatives to the process seem to have stayed on course thus far. The PBA – well, we know about the PBA.

POWER AND SCRUTINY

One thing to watch out for is whether or not the scoping – the questions the actual EIS attempts to answer – is limited to how to best build an expanded crossing here or if it also asks where the expanded crossing should best be located, or if one is needed at all. We have no regional government, no effective regional planning agency, no university program that objectively tackles this kind of thorny question, but it is the key question in this entire process.

Some critics of the PBA have faulted it for failing to look at alternative sites for a truck crossing. That's unreasonable and perhaps even unfair. Why should the PBA undertake a plan that might conclude it should pretty much go out of business? They have a bridge that, on the whole and from their point of view, works pretty well. It just needs a bit of expansion. It provides jobs for almost an entire town, plus untold numbers of subcontractors and consultants. Why risk throwing all that away? Why give away all that power?

Power is as addictive as heroin and, for some people, as pleasurable as sex. Unlike the former, it's legal, and unlike the latter, you can engage in it all day long when you're out of the house and your spouse won't divorce you if it gets in the papers. In *Godfather II* Michael Corleone famously said, "If anything in life is certain, if history has taught us anything, it is that you can kill anyone." He might equally have said, "If anything in life is certain, if history has taught us anything, it is that nobody gives up power willingly."

The PBA will ask the questions that serve its interests. It will take a good deal of public scrutiny and attention to make sure this EIS also asks the questions that really matter to the rest of us. We're in a better place than we were a year ago, but this surely is no time for complacency.

YOUR GOVERNOR AND MINE

Governor Pataki was in town last week courting the African-American vote for the election a scant 21 months hence. All the polls show he has a solid lead on both of his likely opponents – former US Housing Secretary Andrew Cuomo and New York Comptroller Carl McCall – but he no doubt remembers the time not very long ago when people said Hillary Clinton didn't stand a chance of getting elected senator from New York.

While he was here, the governor said that what happened at the Peace Bridge was a matter for the community to decide.

That, folks, is Albany jive. The stalemate at the Peace Bridge was entirely predicated on Governor Pataki's appointments to the Buffalo and Fort Erie Public Bridge Authority. For most of the time since he's been in office he's controlled all five of the American PBA appointments; for the past two years he's controlled four of them.

Buffalo attorney and former Erie County Democratic chief Joe Crangle, who represented the Common Council on the Public Consensus Review Panel, has long said that any time Governor Pataki wanted the mess at the Peace Bridge over all he had to do was order it, and that if the governor had wanted us to have a signature bridge all he had to do was tell his representatives to build us one. "People say, 'What if five Americans on the board vote one way and the five Canadians vote the other?' I say [Crangle said], 'If the five Americans voted one way then the Canadians would vote with them. What else could they do?'" They could vote a stalemate, but Crangle, ever a practical politician, says they're too rational for that. Fort Erie has far more to lose from stalemate than Buffalo does: "Their whole economy depends on the bridge. For Buffalo, it's just one thing among many."

At several public events over the past two years Governor Pataki has said he favored a signature span, but his appointees to the PBA never voted that way until the November shift in position. When Victor Martucci became PBA chairman a year ago he was a fierce advocate for the twin span. It was only because of Judge Eugene Fahey's order that the PBA conduct a full EIS and the pending lawsuit in Federal court that he changed course and got the other members of the PBA to do likewise.

So where is the governor now? It depends whose support he's courting these days. Support for a politician running for office takes two shapes: votes and money. Martucci recently announced his intention to resign from the PBA. We'll get some idea whether the governor is courting money or votes when he names Martucci's replacement.

THE DETROIT GANG

The people who run Detroit's Ambassador Bridge who want to set up a privately-owned and operated bridge here have set up an operation they're calling the "Ambassador Niagara Signature Bridge Group." It's got very pretty stationery with a white single-pylon cable-stayed bridge on a gold field as their logo. They've got one of the areas top PR firms – Paragon – handling publicity, and they hired James B. Kane, who headed Pat Moynihan's Buffalo office these past several years, as "regional director."

Kane was instrumental a year ago in bringing them here to perform for the Public Consensus Review Panel. At that time, he helped them meet with just about every public official connected with cross-border trade in this area and every journalist writing about it – presumably on Senator Moynihan's behalf. He is very well connected here in both political and business communities. He has declined to give interviews about what he's doing for Ambassador now and what Ambassador is up to in Buffalo, but promises a press conference later this month in which all will be revealed and all questions will be answered.

The Ambassador group has also announced that it has retained Hiscock and Barclay, a Buffalo law firm, to represent it in local affairs, and that the firm's managing director, Mark McNamara will head up the legal team. McNamara has represented Ambassador here for the past year, so I don't know what's newsy about that, unless it's to say, "We don't just have a local *lawyer*, we've got a local *law firm* working for us. We're serious about this."

Have no doubt: they are serious. They're pouring a fortune into their Buffalo operation – neither MacNamara nor Kane nor Paragon comes cheap, and neither does the office Ambassador has set up in Olympic Towers. And this is only the start: they'll soon begin their own environmental impact study, totally separate from the one done by the PBA.

The PBA's study will presumably study all crossing options at or near the present site, but it's hard to believe they'll take a serious look at anything that doesn't land just about where the present bridge lands. Ambassador seems to be assuming that the best way to handle the congestion is to split truck and auto traffic, to leave auto traffic at the current site and move truck traffic to a new site adjacent to the International Bridge (the railroad bridge) a mile or so down river.

If that were to happen it would accomplish two things: it would leave the PBA with the low rent part of border traffic and a bridge getting ever more expensive to maintain, and it would destroy the Park Nobody Knows on Squaw Island, part of which would provide several much-needed ball fields, and a huge segment of which has some of the city's most beautiful park land.

The Ambassador group got a huge windfall when gambling went into Windsor, Ontario – their revenues from their Detroit bridge operation surged immediately as auto traffic expanded. But they're businessmen, they're not gamblers. If they're poised to pour a huge amount of money into a war with the Public Bridge Authority, it's because they expect to be able to take a much huger amount of money out of here.

ONE DOWN

The Ambassador Niagara Signature Bridge Group has claimed its first Buffalo casualty: as soon as Ambassador announced that it was being represented by Hiscock and Barlcay, Robert Kresse, who has an "of counsel" relationship with that firm, resigned from the Common Council's Binational Peace Bridge Task Force. He didn't want to have or give the appearance of a conflict of interest.

I don't understand the legal niceties here – Mark McNamara has been sitting in on task force meetings as Ambassador's Buffalo attorney for some time now and nobody was bothered by the fact that Bob Kresse is of counsel at that firm and a member of the task force. I don't know anyone sane who would question Bob Kresse's rectitude in this or any other matter. Maybe it's a sign of that rectitude that he felt the expanded role the firm would be playing in Ambassador's Buffalo affairs required him to withdraw from active involvement in Peace Bridge matters.

Kresse is off the panel but Jim Kane, who previously sat on the task force as a representative of Senator Moynihan, is now sitting on the task force as a representative of the Ambassador Bridge people in Detroit. When he worked for Moynihan, Kane was representing the community – us. Now he's representing reclusive Grosse Point, Michigan, businessman Manuel Maroun. That's not a conflict of interest, that's an *interest* plain and simple. No one else on the task force represents a single businessman; every other member of the task force represents a community group or public agency.

I have no problem with Jim Kane being on the panel: he's up front about whom he represents and why he's there and it's surely better having a representative of the Detroit gang in a place where we can find him than in a place where we've always got to be ferreting him out. What's unfortunate is the other part – Bob Kresse's fine sense of legal ethics that forces him to withdraw from this important public conversation.

His resignation from the task force is huge loss to the community. He is knowledgeable, smart, persistent, and thoughtful. The Public Consensus Review Panel wouldn't have done the work it did were it not for him and the Common Council's Task Force wouldn't be as useful as it now is were it not for him. No doubt Ambassador will be a good client for Hiscock and Barclay and they'll run up a huge number of billable hours before they're done with their phase of the Peace Bridge War. I'm sure they didn't hire Hiscock and Barclay to neutralize one of the most important voices of reason and common sense in this long and absurd affair, but that's the effect of their choice. I sure wish they'd hired some other law firm.

50 Victor Martucci: "You have to be realistic about these things"

Victor Martucci was appointed to the board of the Buffalo and Fort Erie Public Bridge Authority by Governor George Pataki in August 1999. His fellow board members elected him chairman six months later. In late January 2001, he announced that he would resign from the board the following month, at the end of his one-year term as chair. By law, the chair of the PBA is held by an American one year, a Canadian the next, and then an American again. Their recent practice has been for two people to swap the chair and vice-chair back and forth across the border for several years. On February 23, John Lopinski, of Port Colborne, Ontario, who had been PBA chairman three times previously and vice-chair in the intervening years, became chairman of the PBA once more, and Paul J. Koessler, the Buffalo businessman who joined the PBA in November, became vice-chair .

Not longer after Martucci became chair of the PBA, New York Supreme Court Judge Eugene Fahey decided against the PBA in its attempt to segregate the plaza and bridge construction projects, which meant that the PBA would have to do what it had tried assiduously to avoid: look at all the implications of various bridge expansion designs at this location. The City of Buffalo refused to give the PBA the easements it needed to begin construction, which meant nothing was going to happen, whether the PBA did the partial or the full EIS. For a time, Martucci was a fierce defender of the PBA's positions. Then, last November 15, he announced that the PBA would give up its plans to fight Fahey's order in the courts and would instead mount a full-scale environmental impact study. To ensure that study began at the beginning, he said, the PBA would renounce all construction permits it had received from American and Canadian authorities and the process would be fully open, overseen by a joint committee composed of representatives from the PBA, the Town of Fort Erie, and the City of Buffalo. That announcement, and the actions that resulted from it, mooted out all the opposition.

Everything Martucci has done since that announcement has been consonant with what he said that day. The PBA has mounted a complex EIS operation directed by Parsons Transportation executive Vincent "Jake" Lamb. (Parsons is the parent company of DeLeuw/Cather, the PBA's longtime maintenance and engineering firm.) Lamb promises the process will have broad public involvement. The tripartite committee has been meeting regularly. Most recently and perhaps most important for giving the public a sense that the project has integrity, they hired Christian Menn, one of the most respected bridge designers/engineers in the world, to join the process as design consultant. Menn was the engineer who brought agreement to the fractious Charles River Bridge project in Boston, so his appointment at this critical phase of the Peace Bridge War suggests that the PBA is, thus far, keeping its word about an open and fair process.

(Menn's appointment received wide support in the community, most of it immediate, some of it retroactive, e.g., former *Buffalo News* editor Murray B. Light's opinion column in the *Buffalo News* of February 25, nine days after Menn had been named the PBA's premier official design consultant:"The authority should take the next needed step," advised Light, "and name Menn its premier official design consultant.")

Martucci has been willing to talk at length with anyone about his sense of the PBA's mission, position and conduct. Perhaps his relative youth is one reason for his accessibility and flexibility: he's 40, about the same age as many members of the New Millennium Group, one of the PBA's consistent opponents in the Peace Bridge War.

The conversation that follows took place Friday, February 16, at the offices of Collins & Co., the PBA's public relations agency, in downtown Buffalo. We began our conversation with that morning's announcement of Christian Menn as design consultant to the PBA. (Martucci's statements are in Roman type, mine are in italics; the material in brackets are my later glosses.)

One of the things that we had discussed when we began this process and put Jake Lamb in place as the manager of it. If we're going to go back and if we're going to do this process all over again, we're going to do it right. And why shouldn't we think big? Why shouldn't we go out and recruit the best people that we can find to help us through this process? Christian Menn, as you know, is the dean of bridge architects worldwide. His reputation is impeccable. As I understand it, he picks and chooses the projects he works on. So it was a coup for us to convince him to consult on this project. I think it's very exciting that he's going to be a part of the project team.

He's interesting not just for his great design skill or reputation, but for the major project he's currently involved in, which has some similarities to our situation.

The one in Boston. I drive by it every week. We're building condominiums in Boston and so I was aware of the project. From my standpoint, what was important about bringing somebody like him in, in looking at the Boston experience, is that he was able to bridge (no pun intended) the differences that existed in that community. It was a highly-charged emotional issue, just as the Peace Bridge issue is here. He was able to combine function and practicality with aesthetics, and that's what this calls for. And he was able to do that in a way where he made everybody feel like they came away from the process a winner. That's what we're hoping he's going to do here.

I saw in the Buffalo News this morning that he was going to be design consultant but you were going to hire a design firm. Could you explain to me how that's going to work?

Christian Menn is going to work for the Peace Bridge Authority as the Peace Bridge Authority's consultant. We also are going through a procurement process for the EIS process and there will be a bridge engineer/architect that will be hired to handle the functions of the bridge engineer/architect through that process. When options are put on the table to be studied, the bridge architect/engineering firm that is hired through the process is going to be the firm that does the analysis of those different options from an engineering and architectural standpoint.

What we want from Christian Menn is, take a look at the crossing, understand the concerns, the challenges, the needs, the requirements of all these diverse groups, and help us develop a concept that best addresses as many of those concerns as possible – so that it gets us as close to consensus as possible and so that at the end of the process people have a result that they can be proud of, that they can be excited about, and that they can walk away feeling confident that the process was open and fair and we got the best possible system that that process could produce because we've got the preeminent bridge engineer in the world participating with us.

Someone whose current project is both steel and concrete, so to again use the pun, it bridges the two opposing –

Exactly.

And it's true, we're starting from a clean slate now and there are no biases or prejudices toward any one design or any one methodology in terms of construction. This project should yield the best possible system, given our circumstances and the needs and the desires of the communities on both sides of the border.

What do these two relationships, the one with him and the other with the design firm that will be hired, do your relationship with Parsons, or will it be *Parsons?*

Jake is from Parsons Transportation and Jake needs the resources and the technical support that Parsons can provide him to do his job as the overall manager. But Parsons Transportation is not going to do the nuts and bolts design work on the bridge. That's going to be done by the architect/engineer that's hired through the procurement process. Christian Menn is going to be on the conceptual level. Whatever system seems to develop through the process, then it will be left to the engineer/architect that was hired through the procurement process to develop that and design it, make it happen. I guess that's the distinction. Christian Menn is more on the conceptual side and this other architect – although this other architect/engineer is going to have an opportunity to offer concepts as well – they're really going to be there for the nuts and bolts of it.

When I came onto the board I was coming in with an open mind. I became convinced that there certainly was a capacity problem, and that the Peace Bridge Authority went through an extensive and involved environmental review. It didn't make sense to me to throw all that away.

But the defining moments in this whole process were the judge's decision and the reality in the wake of the judge's decisions that the City of Buffalo was not going to provide the easements unless we completed an environmental impact statement. At that point it became a new ballgame.

You have to be realistic about those things, and if we're going to go through process again, we owe it to ourselves, we owe it to the communities on both sides of the border, we owe it to future generations to do it right, not to think small, to think big, and to try to come up with the best possible project that we can, given whatever straits we have to operate with, both financially and given the nature of the border crossing, and the fact that you've got to have a binational agreement on whatever you're going to build. We should do it right, and we're going to do it right.

When you came on, I remember you being very strong for the twin span. Did you change your mind or was it just that the judge's order changed the political reality?

I changed before the judge made his ruling, but the board wasn't there. In my own mind, I could almost envision us ending up in this place because the City had pulled the easements off the table. Without those easements we can't build anything. So the judge's ruling was the most public defining moment in the process, but even if the judge had ruled in our favor, if the City of Buffalo said "We're not going to provide the easements," there was nothing we could do to compel them to provide the easements.

What I quickly learned in my time on the board was that this was not just a very simple transportation project, it was a political issue. There are all kinds of competing interests involved and it was going to require a political solution. The political solution is to go back and do this environmental review and do it right so that the public has confidence that no stone was left unturned and that what ends up getting built is the best possible alternative, given all the competing interests that are involved. Whether it was real or not, the public's perception was that that didn't happen in the previous process. We can argue that till the cows come home, but why bother? That's history. We are where we are today and we've got to do this thing right.

Did you have any difficulty with the Canadians over this shift?

I think that the problem wasn't just with the Canadian members, it was other American members on the board, too. They felt very strongly that they had done the process right and they had commitments from elected officials and from the City on the easements, and that there was plenty of opportunity throughout the environmental review which had taken place over five or six years for people to come forward and make the arguments that they were making, in their view, at the eleventh hour – actually the twelfth hour They didn't see why they should have to go back and do this all over again.

And there were other points of view on the board that view this as an economic development project and they were very concerned that if we had to go back and start from scratch that this process could take five, six, seven years to complete and there would be economic harm done to both sides of the border. Their view of it was not so much that they didn't like the fact that these objections were being raised at the eleventh hour, although that was part of it, they felt like they did the process right, they came up with a solution that the Peace Bridge Authority could afford to build, and that their mission was to provide a safe and efficient border crossing. So in that view, in that context, they did what they were supposed to do.

Sometimes it takes a little while for reality to set in, and for some people to accept the fact that there had to be a political solution, that it more than just a business or a transportation issue, it was a political issue.

The ground shifted over the years, didn't it?

Yes. Think about it: a lot of these folks are business people and they're not used to making decisions and then having to go back and change those decisions because of a changing political environment. They're used to making a decision and moving forward and implementing that decision. There's positives and negatives in that kind of an outlook. I guess the advantage I have is that I've been on both ends of the spectrum. I've been in government and now I'm in the private sector and so I understood the politics of it, I understand the business end of it. So maybe I was able to see things a little differently.

Say more about that.

I graduated from Buffalo State in '82 with a degree in political science. I interned in Jack Kemp's office and worked on Eddie Rutkowski's second campaign in 1983. Then I was the assistant to the director of field operations for Regan-Bush New York in '84. I traveled around New York and set up a Regan-Bush organization in all 63 counties of New York. I went to work for Eddie Rutkowski after that, till he lost, and then I went on to the state senate [working for state senators Walter Floss and John Sheffer] from that. I was a political junkie, I thought I'd be in government forever.

Then I got smart. Something hit me in head: I decided I'd rather be in a business where my future depended on what I did as opposed to somebody else having to get reelected. I got into the advertising business with Bob Davis, who's the Republican chairman now, and then ended up with Marrano-Marc Equity, my first client when I worked for Bob Davis's ad firm.

Our company is an exciting company to be with. When I started at Marrano-Marc Equity in 1995 we were a $24 million company. Last year we grossed $104 million. And we're growing. We're doing condominium conversions in Boston, Massachusetts, we're doing stuff down in Palm Beach Gardens, Florida, we're doing business here.

That really is the primary reason why I had to step down. When I became chairman of the Authority, I had no idea how much time it was going to require. It really does require a lot of time if you're going to do it right. At first, I was able to juggle all the competing interests, but as our company has grown, we've reorganized and struggled with how we're going to deal with our growth. My workload has doubled in just the last two or three months. In addition to traveling to Boston once a week, I've been traveling down to Florida on a regular basis. I've got three young kids – my kids are nine, five and two. Whatever free time I have, I'm going to spend it with my kids. I don't like to do things half-assed, so if I can't spend the time that the Peace Bridge Authority needs, then I'm not doing anybody any favors by being involved. I just can't put the time into it that I have over the last year, so it's time to let somebody else who does have the time and the passion to step in.

It also has a lot to do with timing. If you had told me five months ago that the process would be where it is today, I would have told you that you're nuts. I feel really good about where it is today, with the formation of this partnering group, and with Jake Lamb being in control and running and managing the process, and now having Christian Menn come in as a consultant. I feel really good about walking away now feeling like this is set up to succeed. And I think it will. I'm optimistic about it.

Are there any other reasons why you're leaving?

Absolutely not. In fact, it was a hard decision for me to make. I really enjoyed it (believe it or not) and I felt like I was accomplishing something. I really wanted to see it through to the end, because I've put a lot of time into it and a lot of effort into it, and there was some personal sacrifice. A lot of people called me names, and were thinking that I was the devil incarnate, and sometimes those are difficult things to deal with. I wanted to see it through to the end, but I just couldn't do it. I just don't have the time to do it. And if this thing is going to work it has to be driven by the board of directors, and it has to be driven by the chair and the vice-chair and it requires the time that I just don't have any more.

Could you tell me how you got on the board and how you got to be chairman so fast?

I got on the board because of my relationship with Bob Davis and the governor's office. There was a vacancy created and they called me and asked me if I'd be interested in serving. I said "Yes." I'd told the governor's office that I was interested in serving on a board if something became available. I've always believed that you should do some community service. When the community's good to you and you make a good living, you should put something back. I had made it known that I was interested in doing something. This came along and so I jumped at it and said, "Sure, I'd be happy to."

I kind of got drafted to be chairman, I wasn't actively seeking it at the time. When Brian Lipke made the decision to leave the board, they came to me – Luiz Luiz Kahl and Lou Billitier and the others that were serving with me on the board – and asked me if I would do it. You'll have to ask them why or what they saw in me, but they asked me to do it. I thought it was a neat opportunity to try and make a difference because things were floundering at the time, so I jumped at it.

Do you have any sense of where things went wrong in all this?

It's hard for me say because I honestly didn't know a heck of a lot about it other than what I'd read in the newspaper until I got on the board and really dug into it. I went back and I read a lot of the documents that were put together through the environmental review process over the years. I think a lot of things happened.

I think that the board did what the board really believed was the right thing to do. That is, they felt their mission was to provide safe and efficient transportation across the border, they felt very strongly that there was a capacity problem, they knew that they were limited in what they could afford to do. And they felt like they went through a process that led to a solution that they could afford to build.

When I went out and met with the New Millennium folks, with Jeff [Belt] and Bill Banas, and Mark Mitskovski, I said, "You know, try to put yourself in their shoes for a minute. They really didn't have any sinister motives. Put yourself in their shoes. Could you imagine that they went through this process and they came out and they said, 'Okay, here's what we're going to do. We've got a capacity problem. And we're going to build a new six-lane cable-stayed bridge. And in order to do that we're going to tear down the historic Peace Bridge. And we don't think that the American plaza works. We want to move that plaza to the north, and in order to do that, we're going to take about 300 homes and 100 businesses. And, oh, by the way, we're going to need about $150,000,000 of taxpayers' money to make it work.' What do you think the reaction of the public would have been to that?"

They did what they thought was in their charge to do. There's plenty of blame to go around. A lot of these folks that got very active and very vocal at the end were not very active and very vocal early on. For whatever reason. But they weren't. Had they been active and vocal during the initial process, then maybe we would have ended up with a different result. So I don't think it's fair to put it all on the Peace Bridge Authority. I think there were plenty of mistakes to go around.

But that's, as I said, history, and I don't like to dwell on that. I like to look at the future and I feel very confident and very optimistic that we've got a process in place now that the board is committed to that's going to work, that's going to lead to a result that the community can be proud of and be confident in. It will be a legitimate solution because it will have been done in the right way. At the end of the day, I think if people can walk away from the process feeling that way, whether they got what they wanted or not, I think that's what's important.

Any difference between working with the PBA and working in private industry?

In our business, we don't have a board of directors. We're a privately held company, so it's a much more autocratic type of organization. Here, not only do you have a board of directors, but you have two officers on equal footing *[the Authority has two General Managers – Earl Rowe, a Canadian, and Stephen F. Mayer, an American]*, and you have a binational board of directors that doesn't always have the same interest. That becomes a difficult thing to manage.

But if everybody respects each other and respects their opinions and you're not afraid to have open and honest debate and your officers and your staff gives you good objective advice, then you make the best decisions you can make. As long as you don't allow it to become personal and you don't take this view that 'my view is the right view and the hell with the rest of you.' [If you did that] you'd never get anything done over there. Most successful companies don't have two CEOs. But by the virtue of the fact that it's a binational agency, you have to make that work.

I had a real short learning curve. I had a lot to digest and a lot to understand, and you know, you have to learn things for yourself. Because it's human nature, people are going to tell you something, and what they tell is going to be – I don't want to use the word 'tainted' because that's a negative word – it's going to have the wrong personal bias interjected into it. So at some point you have to be able to sift through the bias and look at what's objective and what's fact and then come to your own conclusions. That took time. It took time for me to understand what was fact on the engineering end of it and the economic end of it and whether we're doing as good a job as we can do managing traffic with what we have. Then there's the politics of it. As I said before, it was a political issue and it required a political solution, so I had to understand who the players were and what their interests were and all of that took time. But I think once I was able to wrap my arms around it and understand what was going on and came to the conclusion myself that I thought the best path for us to go down was to do this environmental review and do it right, then it just was a matter of time before I was able to talk to people on both sides of the border and convince them that that was the direction we should go in.

Can you talk a little bit about the Detroit bridge group, what they're up to, how they relate to you.

I don't know what's going to happen with them. They're going to have the same opportunity as anybody else to come into the process and put an option or a solution on the table to be studied. I think the process will determine whether or not their proposal has any merit. Beyond that, I'd feel uncomfortable commenting on it because if this is really going to be a true, fair and open process, we can't go into it with a bias one way or the other. I don't want to say I like what they're saying, I don't want to say I don't like what they're saying. It's not for me to take that position today. But I can promise them that they'll get a fair hearing, like everybody else will.

What about the involvement of various politicians in this – Moynihan, Schumer – is that problematic for you? Do you feel they've dealt with you fairly reasonably or –

I never had a conversation with Senator Moynihan and I never had a conversation with anybody on his staff. That's the truth.

Jim Kane? [Before becoming the Buffalo representative of the Detroit Bridge Company, Kane headed Moynihan's Buffalo office.]

Jim Kane never called me, never asked to talk to me. This is what mystified me because here I am, a guy coming in with an open mind and a clean slate. Jeff Belt, too. Joel Giambra introduced me to Jeff. I said, "I want to hear your side of the story. I have no preconceived notion, I have no predisposition to any one side or the other. Tell me why you're right." And I never heard from him again. Chuck Schumer called me. I had two or three conversations with him personally, but nothing from Moynihan's people.

So you wonder if they're really out there trying to help or if there's some other agenda. Because I don't think I ever gave anybody the impression that I wasn't open to listening or to talking to anybody. In fact, when I became chairman, I reached out to a lot of these folks. I met with Sam Hoyt, I met with the New Millennium Group, I met with the Episcopal Church Home people, I had conversation with the Olmsted people. I went and reached out to those folks, but they were reaching out to me too, so it was mutual. But I never had any contact with Senator Moynihan's office.

Did you have any relations with Alan Gandell? [Gandell recently resigned his position as head of the Niagara Falls Bridge Commission.]

Never met him, never met him.

Did you ever talk to anybody up there?

We had a meeting with those folks in December, I think. But that was after Alan Gandell left. We were talking about things that we could do even though we're acknowledged competitors, things that we could do together to help manage the border crossings on a regional basis, if we could share technology and make sure that we're communicating with each other. Like if we're going to be closing down a lane to do some type of repair work that we notify them so they're preparing from a staffing perspective to handle the increase in traffic. Things that we should be doing. This is a region and we have a common interest even though we're competitors, so we sat down and talked about those things. But prior to that, I had never any contact with those folks.

Some people, as you know, have been suggesting that a possible solution to some of the border problems would be to shift the truck traffic out of Buffalo. How does that seem from your point of view?

Well, again, without saying anything that biases the process, I think from a logistics standpoint and from an operational standpoint, and an environmental standpoint, it makes all the sense in the world. It just doesn't make sense to have the functions split the way they're split. I think that's part of the problem, why you have the congestion that you have. If we were able to move those functions over to the Canadian side where you have a much bigger plaza, that makes sense to me.

Have there been any steps towards setting up an international zone?

That was where we were heading before the Federal lawsuit was filed. We were in the process of trying to settle the Fahey decision, and what we were looking at doing was moving the operations over to Canada, which would have satisfied the Olmsted Parks Conservancy's interest of reclaiming parkland. Then we would have been moving our operations off the plaza. One of the scenarios we were looking at was taking over the Episcopal Church Home for our administrative building, and that would satisfy what the Episcopal Church Home wanted to do. But when the Federal lawsuit came down, that put all that to a halt.

So right now, there's nothing in the works in terms of an international zone, because I don't know if the process is going to determine if that's the best solution. We're starting with a clean slate and we've got to get through the scoping process. Once you get through that scoping process and you get to the draft EIS stage, I think you're going to see some options begin to crystalize. And then that's when you need to start doing the legwork to see what's practical and what's possible and what's not.

One of the things I've had a hard time understanding is the relationship of the PBA to the government of Fort Erie. It seemed that the relationship on that side is very different from the relationship of the PBA to Buffalo. And it seemed that a lot of the problems resulted from two very different conversations going on on the two sides of the river.

That may be.

I think what's important now is that for the first time you've got a mayor and a county executive that are working together and they're working together with the governor, so now you've got the three key players on the American side and they all share the same agenda. And for the first time we're getting clear direction at the Authority as to what our elected officials want us to do. We didn't have that before.

Sam Hoyt said to me, "Maybe this is just my perception and it's not reality, but I gotta say it, the perception is that the Canadians run things over there."

I said, "Sam, you know what? To a certain degree that's true but let me tell you why." I said, "Look, the Canadians have a master plan. They have a regional economic development plan, they know how the Peace Bridge fits into that plan. Their board members get very clear direction as to what their objectives are and what their goals are and how the Peace Bridge fits into that. We weren't getting that. We're getting it now, but we weren't getting that. So eventually, decisions have to be made, and because the Canadians have their act together and we don't, the decision is made and it looks like it's built towards the Canadians." Sam thought about it and said, "That's a valid point. I think you're absolutely right."

Now we've got the mayor and the county executive and the governor all working together with a common agenda and for the first time they're giving clear direction as to how the Peace Bridge fits in to what the mayor and the county executive envision, and what the governor envisions for regional economic development plan. And clearly this process is in the best interest of what Tony Masiello and Joel Giambra and George Pataki want to accomplish. All of that will factor into the process. If we'd had that two or three years ago, maybe that perception wouldn't have been there.

What is it that the governor is saying to you now that was not being said before?

It's not that the governor wasn't saying anything. The governor really was taking a hands-off approach because there was no clear consensus from the community. The governor's philosophy has always been a bottom-up type philosophy of government that home rule should rule and the locals should decide what it is they want, that it's not the state's place to impose on them. The problem was that the locals couldn't agree. So now that you've got the mayor and the county executive and the governor all pulling in the same direction – and Jim Pitts, too. Jim Pitts is being very cooperative in this whole process with the mayor. So even though you don't have a monolith in this city – you've got to deal with the administration and the council – it's been much easier to deal with in the last few months because Jim Pitts and Tony Masiello put whatever political differences they have aside and they're working together on this thing. I've seen that in our partnering discussions. So that's going to help.

Do you have a time sense for what happens next?

I think if everybody works together, like they are now, and I think this partnering group could really ensure that that happens –

By 'partnering group' you mean –

Fort Erie, the City of Buffalo, and the Peace Bridge Authority. The *Public* Bridge Authority.

I was waiting for the end to remind you of that.

See, I read your stuff.

I think it could be done in 2 ½ years. I really don't see how it's going to be done quicker than that. A lot of the stuff that was done in the Public Consensus Review Panel process is going to have to be updated. The engineering analysis that was done through that process was not very detailed and not to the level that the EIS process requires. The only real hard look was our plan because we had bid it out. I mean, that was the only one of the options that was studied through that process, that you could say with any certainty, "This is what it was going to cost to build and this is what it was going to take to build." The others were very superficial looks because of time and because of money. That was one of the fallacies that came out of the process, that they could say with any degree of certainty that this is cheaper and this can be built faster. That really wasn't the case. But who knows, that may end up being the case after you go through this process, and if it is, why wouldn't we build it? I've always said that. If it can be done faster, cheaper and better, that's what we ought to do. We shouldn't be afraid of that.

My hope is that everybody who has an interest in this process is going to respect everyone else's opinions, and not to assume that there's any hidden agenda or any sinister motives. I want people to feel confident in knowing that the Peace Bridge Authority is going to do this thing right. We've done some things over the last few weeks that I think should lead people to that conclusion. I understand: Don't only tell me in words, show me in deeds. Hiring Jake to manage the process – Jake Lamb is somebody that is respected by everybody on both sides of the border. Bringing in a guy like Christian Menn, who brings world-class credentials to this project. That should show the community that we're serious about doing this thing and doing it right.

We're going to strive to come up with the best possible bridge system that this process can produce and that people can feel confident and proud of at the end of the process. Everybody's going to have their day. Everybody's going to have the opportunity to say their piece. My hope is that it's done in a civil way and in a respectful way and everybody respects everybody else's opinions and when decisions are made, that people rally behind the decisions that are made.

Because as a community, I think that we need to mature that way. I think once we go through the process, we need to understand that the process is over and it's time to build a bridge. *If* the process says we should build a bridge. It may say that we don't even need to build a bridge. Who knows?

One suggestion has been to change the composition of the board so that there's some local Buffalo representation on it.

I think that was largely the result of the mistrust and the dissatisfaction that built up over a period of time because of how the whole process degenerated. But I think we've moved beyond that now. I really think if you talk to the mayor and to Jim Pitts and to others, there's a better comfort level and a higher degree of trust now between the Peace Bridge Authority and elected officials on this side of the border. I think if there is. I think it eliminates the necessity for doing something like that. I just think it's a two-way street.

I said this to Sam Hoyt. I picked up the phone and called Sam. I went to one of his fund raisers. He was shocked when I showed up at his fund raiser. I said, "Sam, I read about what you have to say about me in the paper and letters that are going back and forth. Why don't you ever call me? You and I are contemporaries. We're the same generation. Maybe we have a difference of opinion on this thing, maybe we don't. How do we know? We've never bothered to sit down and talk." So I took him out. We went over to Pilot Field one day and had a hot dog and watched a couple of innings of a baseball game and found out we agreed on more than we disagreed on.

I think that's what has to happen through this process. You can't get into this bunker mentality like it's us against them. It happens on both sides: it wasn't just the board on the Peace Bridge, it was on the other side too. People have to reach across the table and sit down and listen to each other and find areas where you can agree instead of just constantly sniping at each other. And that's what I tried to do in my year as chairman on the board.

That mistrust was a key factor.

Sure, sure it was. And that kind of thing really snowballs out of control because then it becomes personal and people get their guard up and their egos up and eventually they're not talking to each other, they're talking at each other, and they're doing it through the press. What gets accomplished when you do that?

And they're also looking for the hidden meaning in everything.

I understand that to a certain degree. You're never going to eliminate that. It's human nature. I know that. And I guess maybe that's why it didn't get under my skin as much maybe it did others. But, after I had lunch with Jeff Belt and Bill Banas, we all said, "Why didn't we do this before? We agree on more than we disagree on." I don't know why things get out of control the way they do.

51 THE SENATOR'S LETTER & THE GOVERNOR'S CHOICE

Senator Charles Schumer, in a February 20 letter to New York Governor George Pataki, said he was concerned about the impact Victor Martucci's resignation would have on the bridge development process. He asked the governor to name as Martucci's replacement someone who had served on last year's Public Consensus Review Panel,

> someone who will continue to make public participation a cornerstone of the Peace Bridge project....
>
> After years of gridlock and delay, the Authority is finally moving forward with a full environmental impact statement for the Peace Bridge and its accompanying plaza, and is, at long last, making a concerted effort to include the public in its efforts to develop a plan.
>
> Now that the process is beginning anew, it is crucial that you bridge the gap that will be left by Mr. Martucci's departure by choosing a replacement who will keep the project moving steadily towards completion. In keeping with the spirit of cooperation and inclusion that developed during Mr. Martucci's tenure, I urge you to consider choosing a member of the PCRP to replace him when he leaves his post at the end of this month.
>
> By including representatives from the City of Buffalo, the County of Erie, the private sector, and other interested members of the community, the PCRP played an important role in moving this process out of gridlock and forward in a direction of real progress. Inviting a member of the PCRP to join the Authority would improve the region's ability to capitalize on the commerce and tourism opportunities of this binational gateway, while earning the Authority the much-needed support of the public it serves.
>
> By filling the vacancy with someone dedicated to an open process and with the expertise needed to help complete this project expeditiously, you will send a signal to the

> people of Western New York that their voices will be heard as we take another step towards building the bridge that will best serve the people of Western New York and Southern Ontario in this new century.

I assume that Senator Schumer has in mind someone who possesses a great wealth of information about all technical, social, political and historical aspects of the project and who has no direct financial interest in the outcome and no alliances with any organization or groups with interests inimical to the community's. That could be someone like the New Millennium Group's Jeff Belt.

I also assume the senator wouldn't want the governor to name a representative of an organization that relentlessly opposed both an environmental impact study and serious examination of the feasibility of a signature bridge, and has yet to endorse the collaborative process now under way – an organization like the Buffalo Niagara Partnership, say.

At the PCRP's final meeting on March 30, 2000, the Partnership's representative, Natalie Harder, moved that the PCRP recommend immediate construction of a steel twin span and that it come out *against* a full environmental impact study. Her motion failed for lack of a second. The PCRP then voted on Edward Cosgrove's motion that advocated a signature bridge, a full environmental impact study, and establishment of an oversight commission to ensure that the work was done correctly. Harder, as the Buffalo Niagara Partnership representative, cast the single negative vote. The Partnership's director, Andrew Rudnick, has argued against a cable-stayed signature bridge for many reasons, one of them being the danger such bridges pose to "seagulls, ducks, geese, what have you."

The Buffalo and Fort Erie Public Bridge Authority has moved two-thirds of the way toward Cosgrove's motion (they're doing the full EIS and they've set up a "partnering group" to oversee the process) and they might very well come to adopt the signature bridge as well. Victor Martucci says the table has been cleared of all prior designs and the PBA will give fair and equal consideration to all options. It's difficult to imagine that a steel twin span will beat out a modern signature bridge in a truly open process. Andrew Rudnick or another one of his spokespersons would be an inappropriate and retrogressive appointment to the PBA. Jeff Belt or someone like him (is there anyone like Jeff Belt?) would be a spectacular step into the future.

Martucci says in his *Artvoice* interview that, "The governor's philosophy has always been a bottom-up type philosophy of government that home rule should rule and the locals should decide what it is they want, that it's not the state's place to impose on them." Governor Pataki's response to Senator Schumer, his choice of Martucci's successor on the Buffalo and Fort Erie Public Bridge Authority, will give us a clear indication of whether or not that is true.

52 PARTNERS, PAPERS, WORRIES & TRUSTS

Some news, a few opinions, & more questions than we'd like.

THE PARTNERING GROUP

The Partnering Group held its first public meeting at Buffalo and Fort Erie Public Bridge Authority's Peace Bridge Plaza office in the early evening of Wednesday, February 28.

The Partnering Group is composed of the mayors of Buffalo and Fort Erie, the president of the Buffalo Common Council, the chair of the Fort Erie Town Council, and the chair and vice-chair of the Public Bridge Authority (PBA). The Public Bridge Authority set up the Partnering Group to oversee the environmental impact study New York Supreme Court Judge Eugene Fahey said was necessary before any construction could begin on a new bridge or plaza. The Partnering Group is one element in the PBA's attempt to get the public on this side of the river to start trusting it once again.

Only half of the Partnering Group made it to the February 28 meeting: PBA chairman John Lopinski, Buffalo mayor Anthony Masiello, and PBA vice chairman Paul Koessler. Buffalo Common Council president James Pitts was ill and neither of the Canadian politicians showed up or sent a representative.

It wasn't so much a meeting as a press conference, since the three Group members there didn't talk to each other but rather to reporters from two newspapers, two tv stations, and one radio station, who comprised most of the audience, if you exclude assistants to Mayor Masiello (3), public relations representatives of the PBA (2), other PBA board members (2), PBA staffers (2), county legislator Al DeBenedetti, and what I supposed were a handful of concerned citizens (4).

John Lopinski, who only a few days earlier had replaced Victor Martucci as PBA chair, thereby occupying that position for the fourth time, opened the meeting. Mayor Masiello made welcoming remarks and expressed his pleasure at the appointment of Swiss bridge-builder Christian Menn as the PBA's chief design consultant. Lopinski and Masiello both said the absentees were probably stuck in traffic somewhere. No one in the room mentioned that there was only one place between Fort Erie and Buffalo where anyone ever got stuck in traffic, perhaps because you could look out the windows of the adjoining room and see that bridge traffic was moving briskly that evening.

After the chairman and the mayor finished their introductory remarks, Vincent "Jake" Lamb, the former Parsons Transportation vice president who is directing the environmental impact study of the bridge expansion project, announced the names of the consultant firms hired to handle key aspects of the project, after which he introduced Christian Menn, who spoke briefly.

A MIXED BAG OF CONSULTANTS

Lamb said that the environmental issues consultant would be Buffalo's Ecology & Environment and the traffic consultant would be Wilbur Smith Associates of Latham, NY and New Haven, CT. He said there would be two bridge designer/architect consultant teams, one a joint venture headed by Modjeski & Masters (Poughkeepsie, NY) and Buckland & Taylor (Vancouver), the other by the Figg Engineering Group (Tallahassee, FL). The two designer/architect teams had exactly the same role – to "develop and assess bridge concepts (locations & types), bridge aesthetics, bridge cost estimates, input to and participation in Public Involvement activities." When asked if the two teams were redundant, Lamb said he expected the actual work the groups did would vary, according to the particular areas of expertise each had to offer.

Eugene Figg is well-known to anyone who has followed the Peace Bridge War. He has built several award-winning cable-stayed and arched concrete bridges and he is presently working on the Charles River Bridge in Boston, which was designed by Christian Menn. Figg has visited Buffalo many times over the past three years expressing his interest in the Peace Bridge expansion job, and he has courted just about every politician, political group, community group, corporate interest and newspaper that might influence the decision.

Buckland & Taylor is well-known to anyone who followed the work of the Public Consensus Review Panel last year. They were hired by the PBA as part of the PBA's last-ditch attempt to keep New York Supreme Court Judge Eugene Fahey from issuing an order forcing them to do an environmental impact study. The Buckland & Taylor engineers imposed delay after delay on the process and finally "decided" that a steel twin span was the only viable option.

The four consultants listed a wide range of subconsultants as parts of their teams. Ecology & Environment, for example, included a recognized expert on Olmsted Parks, Pressley Associates of Cambridge, MA, and Terry Yonker of the Buffalo Ornithological Society. The joint venture listed seven subsidiary organizations and Figg listed eight.

I think it is inconceivable that Figg would have been hired a year ago by the PBA for anything but window dressing, and the combination of him with Christian Menn is promising. But we shouldn't forget the this-gun-for-hire role that Buckland & Taylor played for the PBA in the Public Consensus Review Panel process.

Will the Buckland & Taylor engineers and designers feel the need to honor and justify their firm's previous performance for the PBA by insisting on a steel twin span once again or will they be able to say, "The last time we were in Buffalo we were representing a client who wanted us to justify a certain conclusion they'd already reached and we did what the client wanted, but this time the client is asking us to do what's best for the site and the community so we're going to really give it an objective look"?

Consultants are always pulled by two masters – the truth and the desires of the people paying the bills. When the consultants are lucky, the people paying the bills want only what the consultants think is the truth. When the people paying the bills have a "truth" they want delivered, the work of consultants can get sticky. We know how Buckland & Taylor performed last time. What will they do this time?

ALL THE NEWS THAT FITS (WHAT THE *NEWS* WANTS YOU TO KNOW)

In editorials last week, the *Buffalo News* praised the PBA for hiring Figg and Menn. They're right: the PBA should be praised for both appointments.

Several people have said to me they've noticed improvement lately in how the *News* covers Peace Bridge matters: more facts, less stuff churned out by the PBA's press agents, no more of those beastly editorials telling people in the community we ought to shut up and accept decisions made for us by the Buffalo Niagara Partnership and the rest of our betters. I agree. I suspect it's primarily a function of the fact that the PBA is no longer working behind closed doors or at war with the community, so there is no longer as much need for slanted coverage.

But you still must read *Buffalo News* articles and editorials about Peace Bridge politics with care. In a March 1 article on the Partnering Group meeting, for example, *News* reporter Patrick Lakamp discussed at length the appointment and qualifications of Menn and Figg, but mentioned only in passing the appointment of Buckland &

Taylor, and he nowhere mentioned the delaying and distorting role Buckland & Taylor had played in the PCRP process. In a recap article published on page 1 the following Sunday, March 4, Lakamp again discussed Menn and Figg, and that time he didn't name Buckland & Taylor at all.

Lakamp had been there when Buckland & Taylor engineers delayed the PCRP process and, at the end, when they fought diligently for the steel twin span. I saw him. He was there last week when Jake Lamb said Buckland & Taylor was on the job in exactly the same capacity as Figg. I saw him then, too. Was the double omission mere oversight or continuation of the old style of *Buffalo News* Peace Bridge coverage?

More important, was this Lakamp's omission or his editors' cuts? I would bet on the latter. The *Buffalo News* has long had what can most kindly be called a strange editorial attitude toward anything connected with the Peace Bridge.

If you write a letter to the editor of the *News* about the Public Bridge Authority, for example, the *News* will in all likelihood change what you wrote to "Peace Bridge Authority." Lakamp's two articles last week continued that deliberate policy of using an incorrect and misleading name for the Buffalo and Fort Erie Public Bridge Authority: the phrase "Peace Bridge Authority" appears eight times, "Peace Authority" once, and the single word "authority" forty-two times. The correct name, "Public Bridge Authority" did not appear once.

My consistent notice of this misnaming by the *Buffalo News* isn't English professor quibbling. As I've frequently pointed out here, the Public Bridge Authority took, to calling itself "Peace Bridge Authority," which exists nowhere in law, as a deliberate part of its plan to distance itself from public scrutiny. I can understand why the PBA, anxious to avoid public scrutiny, used any device at its disposal to get us to stop looking at it closely. But why would the city's only daily newspaper, ostensibly a purveyor of facts and advertisements, decide (as one of their editors told me they did) to use a misleading name for a public agency?

Beats me. It really does.

This resonates beyond the Peace Bridge. I don't know as much about other current civic issues as I know about the Peace Bridge, but I do know a lot about the Peace Bridge and I know how consistently distorted the *Buffalo News* coverage of that story has been and continues to be.

If they're so willing to mislead us about this construction project on Buffalo's West Side, if they're so willing to leave out of their news coverage and editorial commentary key facts we know they have, how, in other key civic issues, can we know when they're telling us all the news that fits or all the news that fits whoever is beating the drum to which they're marching that day?

Given what we know about the activist role they played while pretending to be neutral observers and reporters in the Peace Bridge War, why should we trust them about any other project with big bucks at stake? The PBA seems to be working very hard to win our trust these days, and hooray for them. But with the *Buffalo News*, it's business as usual. Boo.

THE OTHER PARTNERSHIP

One of our far-flung correspondents reports that the Buffalo Niagara Partnership, the only civic organization that opposed a signature span built primarily by local labor in last year's Public Consensus Review Panel, last week assigned an employee to work on "the Peace Bridge project." Natalie Harder, their former development director, left that job not long after the PCRP rejected the PBA's and BNP's attempt to get a steel twin span built. Why does the Partnership once again see the need to assign a staff member to the Bridge? Whose interests are they protecting or pursuing this time?

A SURPRISING APPOINTMENT

Speaking of the Buffalo Niagara Partnership, Andrew Rudnick, that organization's president and CEO, a dedicated and relentless steel-supporter and signature bridge opponent, was recently named to the board of the Community Foundation of Greater Buffalo. The Community Foundation was one of the principal funders of the Public Consensus Review Panel, the group that, during the Peace Bridge War, helped clarify and focus public opinion and which made accessible a huge amount of information about the PBA's plans and the implications of those plans for the region.

The Community Foundation has eleven board members, four appointed by banks (HSBC, M&T, Fleet, Key), three by three different judges representing different layers of the judiciary, one by the mayor, and three by the Foundation's own board. What is curious about Rudnick's appointment is that it wasn't made by his long-time patron Robert Wilmers, president of the M&T bank, but rather that he was named by HSBC.

One former Community Foundation member told me he was really surprised at the source of Rudnick's appointment and "sick, really sick" at the fact of it.

Is this simply a sudden fit of public service activity on Rudnick's part or was he appointed to that position by the moneymen and powerbrokers to make sure that the Community Foundation didn't interfere with any other big-profit public works ventures? This bears watching.

SHOULD WE WORRY ABOUT CHRISTIAN MENN?

Christian Menn is one of the world's most highly regarded bridge-makers. Jake Lamb shopped him all over Buffalo during his three-day visit here last week. Menn met with the Partnering Group, the Bridge Action committee of the New Millennium Group, political officials, engineering and architecture students, and more. Menn proved himself smart, experienced, and already savvy about many of the interests involved in the Peace Bridge War.

He said he thought this crossing deserved an important bridge and he also said that the current bridge was worth preserving, even though the Parker truss is an eyesore. He said he thinks it important that a bridge builder work within realistic boundaries, that he honor, for example, budgetary constraints. He loves the work of Santiago Calatrava (who studied with him) but, he says, Calatrava doesn't think about budget. No more than ten percent of a bridge's budget, he said, should go into aesthetic aspects.

The two questions I heard again and again after his visit were, "What did they tell him his job here was?" and "What happens if he comes up with a twin span?"

Lamb says that Menn was asked to consider the conditions and come up with the best design suggestions he could, nothing more. Did members of the PBA board or did Stephen Mayer or Earl Rowe tell him that what *they* really wanted? We'll never know.

All we know for sure is what Menn said publicly while he was here and the quality of work he has done in the past. He has not, in James Pitt's felicitous phrase, built any ugly bridges. If you look at the bridges Menn has designed and built (such as the Ganter bridge in the canton of Valais, the Reichenau bridge over the Rhine, the Felsenau viaduct over the Aare in Berne), and bridges he has designed for others to build (such as the Charles River bridge now under construction in Boston), you'll see that his sense of aesthetics goes far beyond ten percent of decorative costs.

The day after the Partnering Group meeting, he told Diane Christian and me that one of the reasons he likes the present bridge is that it comes from "the golden age of bridge design," a time that produced the George Washington Bridge and the Golden Gate Bridge, two of his favorites. He loves the arches of the Peace Bridge. "I would like to maintain the bridge," he said, but "the new one should be an extremely new and elegant bridge." He sees no reason to copy the present bridge: if you tried to imitate the arches of the old bridge, "you would come to a confusion." Nearly a century of design separates the existing bridge and whatever new bridge goes up. If the old bridge is maintained and a new bridge built near it, he said, the new bridge should reflect the cutting-edge of 21st century design, not the current bridge's early 20th century design. "I'm thinking of a cable-stayed bridge with relatively small spans," he told us.

How committed is he to preserving the old bridge, to designing a new bridge that complements it from a 21st century point of view? "That's what I'm thinking today," he said. He told us that before he came here last week he'd read everything he had been sent and could find about the old bridge. He said everything he'd decided before his trip had been revised once he visited the site. It was entirely possible that his ideas of that day would change and change again the longer he was involved in the process and the more involved he became with the communities on either side of the river. "It's just starting," he said.

As things stand now, Christian Menn isn't going to design the Buffalo-Fort Erie bridge. He's going to tell the PBA if he thinks what the actual designers come up with is a good idea or not, and he's going to offer improvements to their ideas if he thinks any are needed. But, given his position in the structure of the EIS and the great hoopla his involvement has already engendered in Buffalo, what he suggests will have great weight with the designers and the decision-makers.

There are key issues I don't think Christian Menn has yet thought through. One is, how do you weigh the aesthetic value of the sweet old bridge (less the ugly Parker truss) against the several years of disruption of life on Buffalo's West Side rehabilitating that old bridge will require once the new bridge is built? If cost is a key factor (as Menn insists it must be), how will he factor in the long-term costs of a high-maintenance geriatric steel bridge alongside a low-maintenance concrete bridge?

Menn is an engineer. With rare exceptions, engineers are practical people, they're not dreamers. You give engineers a problem and they solve it for you. You don't go to engineers with questions like "How should the world be?" You ask engineers "How can I make this happen?" But, within that restriction, there are engineers who try to make things happen in the most elegant way possible. Based on what I know of his work, I think Menn is one of those engineers. It's really important that, during the many visits he will make to this area in the next few years, people here let him know how they feel about the impact on their lives of long-term rehabilitation and continuing heavy maintenance of the old bridge. He's got to be reminded that a bridge has to fit more than the physical environment and the PBA's ostensible budget.

CAN WE TRUST JAKE LAMB?

Jake Lamb is presently the key person driving the environmental impact study. The EIS will feed directly into whatever design is actually built. Lamb is coordinating the consultant groups, he set up the complex organization by which all the information will be acquired and through which all the information will filter and be managed. Since he took over the preliminary work for the EIS, the PBA's two general managers, Earl Rowe and Stephen Mayer, both of them relentless and dedicated steel twin span advocates, have been nearly invisible and almost totally silent. Jake Lamb used to be a strong advocate of the steel twin span; now he says he's a strong advocate of a full and open process.

Jeff Belt, president of the New Millennium Group and one of the key signature bridge advocates, says he finds "accumulating evidence convincing me to believe that Jake has had an epiphany over this project. Prior to Judge Fahey's ruling, Jake fought doggedly to drive his client's project through – as I am sure he has learned a civil engineer must do. After the Judge's ruling, Jake's practical mind instructed him to abandon the old plan that had failed and move quickly toward a new plan, as a civil engineer would know he must do. Jake is now speaking and doing the language of civic leadership. He is and has been all along, a civil engineer! I am glad to learn this and more confident that our Peace Bridge project will be a great success."

Jeff may be right, though there might be another way to put it: before Judge Fahey's ruling and former PBA chairman Victor Martucci's consequent decision to abandon warfare and embrace the environmental impact study process, Jake Lamb had been hired by the PBA to help enact its decision to construct a steel twin span. He did that vigorously. Now he's been charged with examining *all* the ways bridge capacity might be expanded here, or if it should be expanded here at all, and he's doing that with equal vigor – and with a great deal more power. His earlier assignment made him answerable to a small group of people meeting in closed-door sessions down at Peace Bridge Plaza. His present assignment makes him answerable to the community at large. Thus far, he seems to have taken that new assignment very seriously.

CAN WE TRUST THE PBA?

A lot more than we could six months ago.

But this is no time for anybody on the signature side to relax: the snakes are still in the grass. Even so, there have been real changes. I don't believe that Paul Koessler, the newest member of the board and Lopinski's successor as chairman a year from now, has any allegiance to the old PBA way of thinking and planning that caused so much distrust and engendered so many lousy ideas. Koessler seems to have a good sense of what went wrong and a realistic appreciation of what a signature bridge might do for this community. If Governor Pataki's replacement for Victor Martucci has any sense of the needs of this community, the old gridlock will not easily reestablish itself.

I've heard that the Fort Erie politicians aren't the least bit happy about the current state of affairs (which may be why they weren't at the February 28 Partnering Group meeting) and that several members of the PBA were dragged to it either kicking and screaming or sulking and muttering. But they *have* been dragged and most of them are realists. Joe Crangle was right when he said that if the Americans on the board ever got their act together, if they ever decided not to walk in lockstep with the steel twin spanners, then things would turn around quickly, and for the better.

I think of a safecracker I knew in Texas 35 years ago. He'd quit safecracking and become a check forger instead. I asked him which he preferred, safes or checks, and he said safes by far. So why had he changed? I thought he'd tell me why forgery was more interesting or profitable or aesthetically satisfying than peeling, punching and blowing safes. No, none of that. He told me that he had gotten older and the physical aspect of the work just got to be too hard for him: he couldn't climb up on roofs carrying all his tools, cut holes in the roofs, lower himself down on ropes the way he used to. Did he miss safes? Every day. Would he ever crack another one? No, that part of his life was over. It was necessary, he said, to be realistic about these things.

And maybe likewise the PBA. They may miss their steel bridge, but because of Mayor Masiello's firm refusal to issue the easements and Judge Fahey's EIS order, they can't jam it down our throats and they know it. They're practical. If they can't build the bridge they want, they'll build the bridge they can. If it turns out to be a bridge we can all look at with pride, then lucky all of us. They may not all be doing it for the reason some of us would have preferred, but they seem to be doing it, and that, for everyone but the kamikaze moralists among us, is finally what really matters.

53. THE PBA THROWS A PARTY: TALK, FOOD, FURIES AND SCORPIONS

A PUZZLING ADDRESS

The Buffalo and Fort Erie Public Bridge Authority had an event in the Mary Seaton Room at Kleinhans Music Hall on March 22.

The invitation was to the "Peace Bridge Strategic Gateway Address." I have no idea what that means and neither, I suspect, does anyone else. At some point there may have been a plan to deliver real information from the recently-hired technical consultants, but that's not how it turned out. What it turned out to be was 47 minutes of mostly-innocuous feel-good speechifying by PBA officers and Buffalo and Fort Erie politicians, interrupted for a short, very well-made high-tech video, the point of which seemed to be very much at odds with the speeches.

Followed by a party.

The party was catered by Oliver's. There were two bars, each serving six American and six Canadian wines; two meat tables, each with a large turkey breast, a large chunk of roast beef, rolls, horseradish sauce and another sauce I didn't try; two pasta tables, each with two kinds of pasta; and, in the middle of it all, a huge table overflowing with a grand variety of pastry treats. For those who got stuck in conversation before their appetites were sated, a squad of waiters constantly patrolled the room, carrying trays of chicken satay, little shrimp thingys, and other delights.

The feeling was friendly, easy. People chatted, said hello, milled around, talked about the past, introduced people to people they hadn't known. It was like the war was over, and maybe that's all it really was: the PBA trying to say the war is over, have some turkey, try this wine, look at that pastry, oh boy!

TAKING ATTENDANCE

I saw many familiar faces from both sides of the Peace Bridge War and many faces I'd never seen before, nearly 200 in all.

Most members of the Public Consensus Review Panel were

there, as were many members of the New Millennium Group. There were mayors and other politicians from both sides of the border. Six members of the Buffalo and Niagara Public Bridge Authority were there. Scott Sroka from Senator Schumer's office was there, as were longtime PBA attorney Arnold Gardner and Episcopal Church Home and Olmsted Parks Conservancy attorney Guy Agostinelli. Architect Clint Brown was there, and so was Community Foundation head Gail Johnstone. Peace Bridge general managers Earl Rowe and Steven Mayer were there. There were reporters galore.

A good number of people were identified to me as Canadians and there were others I assumed were Canadians because of the things they wore in their lapels. Canadians never took part in any of the Public Consensus Review Panel work, save for one evening when Fort Erie Mayor Wayne Redekop came over to tell the Panel how much the old Peace Bridge meant to him and a representative of the Toronto City Council drove down to tell them that the Toronto City Council had voted that the Americans should stop shilly-shallying around and get on with the twin span construction project.

Vincent "Jake" Lamb, who is managing the environmental impact process and who has become the PBA's point man for public encounters, was home in New Jersey with the flu. No mention was made of bridge-builder Eugene Figg, or any of the other recently-appointed consultants to the process, some of whom I'd previously heard were to be part of the evening's program.

The newest member of the PBA, Colleen DePirro, president of the Amherst Chamber of Commerce, appointed to replace the late Louis Billitier, was not there. I would have liked to have asked her why she thinks Pataki appointed her. She was quoted in the *Buffalo News* as saying she had no conversations with Pataki about the appointment and had no opinions on the bridge. To live in this area, be active in public life, and have no opinions about the bridge must be like having no opinions on smoking in restaurants or local taxes. How can you not? And if you managed to survive living in this area the past few years without having any opinions about the Peace Bridge, why would you want to serve on the board that administers it?

Canadian Consul-general Mark Romoff wasn't there either, which was surprising. During the two or three years of the Peace Bridge War he rarely turned out for the contentious events, but he was always there for the pleasant announcements.

There were two other notable absences, both of whom had worked assiduously for a steel twin span that would sacrifice Buffalo community interests in favor of Canadian manufacturing and trucking demands: Rep. John LaFalce and Buffalo Niagara Partnership president Andrew Rudnick. Perhaps they had pressing interests elsewhere. Or maybe they didn't want to be in a room where all kinds of people could come up to them and ask why they had tried so hard to prevent what seems to be going on from coming about.

SPEECHIFYING

The two bars were set up, but not serving yet, and the several other tables were obviously ready to dispense food, which hadn't arrived yet, at 5:00 p.m., the time the invitations said the evening's events were scheduled to start. Nobody began moving toward the folding chairs in the front of the room until nearly 5:30.

The speeches were all mercifully brief, and three of the listed presenters didn't show up at all. There was nothing in any of the speeches you had to worry about or think about. Mayor Masiello was quite good. He was followed by Fort Erie Mayor Wayne Redekop, who indulged himself in a lot of hockey metaphoring. Redekop was followed by Erie County Executive Joel Giambria, who said good things about involving the community in public works decisions. Giambra also noted that Redekop, for all his hockey metaphoring, hasn't mentioned that the Sabres had clobbered the Mapleleafs a few nights earlier.

REDEKOP

PBA chairman John Lopinski introduced Redekop. He said that among Redekop's achievements was successfully negotiating slot machines for the Fort Erie racetrack. That wasn't a peripheral vita entry: much of the passenger traffic on the Peace Bridge consists of Americans crossing over to gamble. Both the PBA and Fort Erie profit from those slots.

Redekop said he was surprised that Lopinski hadn't mentioned that he'd also been a hockey player. His entire speech percolated with sports metaphors:

> I'd like to emphasize that Mayor Masiello and the city of Buffalo, the directors of the Peace Bridge Authority, and the members of council of the town of Fort Erie and me are all members of one team for this particular game.... "Think of us as players in what is perhaps the most important game of our career.... We have a good team....Our preferred team is only temporary unless our team can agree on a winning strategy. I doubt that any team has ever made it to the finals without having one strat-

> egy that all can work together....As we continue through the final minutes of what is an important game to all of us, let's follow some simple rules. Let's play as a team. Let's agree to think like winners. Let us agree to respect the value of other views. Let's agree to turn our difficulties into opportunities. And let's use our heads and move forward together. Finally, as that great Canadian social commentator Red Green says, "Let's keep our sticks on the ice."

The hockey metaphors got old almost immediately, but who cared? It was fluff at a border party. So what if the mayor of Fort Erie didn't have anything but platitudes to utter? Politicians issue platitudes the way idling diesel trucks issue carcinogens: it's in the way they're built.

It turned out that Redekop wasn't just platitudinous, he was also duplicitous. As soon as the party was over, he told reporters from WBFO and the *Buffalo News* that he had no intention of going along with any plan that did not keep the current bridge fully operational and that put a new bridge anywhere but south of the current bridge.

The reason all construction had to be south of the present bridge, he said, is to protect archeological sites underground to the north. I remain unconvinced. Possible archeological artefacts didn't stop construction when they built the new duty-free shop or the Commercial Vehicle Processing Center: they just built right on top of possible sites with (I've been told) a concrete shield between the buildings and the ground to preserve whatever was there. Why couldn't the same procedure be followed a hundred yards to the north?

Redekop's preconditions were disturbingly close to Buffalo Mayor Masiello's February 1 proposal that we keep the old bridge and build a new one to the south of it. Is Redekop simply jumping on Masiello's bandwagon? Are the two in collusion, trying to force the pending environmental impact study into a foregone conclusion?

I don't know, but I think we're past the point where a Fort Erie official stomping his feet and holding his breath until his face turns blue is going to hold up this project. Central command in Ottawa just isn't going to stand for that kind of petulance. They are desperate to have a clear channel for their truck traffic and I can't believe they'll let Redekop block it.

Maybe this is Fort Erie trying to squeeze out a few more bonus bucks before it all goes equitable. As long as the PBA was at war with Buffalo, Fort Erie could get rich by making nice and sucking on the big sugartit. But if the PBA and Buffalo are really working together for a solution, then Fort Erie has no special edge and the PBA has no special need to buy it off.

Especially if, as I was told while waiting for a refill of the Long Island Paumanok merlot, the PBA is about to do a management audit, something it has assiduously avoided and which its critics have pointed to as evidence of its fear of sunshine. A financial audit checks on your paper trails, but it doesn't look at reasons for anything: your books say you paid so many dollars to a certain supplier and your records show a cancelled check and an invoice, and the accountant is happy. In a management audit, you're asked what did you get for that money and how did you decide on that supplier and did you need it in the first place? If that phase of the PBA's operations go public, it truly will be a new day down at Peace Bridge Plaza.

THE VIDEO

The speeches, as I said, were nice-nice. The focal point of the 47-minute session was a video, in which a sexy female voice told us that the Peace Bridge "is the backbone of our region's unique binational character and lifestyle" and "our gateway to success." She talked about all the tourists it moves both ways, how it provides thousands of jobs, how it can provide thousands more. While that was going on the music was variously inspirational and dramatic. Then it all changed, and so did her voice:

> But traffic volume is reaching capacity...NOW! [*The music got sinister*] For every traveler, there are tourist operations waiting to serve. For every truck, a retailer waiting to sell. For industry needing parts JUST IN TIME missed connections, or production halts spoiled shipments leave frustrated customers on both sides tempted to reroute, move or built their business ELSEWHERE. The Peace Bridge is our link to success and progress! [*Happy dance music again*]

Her voice was all over the place in that bit: passionate, fearful, sexy, assertive, hesitant, soft, loud, fast, slow, and the effects soundtrack was equally busy. It was quite something.

Then she told an old lie: "The Buffalo and Fort Erie Public Bridge Authority is a partner in progress to its customers and neighboring community, a binational organization entirely funded by bridge tolls, duty free revenues, and tenant leases." Give a rest! The PBA is also funded by huge tax exemptions (which is getting tax money, only it's located in a different page of the account book), disproportionately greater lease payments from the American than the Canadian government (paid by tax dollars), and outright grants from the American government (out of tax dollars).

What came next seemed more appropriate to the state of things a year ago, when the PBA was running its very expensive ad campaign on Buffalo television and radio stations designed to pressure the Public Consensus Review Panel, the City of Buffalo, and Judge Eugene Fahey to let them build the steel bridge immediately:

> The Peace Bridge shares that vision with great hopes and plans, and all eyes are on us, waiting. What rests in our hands is the fate of a legacy, the future of our link to success. We all need to focus on that link. Find a gateway solution NOW and take pride in the progress it will bring to both sides for generations to come.

The video was very well made, it had (as they say in the trade) high production values, with MTV-style rapid cutting, a punched-up digital soundtrack, nice archival footage of the 1927 opening of the bridge, and swooping aerial shots. It really moved.

The only thing about the video I couldn't figure out was its point. The EIS is underway and it's not going to be stopped, so there will be no rush to judgment. Everyone in the room knew that the bridge served tourists and truckers, so there was no need to tell them that. Maybe they has commissioned before then-PBA chairman Victor Martucci talked the board into giving up the fight and doing the EIS and they figured they might as well get some use out of it. If that's all it was, fine, nothing to worry about. If the video reflects a split in the PBA – some of them comfortable with the EIS and the open process, others wanting to go back to the old days of three months ago and the policy of pressure and mislead – then we've got a problem.

FOLKTALES

Afterwards, I heard two things, again and again. The first was some variant of, "Did you notice how every one of the speakers used the term 'Peace Bridge' and never the term 'Public Bridge.' They still can't face that they're a public agency." The second was more immediate: "What's this about? Do you trust them? What's really going on?"

I had no better answers than anyone else, but after a while I thought about a Greek play and an African folktale.

The play is “The Eumenides,” the third part of Aeschylus’s *Oresteia.* In it, the Furies, who were mean and nasty and vile are convinced by Athena, goddess of wisdom, to be really helpful instead. They had been living as outcasts, forever in a state of hatred and always engaging in modes of destruction, but Athena makes them an offer they cannot refuse, after which instead of punishing Athens they decide to defend it. Their language in the play changes from anger, defiance and belligerence to words of harmony, peace, mercy, love. The ending of the play is optimistic: the Eumenides continue to be powerful, but now they are powerful as part of the community rather than in opposition to it.

The African folktale is the one about the frog and the scorpion.

> The scorpion wants to get to the other side of the river and asks the frog if he can ride on the frog’s back.
>
> “No way,” the frog says. “I know about you. You’re deadly.”
>
> “Hey, I’m asking you for a favor, to get me across the river. If I were to bite you, we’d both drown. Why would I do that? Trust me. Be a pal.”
>
> That makes sense to the frog, and he’s a good guy, so he holds still while the scorpion clambers up on his back, then he eases into the water and starts swimming to the other side.
>
> About halfway across the river the scorpion does it: zap! That poisonous tail cuts deep, midway down the frog’s spine. Before the tip of that poisonous tail has receded, the frog can no longer control his legs.
>
> “Why did you do that?” the dying frog asked. “Now we’re both going to drown.”
>
> “I’m a scorpion,” the scorpion said, “and that’s what scorpions do.”
>
> “But you said you wouldn’t bite me.”
>
> “I’m a *scorpion,*” the scorpion said, “you knew that when you met me.”

The great thing about examples from literature and folktales and proverbs is, you can find one for every occasion. Which story is the right one for us right now? What do we have here? Furies entering a new life, now friends of the city? Or the scorpions we know so well? Is it Jake Lamb, running a process that is truly open and intelligent? Is it Wayne Redekop saying from the lectern that this is a time for us all to act as a team but really meaning that the team is playing in a game with the score fixed before the opening whistle?

Stay tuned.

54. VINCENT "JAKE" LAMB: POINT MAN AT THE PEACE BRIDGE (I)

Canada's Minister of Transport recently told a reporter that Buffalo's environmental and aesthetic concerns in the Peace Bridge expansion project mattered not a whit in Canada's mega-trade plans. The mayor of Fort Erie said that he'd cooperate in the design process currently going on, so long as it resulted in a new bridge located at a site he'd already selected and didn't interfere with continued operation of the current bridge, and the mayor of Buffalo proposed a bridge alignment and design plan that was only a slight modification of an idea he had three years ago. This was at a time when both mayors had joined with the PBA in a decision-making group ostensibly open to all suggestions about design, alignment, and location.

But, I think, political posturing, preening, and pontificating matters very little now. The mayors of Buffalo and Fort Erie can suggest, but they cannot control the design process. Canadian government officials can jabber all they like about the inappropriateness of Buffalo interfering with their traffic plans, but the decision about what bridge will cross the Niagara River between Fort Erie, Ontario, and Buffalo, New York, will be made after a public process that takes place right here, not behind closed doors in Ottawa.

The final choice of design will be made by the Buffalo and Fort Erie Public Bridge Authority, the PBA. About eight years ago, the PBA came up with the idea of twinning the current steel bridge and in the process occupying a bit more of the pitiful remaining fragments of Frederick Law Olmsted's Front Park, which their present operations have made all but unusable anyway. The PBA's professional staff and its board of five Canadians (all appointed by the Minister of Transport) and five Americans (two appointed directly by the governor and three serving ex officio) expected the expansion would occur without impediment, as nearly all the previous Peace Bridge expansion projects. There might be some opposition, but money would be spent, the opposition would fade away, and earth would be turned right on schedule.

Not this time. There was a huge amount of community opposition on the Buffalo side of the river, so much of it that the mayor of

Buffalo and the Erie County Executive eventually joined the chorus. Buffalo Mayor Anthony Masiello refused to give the PBA the easements it needed to begin construction. Two community organizations – the Episcopal Church Home and the Olmsted Conservancy – filed suit in state court asking that the PBA be forced to do a full environmental impact study of its proposed bridge and plaza projects, a study the PBA and its staff desperately tried to avoid. With the help of two local foundations, the county and city set up the Public Consensus Review Panel, a broadly-based community and citizens' group that considered alternative ways of expanding border traffic and the social and environmental consequences of the apparent range of choices. The PBA refused to take part in the panel until it seemed Judge Eugene Fahey was about to rule against its plans, at which point it agreed to underwrite part of a team of engineers to consider design possibilities. The PBA and its engineers delayed the process for two months, then grudgingly took part in it. The PBA itself never sat at the PCRP table. The engineers it hired never did anything but advocate the PBA's party line.

It was all resolved, finally, not by reason and consensus, but by Judge Fahey, who ruled that the PBA couldn't turn a shovelful of earth until it undertook a full environmental impact study, one that examined all aspects of both the bridge and plaza construction projects. Judge Fahey told the PBA that they'd have to let the rest of us in on the process.

Judge Fahey's decision was followed by a major change in attitude within the PBA itself, at least on the American side. While the Canadians held firm to the idea of a steel twin span, PBA chairman Victor Martucci, an astute businessman, realized that strutting and blathering were not going to undo Judge Fahey's ruling. A new set of attorneys advised him that appealing the ruling was likely to go nowhere. Martucci is obsessively driven to do things well: when he opposed the EIS, he fought it absolutely; when Judge Fahey's order made the EIS inevitable, Martucci decided they would do the best EIS possible, and he got the rest of the PBA to go along with him. Soon after Martucci set the current EIS project in motion he left the PBA board. His successor, Paul Koessler, seems to be of similar mind. It was time, both insisted, for the PBA to work with, rather than in opposition to, the community.

Enter Vincent "Jake" Lamb.

Jake Lamb is organizing and running the environmental impact study. He's a former executive vice president and director of Parsons Transportation Group. He's been project manager on major public works projects around the world. His most recent corporate job before coming to Buffalo was managing corporate conflict resolution in both internal organizational/operational issues and external client/project conflict issues. He's been around as a consultant to the PBA for the last two or three years. He always seemed to be at the periphery during hearings and public meetings; he listened a lot, but they never let him say much. He's still listening, and now he's the man doing nearly all the talking for the PBA.

Lamb says that the current process cannot have validity unless there is real, extensive community involvement. This is the exact opposite of the attitude long held by the PBA board, its two general managers, and its primary front man, Andrew Rudnick, president and CEO of the Buffalo-Niagara Partnership (formerly the Chamber of Commerce). Nowadays, Rudnick says nothing about the Peace Bridge, and the two general managers are all but invisible: they no longer give speeches about what we should do, about why the community should shut up, about why they know what's best for us. The PBA itself has left just about all the public engagements to Jake Lamb.

Two years ago, you couldn't get senior staff at the Peace Bridge to take or return telephone calls, let alone come out and talk about what they were up to and listen to community concerns. Lamb has completely reversed that policy. He'll talk to and listen to the concerns of just about anybody. In the past several months he's had meetings with and done presentations for the Common Council's Bi-National Bridge Task Force, Episcopal Church Home, Olmsted Parks Conservancy, Mayor Anthony Masiello, Erie County Executive Joel Giambra, New Millennium Group, Senator Charles Schumer, Congressman John LaFalce, Congressman Jack Quinn, Canadian Consul-General Mark Romoff, some of Senator Hillary Clinton's staff,* Fort Erie Mayor Wayne Redekop, Mike Fitzpatrick (Ironworkers Union), Welland/Pelham Chamber of Commerce, and many other individuals and groups.

(Lamb has met with every major political figure in the region except Hillary Clinton. She continues to maintain a policy of dead silence on the Peace Bridge issue. She never talks about it, ignores questions from reporters about it and, when pressed, responds with bland platitudes about the importance of trade. I assume that's because of her friendship with and political debts to Rep. John LaFalce, a fierce and unrelenting proponent of the Canadian steel twin span design. So far as I know, LaFalce remains the only local political figure of importance who never embraced the PBA's decision to abandon its earlier plans and undertake the full environmental impact study. Andrew Rudnick, another steel bridge advocate, has also maintained silence since the PBA changed direction, but I count him as a business activist rather than a political figure. Clinton's help will be important when the PBA makes its actual application for federal transportation funds and we can only hope that she gets free enough of her impeachment-days debts to John LaFalce to help us when that time comes.)

It is possible that when the environmental impact study is finished the Public Bridge Authority will examine the suggested alternatives and conclude that the only viable plan would be the one they came up with nearly a decade ago, a steel twin span alongside the aging twin span now in place. From all I hear, the Canadian members of the board would be perfectly happy with such a decision.

The only thing that can stop the old foolishness from surfacing once again is a set of recommendations from the environmental impact study so well-founded and well-reasoned that the forces of avarice and expedience have to give way to common sense and human decency. They almost got away with it a year ago, but Mayor Masiello's refusal to give the easement they needed to begin construction and Judge Fahey's insistence that they obey U.S. environmental law stopped them.

They haven't gone away. The forces of avarice and expedience never go away. The most important person between us and them right now is Jake Lamb.

Senator Charles Schumer, an ardent and persistent advocate of a rational bridge project, recently became concerned about what seemed to him lack of movement in the EIS process, and by an article by Patrick Lakamp in the Buffalo News quoting Lamb as saying the PBA needed $150 million in Federal funds to carry out the project. Why, Schumer asked, had nothing happened on the EIS in the year since Judge Fahey's order became effective, and how could Lamb say they would need $150 million in outside money when the project scoping – the stage where they detail what problem they're trying to solve and what the possible solutions are and what those solutions might cost – hadn't even begun, let alone the EIS itself? Did Lamb (hence the PBA) have a design in mind already? Was the process just underway smoke and mirrors, a sham?

The Peace Bridge Action Group of the New Millennium Group, a community organization that was critical in stopping the ugly bridge plan, sprung to Lamb's defense. As far as they're concerned, Lamb has been doing a huge amount of preparatory groundwork, getting the parts in place so when the scoping and EIS start they'll have enough community involvement and professional competency so the results will have validity.

I've been writing about Peace Bridge affairs since August 1998, and I've several times told you about people or processes I thought you should mistrust or at least watch very carefully. The NMG trusts Jake Lamb, and I think they are right to do so. Jake Lamb is one of those people who is, I believe, just what he seems: a highly skilled professional honestly dedicated to doing a good job.

What follows is a transcription of a telephone conversation we had on June 22, and an email he sent June 10 in response to some questions I'd sent him about the issues raised by Senator Schumer. Next week, we'll publish extensive selections from our May 31 and June 1 conversation in which Lamb describes the entire scoping and EIS process, and discusses the reasons for his key choices.

I. Money Matters (an email letter)

Frankly I, did not intend to raise the issue of money and financial aid at this point in the process. But I was reminded in meetings with congressional representatives that *now* is the time to start application processes to obtain ISTEA 2003 [Intermodal Surface Transportation Efficiency Act] funding, and we don't know when the next opportunity will come. So I was (am) in a situation where I am asking for funding in advance of definition (thru the Environmental Process) of a specific plan for the Peace bridge expansion. But I do know, based on the PCRP consultants' report, that all of the alternatives identified in that review that would produce a U.S. plaza plan that restores (partially or in whole) Front park and Fort Porter, would cost more than the PBA's funding capacity....

I believe the work of the PCRP is of value and I am using their work products to try to get some early (consistent with the ISTEA budgeting cycle) political support for money to be in position to *achieve financial feasibility for project alternatives that may be beyond the financial capability of the PBA.* As I have said to Patrick Lakamp and to you, an independent evaluation will be made of the PBA's financial capability (this evaluation will consider toll increases) and the definition of the expansion plan will be achieved thru the environmental process. There is no pre-determined plan which is the basis for my initiative to get ISTEA 2003 money.

You don't need a final plan to support a request for funding. You should have some substantive supporting information and I think we have that in the work produced by the PCRP. And I remind you that the PBA received the approximately $22 million U.S. federal funding commitment for the connecting roads (plaza to/from I190) at some stage well before the connecting roads plan was developed in the U.S. plaza EIS.

At some point in the environmental process financial feasibility will become a decisive factor. I am trying to reduce the risk that financial feasibility may ultimately become the deciding factor in selection of the preferred alternative. I would prefer *eliminating* the financial factor in the selection process. It is toward that end that I am raising this issue now. I want to find out now what the potential is for obtaining federal aid and when we need to formally apply for it, and the potential commitment of local, regional, and state leaders to support aid for the project.

Usually there is a cycle of several years between ISTEA appropriations. I did not want to chance waiting and then find that after we develop thru pubic involvement a favored plan which, based on PCRP's work, would probably exceed PBA's capacity, that there is no available funding source and/or that there is a long wait for new funding. Then the question will be "Jake, why didn't you ask earlier?

Now the political leaders may respond with, "See us later and we will get the necessary funding." Okay, if that is the best I can do now I'll accept that – provided their commitment is made public.

If the PCRP estimates are accurate it seems that the delta cost [the difference between what the PBA can afford and what the project seems likely to cost] is also driven to some extent by the objective to restore Front park and Fort Porter. If that is the case there is some logic in U.S. financial sources assuming a greater role in helping to achieve this quality-of-life-enhancing objective. So in view of the urgency to tap into ISTEA2003 I approached the American side first.

I believe funding discussion at this stage may have the additional benefit of building political support on both sides of the border for the "Shared Border Management Concept" (moving U.S. plaza operations to Canada to the extent possible) because there are preliminary indications (to be developed and confirmed thru the Process) that this Concept (which would be common to all U.S. Plaza alternatives) would reduce the cost of the project, and therefore the amount of additional money needed, and could shift some project plaza work to the Canadian side, thereby supporting/bolstering the logic for Canadian financial aid .

II. WHERE THINGS ARE (A PHONE CONVERSATION)

The Buffalo News *quoted you as saying you want $150 million in federal funds for the bridge project. But the project scoping hasn't been done and the environmental impact study hasn't started, so how did that number come about?*

Actually, the number was 130 to 150. The way the number came about was using the alternatives that were identified in the Public Consensus Review Panel report and the estimates that their team of consultants came up with. If you add the bridge and the plaza together for the combinations that they developed and compared those estimates of costs – they were 1999 dollars – to the estimated capability of the Peace Bridge Authority to fund new capital construction, which is estimated at about $180 million, if I take the difference between that $180 million and the estimates, and there's some rounding, the range is $130 to $150. It depends on which alternative you might end up with.

I think a couple of things are clear from the Public Consensus Review Panel report. One, that if there was a consensus, that there was a consensus to restore Front Park and putting in a plaza at a different location, into a residential area. I think that was clear. The other thing that was clear was that these estimates confirm, when you compare them to the Peace Bridge Authority's capability, that some additional funding would be necessary to achieve the objective of building one of the those alternatives.

It's just not a matter of being close to the Authority's capability. The difference was so substantial that I felt it necessary to point this out at this stage of the game, this stage of the process, so that people would understand going in that the reality of the situation is, if we just looked at a project the Peace Bridge Authority could afford with their own money, without any outside help, it would mean there would be some restriction on how far we could go with alternatives. We would look at these other alternatives as well and estimate the values of them. But given fact that the preliminary work is underway now for getting money authorized and allocated to various projects under ISTEA-2003, this seemed to be the right time–and it is the right time – to raise the issue so that financial planning can go along concurrently with the engineering evaluations and the environmental evaluations so that we're not restricted from picking an alternative that achieves those objectives.

That, basically is why I raised the issue at this point. I know it's causing some concern in different places, but I felt better to do it now than to do it later and have somebody say, "Well, why didn't you tell us before." It's quite clear from the information, and this is valuable information that the Public Consensus Review Panel produced, that there's definitely going to need to be some kind of additional financial help and assistance to get the kind of plan that I think the community is indicating that it wants.

This is talking about the whole system, right?

Yes. This is not bridge alone. As a matter of fact, people say, "The bridge doesn't cost that much." No, it isn't in the bridge. It's in the plaza and the right of way and the relocations that are required to move the plaza. It's the construction and the acquisition of right of way, and relocation of people that would accompany going into those residential areas.

All these numbers are subject to confirmation. Further detailing will be done with them, and maybe another set of numbers is going to come up. But I think the trend there is very clear. It's quite obvious that some additional funding would be required and I wanted start that process.

What happened in your meeting yesterday with Senator Schumer?

Paul Koessler went with me and we met with Schumer's staff and with Schumer himself. He was very gracious to meet with us, we were pleased to have the meeting. We agreed that we want to work together toward constructing the best bridge and plaza design for the U.S. and Canadian communities, and that public input and participation is a key component if this project is going to succeed. So we're on the same page with respect to the process.

We also discussed the issue of shared border management, which as you know would involve transferring some or most of the U.S. Customs operations to the Canadian side of the Peace Bridge. This, of course, would have a very positive impact on the U.S. side in that it would minimize the amount of land the Authority would have to acquire. The increased efficiencies brought about by such an arrangement could decrease the total cost of the project substantially. We agreed with the senator that we would work with him and his staff to explore that shared border management customs operation further.

He also reiterated his concern that the formal environmental impact statement had not yet begun. It actually formally begins when we issue a notice of intent and we haven't done that yet. He's anxious for it to move and we agreed: we're anxious for it to move, too and we're working toward that end.

It was a positive meeting and it's one that indicates support by the senator which is going to be extremely important throughout the process.

Have you ever had any conversations with Hillary Clinton and her staff about this?

We've been around to all of them. I have not met personally with

Hillary Clinton. We have met with staff. We did not bring up this issue of 130-150 because we were not focusing on that at the time. We did talk to them about shared border management and we did talk to them about the prospect of eventually needing some financial assistance, but it was in addition to everything else that we were talking about. We really didn't focus on it.

I didn't focus on it with Patrick Lakamp either. I went through the entire process and he kind of focused in on that part.

Lakamp wrote in his June 8 piece in the Buffalo News *that, "The authority does not intend to ask Canada for monetary aid, Lamb said, in part because the pressure for an enhanced design came from Americans."*

That's not a conclusion that's appropriate at this particular point in time. We certainly haven't ruled out asking Canada for money. My view on it is, I want to see some movement on the American side toward two things in the political world. Support the idea of providing some sort of financial assistance to the project, that's number one. Number two, the idea of moving together with Canada in some cooperative effort to help solve our local problem here. I'm not talking about the full border now. With respect to moving some of these operations to Canada. This isn't something the project can do without outside support from both the public and the political worlds.

If we get that kind of support, we would definitely discuss the financial situation with the Canadian authorities and I think under some situations we would discuss the possibility of getting some Canadian assistance to make the project a reality to achieve these gateway objectives that we all know have been put on the table by various components, especially on the American side, but on the Canadian side as well.

So that's not ruled out at all. There's some logic for it, for the Canadians to participate along with the Americans, but basically I'd like to see some movement on the American side on both of those scores.

Do you have any idea when the letter of intent will occur?

We've been working toward that. We've been doing a lot of preparatory work and just recently I've finalized my thinking with respect to these collaborative workshops on the Binational Civic Advisory Committee. I have a good feeling now for how they're going to participate, for how we're going to get the public involved in these workshops to help bring us to consensus on conclusions and recommendations with respect to selecting as Senator Schumer puts it, the best bridge and plaza plan.

With that in place we are planning a meeting with agencies the early part of July. Immediately following that will be a notice of intent. I would expect the notice of intent would be published by the middle of July or the latter part of July. Of this year, of course.

We would follow that with public information meetings on both sides of the border to be followed by the formal public scoping meetings, which are required by the procedure. The public information meetings that I'm talking about are not required by the procedure but we want to do that to help inform the public of how they're going to participate in the project and what the scoping process is about and how to specifically participate in the scoping process.

Following the public scoping meetings, we intend to have workshops to help the public respond to the issues that are raised during the scoping. In other words, we want to help facilitate taking their ideas and thoughts and suggestions onto a piece of paper with respect maybe to alternatives or to discuss even the process that we're using so that it's clear to them what we're doing, so they're in a much better position to have informed input to the scoping process. We'll take that information, then, and revise our draft scoping document. That would basically mean putting everything up on the table, not taking anything off the table.

The whole idea of the scoping is to get all the issues identified and defined, and all the alternatives that the public is interested in considering identified and defined. We don't do any elimination of anything during that process. We start that after the final scoping document is prepared, which comes at the end of that process.

How long does that take?

Let's follow some scenarios here. If we – and I'm pretty confident that we will – have the notice of intent out in July and we have our formal scoping meeting in August, I would think that the scoping process would be finished not later than the end of September. Then the process of evaluating alternatives would start. That's where we get into more concentrated public collaborative workshops to take that long list of alternatives that's developed during the scoping process, start pruning it down, culling it down. The public's going to be involved in that culling process.

Hopefully, we get down to several reasonable alternatives. The criteria for determining "reasonable alternatives" will be identified in the original scoping document that we will circulate during the public information meetings prior to the public scoping meeting, so you'll know how we're going to evaluate this stuff, what the criteria is, and you – by "you" I mean the public – would apply that criteria when you're comparing one alternative to another.

So it's going to be all up front. We start applying that later, after we've sorted through it and you've had your look at it and you've had your questions and clarifications about it, suggested changes about it. We use that criteria then to start the culling process and ultimately the same criteria would be used to select the preferred alternative.

One of the criteria will be having a doable project, an affordable project. It's in anticipation of where we're going to go here that I talk about money now. I don't have a plan in mind that we're going to end up with. But I do know that if we're going to go in the direction which a lot of people want us to go in and which is pretty well encompassed with the Public Consensus Review Panel report, that it's more than the Authority can afford, at least based on what I understand about what the Authority can afford.

Now, during the process, we're going to have an independent evaluation made to determine if that 180 [million dollars that the PBA can afford] is an appropriate number. It could be more, it could be less. Our traffic engineers are already working on traffic projections. We're going to use those traffic projections and use different scenarios for the toll rates to determine the number. We'll go through that process and have financial people come in to look at the other authorities' net operating revenues and costs and what their debt service is on their outstanding bonds, and through that process we'll get a confirmation of 180 or some other number. That's going to be done by our process, by our group, by our team. We're going to have to bring in a couple of specialists to help us there, but that's what we're going to do.

So when I put out 130, 150, that's not a number that's in concrete. But there's some basis for it, and the basis is the Public Consensus Review Panel report.

The other thing is, as we discussed with Senator Schumer, if we can move operations to Canada. Fortunately, because of the Authority's investment in the Canadian plaza over the past ten years the amount of work that would have to be done over there is not as extensive as what would otherwise be required, or appears not to be as extensive. We haven't worked out all the details. That would be done as part of this process. It definitely looks like it would be less expensive. Certainly we're not going to have to go out and buy a lot of right-of-way, and that would reduce the size of the requirements on the U.S. side and that translates into saving money. And that makes the project a little more feasible, from the financial standpoint.

I'm not saying "Hey, I gotta have 130, 150 million or else." What I am saying is that it's clear – based on the objectives that are known to us, we're not starting here in the dark – that some financial assistance is going to be required. And to the extent that it is required to come up with the best plan, we just putting everybody on notice that it's a reality of this project.

Part of this financial thing could be a staging of construction. In other words, if we came up with a total project, bridge and plaza, you got the possibility of staging the construction to try and stretch out the money requirements so that it's not all done at one time. I don't know whether that's doable or not. It depends on what the form of the final plan is, the preferred plan. But that certainly would be looked at also.

Anything that would help make the project work, meaning the preferred project, the project that's going to come out of the process, we're going to try and achieve. That includes making efforts to get more money from whatever source we can get it from. And it includes efforts to see if there's any possibility of the shared border management plan.

There's a lot of mistrust toward PBA staff but I don't know anybody who doesn't trust you. It's amazing.

They *should* trust me. Because I'm being straight. I've got one agenda. It's the project agenda. I'm not carrying an agenda for the Peace Bridge, I'm not carrying an agenda for the institution. I want to see this project be, as Senator Schumer says, the best bridge and the best plaza. That's where I'm going with this thing. I'm going to do everything I can to achieve that objective.

55. VINCENT "JAKE" LAMB POINT MAN AT THE PEACE BRIDGE (II)

At a June 29 press conference at its Peace Bridge Plaza headquarters, Buffalo and Fort Erie Public Bridge Authority general managers Stephen Mayer and Earl Rowe described a wide range of pending improvements in bridge operations, some of them common-sense and long-overdue, some of them highly sophisticated and impressive. Over the next two years trucks will be tolled on the basis of axles rather than weight, a fourth truck lane will be added to the US plaza, and truck turning patterns will be tuned to work better. They're introducing NEXIS, a CANPASS-type quick entry system for low-risk passenger travelers that will work in both directions and by the end of the year they'll have EZPASS for cars and trucks so nobody will need tokens and hardly anybody will have to stop for change or open windows in the depths of winter. New custom booths on the Canadian side will better utilize the 35 additional customs officers just assigned there. Coordination with the Niagara Bridge commission will be improved. They've set up a virtual traffic model that seems likely to make current operations and planning far more efficient and accurate, and the new team of bridge inspectors they've hired have figured out a way to do their work without shutting down lanes.

All of these improvements, said Vincent "Jake" Lamb," project manager of the Bi-National Integrated Environmental Process, will facilitate the flow of traffic through the Fort Erie-Buffalo corridor, but they won't provide enough capacity to handle the increased truck traffic resulting from NAFTA. If the growth of trade between Canada and the US continues at anything like its present rate, he says, more lanes will be needed, however much efficiency at this crossing is increased.

Before any lanes are added – by widening the current bridge, building a new bridge adjacent to it or as a substitute for it, or building a new bridge near the International Railroad Bridge at Squaw Island or at the southern tip of Grand Island – permits must be acquired from a host of Canadian and US federal, state/provincial, and municipal agencies. Before all those permits can be issued, the

PB A must conduct an environmental impact study. That is the process Lamb is currently managing.

THE PROCESS

The final decision on what bridge will be built, where it will be built, and what changes will be made in Buffalo and Fort Erie roads and parks will be made by the Buffalo and Fort Erie Public Bridge Authority (the PBA). That decision, presumably, will be predicated on the results of an environmental impact study overseen by a "Partnering Group" composed of the PBA, the City of Buffalo, and the Town of Fort Erie.

A few years ago, the PBA operated in almost perfect independence of the two communities at either terminus of the bridge. It bought Fort Erie and it ignored Buffalo. Now, for a variety of political, legal and economic reasons, all that seems to have changed. These days, the PBA is paying attention to community concerns, needs, and interests. Instead of spending a fortune keeping the community at bay, it seems to be spending a fortune getting the community involved. The person running that project for them running all that is Jake Lamb.

The Partnering Group will be advised by the "Bi-National Citizens' Advisory Committee," a broadly-based group which in conception is similar to the Public Consensus Review Panel (PCRP), the group that studied bridge and plaza possibilities last year which was funded by the city, county, Wendt Foundation and Community Foundation. The primary difference is this: the Authority did everything it could to pretend the PCRP didn't exist, it never sent a representative to the table, and whenever it had to send a message it sent it via such surrogates as Buffalo-Niagara Partnership executives Andrew Rudnick and Natalie Harder. Now, the Partnering Group is on record as promising to heed what the citizens have to say.

Before any decisions get made there has to be an environmental impact study, the EIS. The word "environmental" in that term is taken very broadly: air, noise, water, animal life, human life, economic impact, traffic impact, neighborhood impact, and more. The EIS process is a complex matter requiring the services of a large number of specialized consultants. Last time, the PBA tried to avoid doing one at all. This time, they're pouring a huge amount of money into consultant firms and attendant technology.

Before the EIS can begin a scoping document must be prepared. Lamb's staff is presently working on draft scoping document. "The scoping document is a document that we will use during the first public information meetings. It's a roadmap for the process. It will cover such things as project description, purpose and need of the project, goals and objectives and initial listing of alternatives based on information that obviously is available from work that was done before, how we're going to evaluate those alternatives, the involvement of the public in the process, the roles and responsibilities of the various government agencies that have approval jurisdiction, and also the civic advisory committee."

TIMELINE

Lamb says that about July 6 letters will be sent out to an equal number of Canadians and Americans inviting them to be on the Bi-National Civil Advisory Committee. The American political officials came up with their appointments as soon as Lamb requested them; the Canadians have been very sluggish about any actual participation. The first organizational meeting will be about 26 July. In the second week of July, he plans meetings with the American and Canadian agencies to which the EIS will be submitted and which will have to issue the permits before construction can begin. In the latter part of July, they expect to publish a Notice of Intent to do the EIS."That," Lamb says, "is the formal kickoff of the environmental process."

In August and September they will conduct meetings informing the public how the process will work and how public input can influence the decision-making. That will be followed by agency and public scoping meetings and workshops: deciding and discussing what questions the EIS will really answer, what possibilities it will consider.

They will prepare a draft scoping document that will be revised on the basis of the community meetings. In late September or early October, they will issue the final scoping document, the operational charter for the environmental, traffic, engineering, design and other consulting groups actually doing the EIS.

How long, I asked, will all this take?

"Our projection," Lamb said, "is to get this done in 2 years. It's a lot. It's ambitious. When we talked to the agencies, three years has been the best estimate so far. But the public is pretty well informed, the ones that are interested, as to what's going on. And there's been a lot of work that's been done already, certainly by the Peace Bridge Authority on their earlier studies. So there's a lot of stakes in the ground with respect to understanding, for instance, environmental issues, and there's been a lot of work done by the Public Consensus Review Panel. In addition to just technical information that was developed there, there was a broader impact, or a benefit from the standpoint of people being better informed about what was going on. So if we capitalize on all of that, and keep our eyes focused here, I think we can get it done in two years."

"Then what?"

"Well, if we can get this thing done in two years, we'd probably figure two years for the design, permitting and contracting. You've got to figure about three or four years to construct it. Maybe three would be good. That's a pretty realistic timeline. Maybe it's a little optimistic on the front end. That assumes that there's no down time for litigation."

TRAFFIC PROBLEMS

Some work has already begun. "We started in May forecasting the project traffic. We need to do that to size the project and to make a quantitative analysis of the environmental impact. That's underway." They've also begun new aerial photography and mapping. "We need scaled mapping of the area and very recent aerial photography to make quantitative comparisons of the various alternatives. We actually have increased the scope of the mapping in a northerly direction to go up as far as the southern tip of Grand Island because we anticipate that we're going to be reviewing off-site concepts that we believe will be suggested during the scoping process by the public."

The traffic project was stalled for two months because the operators of the Ambassador Bridge in Detroit (who have been spreading a fortune around town pushing their plan to build an entirely new truck bridge privately owned by them) asked a Canadian court for an injunction to keep Earth Tech Canada from working for Wilbur Smith, the firm retained to do the traffic study. Earth Tech, said the Canadian Transit Company in Windsor, did some work for it in Windsor, so working as a traffic consultant down here would be a conflict of interest. [The Ambassador Bridge is operated by the Detroit International Bridge Corporation in Detroit, and the Canadian Transit Company in Windsor, Ontario. Dan Stamper is president of both.]

Lamb reopened the traffic consultant procurement process entirely" When we made the selection it was based on total team." he said, so it would be unfair to let Wilbur Smith reconstitute itself after the fact. They "had new interviews that took place during May. In the meantime, all other activities with respect to the traffic issue stopped." Three firms applies and the Wilbur Smith organization was again hired, but with a different consulting providing the Earth Tech services. "We think they ended up with a stronger team," Lamb said.

The traffic consultants are looking at traffic patterns resulting from various bridge designs and locations the impact of various construction projects on the city while various kinds of new bridge might be built. "One of the first things Wilbur Smith has to do is take the Canadian and American traffic models and develop what I call a 'bridge model,' because there us no bridge model which closes the gap between the two models, and test it for validity to make sure there's compatibility with the Canadian model and the American model.

"That includes the entire Niagara River frontier, so that we would be able to analyze the 'what ifs' of traffic. If we looked at what would happen if we had a non-build alternative, the model might tell us how much traffic would go to other bridge crossings, divert to other locations, and also what the forecast would be here at the Bridge plaza. I wanted to make sure we're capable of analyzing traffic projections if we are looking at what I call an 'off-site alternative.' If someone suggests putting a bridge somewhere else, we still need to take care of our bottleneck at the Peace Bridge facilities, so what's going to happen to the traffic, what would be the traffic projections for the entire frontier? We wanted to make sure we had the capability of doing that because we knew we had to have that capability to respond to public concerns, the public questions that were raised.

"Part of their charter here is to analyze the maintenance and protection of traffic schemes that would be attendant to the various alternatives. For instance, if one plan wanted to use route 198 temporarily, or divert to city streets, we're going to have that analyzed. You need to have those numbers in order to analyze, for instance, air and noise impact. It's a very critical piece. We have authorized them to go ahead."

This study, Lamb said, would also include the potential impact of Ontario's plan to build the Mid-Peninsula Corridor, a new highway parallel to the QEW, ostensibly designed to shift truck traffic off a road primarily designed for passenger cars, but also designed to handle the increasing volume of truck traffic across this border.

Lamb talked at length about some of the key issues in the scoping, EIS and expansion projects. From here on out, we'll give his remarks without quotation marks, my questions in italics. First, his response to my comment that the EIS process seemed massively complex and messy.

COMPLEXITY

Complex projects must be defined in small simple easily understood segments – which are then evaluated within the context of their interdependence. The interconnections and interdependency of all such segments must be understood and managed in creating the mosaic which defines the huge or complex project.

PERMITTING

We're preparing for meetings with the Canadian permitting agencies to assure integration of the Canadian environmental processes with US process. That's extremely important because their process doesn't follow the same kind of calendar sequencing that we use in the American side. The Canadian process basically kicks in, officially or formally starts, when we know we have to get a permit. Their process is a lot more informal than ours. Generally, this is further down the line in terms of the American process.

The American process starts with, "What's the project and why are you doing it? What are the goals and objectives?" which also we are drafting right now.

Basically, the reason we want to meet with them is to make sure we're all on the same page about what we're doing. We're going to meet with the Canadian agencies. And then we're going to have a general, bi-national agency which will bring together the American agencies: New York DEC, FHWA, Coast Guard, all of the agencies that would have any role in the historic preservation. Any agencies that would have a role in approving, or inputting into the process on both sides of the border. We want to bring them together. We're going to go through the process with them, what the project is, the goals and objectives, the purpose. And what we see as their roles and responsibilities. To have them say, "Yes, we confirm that's how we're going to act in this thing." So that's an important step that we want to get ready for.

FREEDOM OF INFORMATION

We're going to act as if we are subject to freedom of information clause. The Peace Bridge Authority are very supportive of all of this, with everything that we're doing. I've met with the attorneys. They're talking about segregating some of the work products. I was in a short meeting yesterday, and I said, "Why don't we just don't bother segregating." It's such a pain to pick out what isn't subject to it. You almost have to be a lawyer to look at every one of these pieces of paper. I don't want something that complicated, I want something simple. We're working it out in advance. And we're working out an availability retrieval system so that the documents can be accessed. You know it takes a little time and thought to do these kinds of things.

THE BI-NATIONAL ADVISORY COMMITTEE

We're preparing specific details for how to organize the bi-national advisory committee and other focus groups.

Who's it going to advise?

It's going to advise the consultant team. It's going to advise the Partnering Group team. It's going to advise the Peace Bridge Authority decision-makers and give input to the agencies, the Federal Highway Administration. They're all interested in involvement. And they're going to be there listening, as well as us. It's for hearing people's ideas and it's kind of a democratic process for presenting, deliberating, opposing ideas. We'd certainly prefer that there be a cross-fertilization, if you will, of diverse interests represented. It will help people better understand these issues and where the sources of those issues are.

We're going to undertake what we call "collaborative workshops to focus on specific decision issues. For instance, the alignment of the bridge. Or the style of the bridge, the type of bridge. Or the plaza alternatives. Or other significant issues that are defined as the process proceeds.

This group would basically get diverse interests working together, and reviewing in detail specific work issues. There would be some sort of voting process. Let me tell you how I think that would work. We would have a series of meetings, for instance, on alignment and things associated with alignment. And the core group would be made up of the bi-national civic advisory group and anybody in the else in the public that wants to come and participate. I don't want to get involved in saying its got to be ten people from here, and two people from there or whatever. I'm very leery of doing that king of thing, it excludes people. You got to have diverse interests represented there or it's not going to work. They have to work towards a solution, not just talk about it: "Hey, let's see what we could do about this and come up with a solution to pass along to the formal, authorized decision-makers."

When I say the formal, authorized decision-makers, I'm talking about the people who have the responsibility to undertake the project, which is basically the Peace Bridge Authority, the city of Buffalo, and the town of Fort Erie. You're not going to get anything done without those three basically agreeing to it. So I want to get the public involved with inputting to that decision.

This bi-national civic advisory group plus interested public could turn out to be 100 people, it could turn out to be 120 people. I don't want to put any restriction on it. I don't want to have somebody at the door and say, "No, I'm sorry, we've already got somebody from your group." That means there's the risk here of slanting. I have to accept that risk from the standpoint of it being a democratic process. It doesn't mean you're going to show up at the last meeting and have a vote. There's going to be a series of meetings to deal with some of these projects. We're going to encourage and do everything we can to facilitate easy participation by people.

I expect that this group will do some self-governing of their own. But the idea is to get them involved, to have this cross-fertilization of diverse interests, and explore and analyze, with the help of the technical team giving them technical input of what we can do and what we can't do, and come up with a solution. We need to have a doable project: it has to be doable from the sense of financial doability, constructability, feasability, and practicality. Those things have to be defined, and we're looking for public input to help us define those things. The whole idea is to come up with a reasoned conclusion.

NATIVE ISSUES

We met with the Natives about the burial issues. When I met with Wayne Hill, we talked and I said, "I am going to ask you to participate in our process. You need to come if you have a concern about what is being planned, and let your viewpoints of your people be known to the public, so it isn't filtered through somebody else. I'm not going to carry that message. I've heard what the message is, but you're the one that has to do the message-carrying."

I said, to Wayne "I want you to know who you can come to to talk about the project and we want to have a continuing dialogue with you as we go through the process so that you know what we're doing, what our thinking is, and what evaluation criteria we're gonna use and we want your input, meaning the people you represent. We want to see them as well and we'll come make presentations to them, we'll find out about what their concerns are, so that they're part of the process.

Ron Williamson was there – the guy that's discovering Indian things. He's really doing a wonderful job there and he participated. There wasn't anybody else there. Nobody from the Peace Bridge comes to these meetings at all. They have given their absolute support and trust in what I'm doing, which I deeply appreciate.

I said "Look, I'll get our bridge engineers to show you what we can do with respect to spans. We may span over your burial sites. We may have to put footings in different locations. But we may be able to alter our ideas based on what your concerns are and based on where these things may be – based on where these burials may be."

He said "OK. We'll take a look at them and we'll consider them."

I almost fell off of the chair. I was ready to see some other reaction to it. I said, "Fine, that's fair enough."

He said, "I'll talk to my people and you come and we'll review what our issues are."

I was encouraged by that, and not from the standpoint that I favor one thing over the other, but from the standpoint that it would mean we'd have a very fair, objective open dialogue about the issue.

Much more efficient at this stage than a lawsuit afterwards.

Exactly. So Ron pointed out that he knew where everything was: every square meter of that place has been dug up on the north side. On the south side, we already had cooperated with Williamson and the First Nation in doing digging and locating the piers in such a way that it was acceptable to them.

I was pleased that they would consider us having air rights over their burial ground. That wouldn't happen with a US cemetery. You wouldn't go anywhere near it. They feel very strongly and they don't want to disturb their spirits. They have allowed some construction to take place over top of the burial grounds as long as that construction did not go into them. Now where would you get this any place else? These people are very fair-minded. This is how their culture works and we respect that culture. And that's fine. Our culture wouldn't go that far. We'd say "Outta here," right?

HISTORICAL SITES

We're going to look at replacement. You know that we have to go through a process on replacement, that there's certain protection afforded to historical structures, historic sites. So it's a hill that has to be climbed. Even changing the arch, we have to clear that with SHPO.

But that's never been declared a historic site.

That doesn't matter. Even if it's declared eligible, which it has been, it's afforded the same protection.

On the Canadian side there is an indicated strong preference to keep the bridge. They met with Christian Menn and they said to him, "We want to keep the bridge." That's what they were telling him.

In any event, we're going to look at it, and we're going to find out what people think about it, what the physical aspects of it are.

OTHER SITES

Basically we have a bridge problem. We have a Peace Bridge problem which is a problem on an existing corridor – the QEW and I-190, a 2000- , 3000-foot section of it, the bridge and plaza. We're not looking at an alternative of changing the system or the corridor, we're looking at taking that bottleneck out of there.

Having said that, we recognize that there are issues being raised of perhaps looking at other sites. Our attitude and approach and policy is that we will look at these other sites. Not within the context of a regional study to determine the optimum location of a potential new crossing somewhere, but we will be responsive to a suggestion in the public forum that would look at, for instance a crossing at the International Bridge. Technically, one could argue that we don't have to do that. But I don't think that's the kind of answer I want to give to he public, because the questions are being asked. That's why we're doing the aerial photography and the mapping and the traffic analyses, so we're able to analyze them to the point of being able to discuss them and evaluate them in the public forum. We will look at specific off-site alternatives as an alternative to what we're talking about here. We have an existing bottleneck there that we have to fix one way or the other anyhow.

Some people say, "If you have more customs people, then you don't need more bridge space." That's not correct. You're still going to need the physical bridge space. Let's say you remove the plazas. You would still not have enough physical capacity on the bridge with three lanes. That's the bottom line. The traffic guys will demonstrate that and they will have these little PacMan displays to run these little ants around to show you what it would look like, so the public can understand. It'll demonstrate what I'm talking about.

If there were no customs immigration or tollbooths, there would still be a problem?

Yes. You'd have a problem at the bridge only. I don't know about the connecting roads – maybe you'd want to improve them too – but basically, if you had no operations on either side, you'd still have to do something with the bridge.

What happens when you take a look at it, and one of these crossings , say leaving cars on the Peace Bridge, putting trucks somewhere else, does make some kind of sense. What do you do with that? Because that's inimical to your interests.

If it's a viable alternative, it's going to take on a life of itself.

It's been suggested to me in a couple of these public meetings that it would make sense to move the trucks. If you're going to move the trucks, you're going to have to have the jurisdictional ability to do that. Those the issues would have to be addressed. Presumably, that means no trucks on the Peace bridge, because if you keep having trucks on the Peace Bridge, you can't move the customs and immigration operations. We have to look and see what the aspects of doing that are.

And we would. If that is a serious alternative that the public wants us to look at, if it comes up in the public scoping session, we'll take a look at it. Our traffic engineers could analyze the "what ifs" with respect to the truck traffic.

MODIFYING THE SCOPING MEMORANDUM

When we put it together, then we modify the draft scoping document or scoping memorandum that we had started with, to reflect issues of substance, or concerns of substance. We're going to document it all. We're going to go through that and pick out and highlight and put it into categories, maybe prioritize these issues that are brought up from the standpoint of how serious they are, and how much concern was expressed.

Then we're going to go through public meetings again to go over or modify that scoping memorandum, which basically is turned into a final scoping memorandum. Now what happens in that final scoping memorandum is, we're going to have a long list of alternatives. In other words, we aren't going to knock any alternatives out during the scoping process. We're going to listen to everybody, especially environmental concerns.

You have to do enough analysis to make sure you basically haven' t missed the big picture, haven't missed something that could eventually turn an alternative into a reasonable one, a practical one. That is going to be a demanding process, requiring a lot of energy, a lot of time.

But we're not presenting an analysis at that point in time. We're just talking about the process, and what we call initial alternatives for screening. That's what I'm talking about. They would survive the scoping process.

After that, how may alternatives go into the next stage?

The draft EIS. Sometimes one goes in as a preferred alternative, sometimes more than one goes in. It depends on what that evaluation and public input is to this evaluation of retained alternatives. Hopefully you end up with a preferred alternative. You don't necessarily have to. You go through the draft EIS process and go through iteration of public information meetings, and, if you have more than one there, to try and really narrow down to one. It's quite possible you could end up with two that are rather similar, or two that are quite a ways apart that would basically satisfy people. In which case you say, "Go ahead applicant, you make the final selection for the preferred alternative."

What goes to the lead agencies? Does the lead agency in the EIS see the two, or does the client?

The lead agency would see the two.

And after that is when the client decides what's going to be done?

The applicant would decide.

This takes us through the end of the environmental impact study. What happens next?

What happens is the applicant decides to go ahead with the project or decides decide not to go ahead with the project. If the applicant decides to go ahead with the project, they would prepare designs and get the permits. A lot of the permits depend on more detailed design than we would develop during this scoping and EIS process, because the amount of work we would do in this process in terms of detailing design would be only sufficient to make a objective comparative evaluation of alternatives and their environmental impact.

I don't anticipate that we're going to get into picking out railings and that kind of thing during the environmental process. But we have to do enough to have comfort about it being what we want. We get into the final design process, and concurrently with that, in a sequential way, you do the permitting. If there's right of way to be acquired, that has to be an early action item, because eventually it means relocation of people. You have to do enough designs to draw your right of way drawings and get that underway. You get the design down, and you go through a tendering process.

PROBLEMS

Do you see any major problems along this very smooth route you've described?

If I've described it as smooth, that was unintentional. I don't see it as necessarily being smooth. I think we're going to have our rough spots.

I mean, are there major, environmental issues, other than the Native burial grounds issue? Are there problem areas that you foresee?

Well, I'm concerned about the burial grounds. I think we have to be very careful there, be very sensitive to the issues raised there. We can't just disregard them. They're serious and I think there's logic to those issues. Especially, comparing it with what our culture would say if similar things were being done on the other side of the river. I think that's an issue.

I think another issue is the bridge itself with respect to historic significance and what I call disposition of the existing bridge. I'm not saying these things can't be worked out, but I think these are focal points.

The other has to do with of course the US plaza and connecting roads. There's definitely a desire in the community, I don't know how broad that desire is but I know it exists, to make some improvements there that are what I call quality-of-life enhancements. They are certainly good objectives for people to have for themselves and their community, and I certainly respect that, and would like to see us try and achieve some of those objectives in conjunction with the bridge. They do cause issues, more serious issues, than would otherwise be encountered because it means we have to move the plaza to a different location. Modify its location somewhat – maybe move completely or partially off its existing location. This means going into the community in terms of moving people. It's really a greater impact from the standpoint of noise and air quality. But right now there's just basically one side of the plaza that's adjacent to a residential area. If you move, depending on how far you swing that arc, you could have three sides of the plaza, which means you have to be thinking in terms of buffer zones. You have to be thinking in terms of how do you mitigate or lessen or decrease the obvious environmental impacts of any transportation facility. There's noise and air pollution from mobile sources. It's unavoidable.

So we're concerned about the environmental impacts with respect to air quality and its associated public health issues. And we're concerned about impact on the community, that includes the potential of an environmental justice issue. We're very sensitive to that. We think it's our job to make sure that that issue is addressed. When I saw the public review consensus panel, and of course I was talking to the engineers as they were going through the process, and we talked about the environmental justice issue and it didn't seem to have as much serious import as I thought it should have had to the people who were involved in the process. I think it was probably that this isn't the time to do that; it should be done later. Which I accept. It's the same as putting, as part of our scoping document, a scheme out there without realizing what all the impacts will be. So I said okay. I accept that. It needs to be looked at later. But it is serious. There needs to be people that have to be moved. It isn't only the people who've been moved – it's what's left of the neighborhood. It's like you live here, and somebody coming though and moving out ten of your neighbors. Is the neighborhood the same? Especially if there's some Chinese Wall there.

The connections, the air quality; public health as I mentioned is an issue. I would like to see park restoration. I would like to see the development of that park to be a flourishing attractive place for people to go to. I'd like to see the state build, as I've suggested here – and I'm not the first one to suggest it, I don't want to take credit for any of this stuff, it's all old stuff – this welcome center, with perhaps some commercial development alongside it. So that it's m ore than a bridge project, more than a transportation project. Not that we would do that as part of this project, and we're not going to analyze that as part of this project, but certainly there are obvious embellishments, things to be done in conjunction with this project.

And of course the money. I'm worried about financing a project that would be most likely to be achievable on a consensus basis, if you follow me. It's going to take money to do this. And I think it's going to be, based on what I know now, and I don't know...we'll have to see what develops here, we'll have to have more detailed estimates and have better information in order to make judgments. And I realize the community has other priorities and other needs, but you know why shouldn't we try to get the money that's needed to do the project in a complete way? Why shouldn't we do that? Why shouldn't we aim for that? And that's going to be extremely important, because in the final cut, when we get down to the end, what you put in that draft has to be doable. It could be doable from the standpoint of staging it, if you can work it out that way. And it's got to be doable in terms of having the money to get it built. Of having a plan to get the money to do it. So that's going to be, again, one of the key issues.

I think that's all of them.

56. THE PEACE BRIDGE NOW

Twice in one week last month Vincent "Jake" Lamb – who is honchoing the environmental impact study Judge Eugene Fahey told the Buffalo and Fort Erie Public Bridge Authority it had to carry out if it had any hope of expanding bridge capacity between Buffalo and Fort Erie – put on exhaustive (and exhausting) informational seminars.

The first was Tuesday night, August 21, at the Leisureplex complex in Fort Erie (built with your toll dollars by the PBA); the second was the following night in the Mary Seaton Room at Kleinhans Music Hall. In both, Lamb described in minute detail the process now going on and the organizational structure carrying it out. When he was done, representatives of the several firms hired to provide studies on or do work involving bridge design, environmental impact, public relations, traffic flow during and after construction, archaeological matters and more talked for five or ten minutes each about their firms' virtues and their projects' goals. Nearly all of them flashed their own PowerPoint slides.

I went to the Fort Erie session. Well over 200 people attended. After the 2½-hour performance by Lamb and his associates and consultants, members of the audience asked questions and made comments. Some commented on bridge design and expansion politics. Many vented about current bridge problems. A former Buffalo politician tried to sell Lamb a computer polling service for sale by a West Coast company he now represents. A retired Canadian customs officer, who had puffed calmly on an unlighted cigar while he waited to get his turn to speak, complained with great heat about the long delays on the American side because so many inspection booths were so often unstaffed; he ended by saying they should just build a bridge now, a non sequitur I thought. Someone else complained about the traffic problems caused when the PBA shifted the American-side duty-free shop to the toll plaza itself. The Buffalo session the next night, I was told, was pretty much the same: the long PowerPoint presentations by Lamb and the consultants, followed by a wide range of current gripes and continuing observations.

Lamb said that the Leisureplex and Kleinhans sessions were

designed to invite questions people in both communities had about the current process. Some of those questions were answered by Lamb and the consultants during those two sessions. Other questions came up in the two weeks after the sessions, after people had time to think about what was said and shown. Here are some of them.

Isn't this the third time we've gone through this process? Didn't the PBA previously have design competitions and a design charette where people came up with ideas?

The Buffalo and Fort Erie Public Bridge Authority indeed had design presentations (one in collaboration with the Buffalo Niagara Partnership and the *Buffalo News*), but this is not in any way the third time we're traveling this ground. It's the first. Last time around, the Public Bridge Authority had made up its mind what it was going to do long before the public got to see anything, and the public sessions were really public relations, not public involvement or participation. A small group of people sought to impose a very narrow vision on the rest of us, ignoring the environmental and social consequences of what they were doing. They ignored Buffalo and bought off Fort Erie. They ignored New York environmental law. This time, because of Judge Eugene Fahey and also because of the Public Consensus Review Panel and the New Millennium Group, the process has been reversed: there will be a huge amount of public involvement and consultation, and the process will be open all the way. No public works project in Erie County has ever had the kind of visibility promised in the current Peace Bridge process.

Jake Lamb works for Parsons, the PBA's long-time engineering consultant. They came up with the previous twin span design. Isn't he in a conflict of interest situation now?

It's not clear whether the twin span idea originated with Parsons or with the PBA itself, and that's history anyway. People who know Lamb well say he's an engineer dedicated to whatever project he's doing now, not to ratifying an idea someone else had once upon a time. In any case, Lamb is isn't doing design; he's managing a process involving design and environmental consultants who represent and possess a far greater range of competency and skills than anything the PBA ever employed previously. Perhaps the question to ask isn't "Is Jake Lamb wedded to the past?" but rather, "Will world-class bridge builders like Christian Menn and Gene

Figg let their names go on a design they think is stupid?" When Figg made his presentations at Lamb's two evening sessions last week he mentioned at least a dozen times the awards his bridges have won. I picture him sleeping with those awards under his pillow. I just can't envision him signing off on lousy design or keeping his mouth shut while a fix goes down.

The process Lamb described is very complex. Isn't this going to take years longer than the twin span plan the PBA had?

No. Although things seem to be moving slowly because of all the public involvement, the timeline is almost exactly the same as the timeline for the completed bridge project the PBA projected back when it thought it could simply impose its will on everyone and everything else. They projected about a decade from when construction began to full six-lane capacity with connecting roads and plazas in place; so does the current team. People who say, "Just build it," are jabbering. Build what? No complete, feasible design was ever prepared. The PBA had an idea of what kind of bridge it wanted to build, but it had no design for a plaza or connecting roads on either side of the river. The first coherent complete plan will be the one to come out of this process.

Why should little Fort Erie have the same voting power in this enterprise as Buffalo?

It would take an international treaty for things to be otherwise. It took specific actions by four governments for the private corporation that built and operated the Peace Bridge to become the public benefit corporation called the Buffalo and Fort Erie Public Bridge Authority: Canada, the United States, the Province of Ontario, and the State of New York. Some structural aspects of the PBA have been modified over the years, but that quadripartite empowerment still obtains. The control of the PBA is vested in five New Yorkers (two appointed by the governor, three serving ex officio) and five Canadians (all appointed by the Canadian federal Minister of Transport in Ottawa, though the individuals appointed have always been from Fort Erie or one of the nearby towns). Buffalo has never had a seat on the PBA or a voice in its decisions, though sometimes Buffalo residents were appointed to it. Before Judge Fahey's decision, that ten-person agency held sham public hearings on decisions that had already been made in private. After Judge Fahey's decision, it decided to open the process totally and to set up a three-member "partnering group" to administer the decision-making process. That group consists of the PBA, representatives of the city of Buffalo and the town of Fort Erie. All three parties have agreed that no bridge will be built that doesn't meet with the approval of all three members of the group, so – if that agreement is honored – Buffalo now has far more power in this than it ever did before.

Why, when they talk about money, does it seem that the US is going to be paying far more for the new bridge than Canada? Shouldn't the PBA pay for everything?

If the money is there, sure it should. If the money isn't there and if Americans want changes on their side of the river (the Canadians argue) the Americans should pay for it.

The PBA says it has and can earn enough money to build a bridge; money to restore Olmsted's Front Park and provide new connectors must come from other sources. This may or may not be true. If the process is as open as Lamb and the PBA says, we'll have a chance to inspect those numbers.

We could say, "Well, we never should have lost the park in the first place," which is true, but that was then and this is now. Canadians aren't going to pay for stupid American decisions 30 and 40 years ago. We could say, "But the PBA has bought off Fort Erie, it's given the town a city hall, a courthouse, a recreation center with two ice skating surfaces, and more," and the Canadians would say, "Yes. The PBA wanted concessions from us and that's what we traded for those concessions. How come you didn't bargain for good things before you gave up one of your most important public spaces? You can't expect us to pay now because you let us get away with murder all these years."

I'm not saying they're right; I'm just telling you what they say.

Keep in mind that for years, Canadians have gotten a better deal out of this crossing than Americans have. Canadian federal agencies pay nothing for office space on bridge plazas that Americans federal agencies pay millions of dollars for, and travelers who pay tolls with Canadian money pay far less than travelers who use American money (the current toll is $2.25 American and $2.75 Canadian, but $2.75 Canadian is worth only $1.78 American). The simple fact is, we share a bridge, but we each have our own interests. The only thing that's different now is, the wheeling and dealing and the payoffs may for the first time be public, not secret as they have been for so many years. We can't undo the past, but we can perhaps work to make sure there is more equity in the future.

Why isn't the PBA looking at other places to put the crossing? Do all those trucks have to come in and out of Buffalo?

No, those trucks don't have to cross here. A new toll road for trucks is about to slice through Ontario and it can connect at Front Park or a little north at the International Bridge, or at the southern tip of Grand Island or it can feed into one of the northern crossings. Jake Lamb says that the PBA's EIS process will consider other crossings, and it will consider no change at all at the current location, but the PBA owns the Peace Bridge and it owns facilities connecting to it on both sides of the river. Given any choice at all, it will expand its options right where it is.

They're not the only game in town. The operators of the Ambassador Bridge between Detroit and Windsor, Ontario, are fairly panting to build a truck bridge connecting the QEW or the new Canadian truck highway with the I190 adjacent to the International Bridge (the iron bridge where trains cross the river a few times a day). They've recently poured a huge amount of money into Buffalo hiring key workers and flacks, such as the former head of Pat Moynihan's Buffalo office and an attorney who specializes in public works environmental law. They said a while back that they were going to do their own environmental impact study, but since then they've been working pretty much in the dark, just like the PBA of old. Their spokesmen talk as if they have all kinds of experience in this sort of project, but so far as I've been able to find out, the firm has never built anything. Its owner is a very rich reclusive Detroit businessman, a heavy Republican contributor with major connections. They're a wild card in all of this.

So where are we?

Not where we were, that's for sure. The PBA may be trying a scam, but if they are, it's going to be hugely expensive and they'll probably get busted at it. Democracy is a genie very difficult to stuff back in the bottle and I think that's what they've unleashed in the public process currently going on. It looks inefficient and slow, but only if you think that the previous imperious assertion of raw power and contempt for community needs was a desirable or useful efficiency.

One reason this is all so slow and klutzy, I think, is no one in any of our earlier major public works projects tried to involve so many members of our several communities in a process certain to impact their lives for years to come. We didn't have a voice in the disastrous suburban relocation of UB, the klutzy design and wretched location of the convention center, the design and location of all those highways that sliced and diced our city. We had no voice in that half-baked Main Street trolley project that took so long to get operational it killed downtown retail business, then literally stopped dead in its tracks before it reached the city limits (in large part because powerful suburban politicians were horrified at the possibility of cheap public transportation providing easy suburban access to people otherwise confined to the inner city).

For years, the PBA fought to keep us out of their private rooms, rooms in which power and money were controlled and parceled. Judge Fahey got us access and the Public Consensus Review Panel and the New Millennium Group told us what to look for once we got inside. Some members of the PBA would like nothing better than to go back to the way things were. Others think their task is to prevent that from ever occurring again. The only thing that will give that new vision a chance of success is continuing pressure from the rest of us on every one of Jake Lamb's panels, committees and consultants.

57. BORDER LAW

SOME BACKGROUND

It was supposed to be simple, like most other major public works construction projects around here: a small group of people meeting in private decided what they wanted to do, how they wanted to do it, who would profit from it, then the construction would go ahead and then everybody would move on to something else; somewhere along the line the public would be informed about what was being done to them, to their environment, to their community.

In this particular case, it was the Buffalo and Fort Erie Public Bridge Authority deciding long before going public with their plans that they would put up a steel bridge alongside the current 75-year-old bridge, using pretty much the same technology and adopting pretty much the same design as the older bridge. They held some sham design hearings, then announced the conclusion they'd reached long before.

Buffalo architect Clint Brown and businessman Jack Cullen said with all the money this was going to cost, the PBA should take a deep breath and rethink the whole thing: this might be an opportunity to build something really beautiful. UB architecture dean Bruno Freschi, working with San Francisco bridge designer T.Y. Lin, came up with the sort of thing Brown and Cullen had been talking about: a design for a beautiful, soaring, cable-stayed, single-pylon bridge made of concrete, using the most modern materials and technologies.

The PBA said, Too late, you should have made your suggestions earlier. People said, How could we, when you'd made up your mind about what you were going to do before the first public meeting was held, and when you dismiss out of hand every suggestion not your own? The PBA said, Too late, too bad.

But something happened this time that was different from the other times the rich and powerful guys had their way with the city. A huge groundswell of opposition developed, and it didn't go away.

The New Millennium Group took on the bridge as a major public interest project. The *Buffalo News* ran editorials saying that a signature bridge would be a better idea than what the PBA had in mind, but the PBA had the responsibility for all this so the New Millennium Group and all the other people asking for something better should just let this one go. Artvoice, a weekly newspaper, took exactly the opposite position: it published dozens of articles by a half-dozen authors challenging the PBA on aesthetic, economic, political, environmental and other grounds. Every time the *Buffalo News* editorial page said "roll over," *Artvoice* said, "stand up." Every time the Buffalo Niagara Partnership (the former Chamber of Commerce) urged immediate construction of the anachronistic steel bridge, the New Millennium Group's Bridge Action Committee urged people to look at the implications of hundreds of thousands more diesel trucks with their engines idling in a residential neighborhood, the foolishness of duplicating an ugly and anachronistic design, and the need for public participation. Andrew Rudnick, the Partnership's CEO, kept saying that citizens should stop interfering in things that were none of their business; Jeff Belt, spokesman for the New Millennium Group kept saying that if citizens didn't interfere the businessmen would romp over them one more time.

The Margaret L. Wendt Foundation, the Community Foundation, the City and the County, contributed money to set up the Public Consensus Review Panel. The PCRP hired consultants and had hearings about all the things the PBA was trying to avoid. The PBA refused to have anything to do with the PCRP until it seemed a legal judgment was going to go against it, at which point it seemed to join but in fact worked very hard to make a mess of the PCRP's work.

The PBA came up with a scam to make the legal part easier than it otherwise would be: they said that the bridge construction work and the consequent work on the US plaza and connecting roads were two separate projects. They got a finding of "no significant impact" from the US Coast Guard, which, they said, meant they did not have to do an environmental impact study [EIS] before starting construction on the bridge. That meant they did not have to talk or listen to any citizen concerns. They did not have to respond to anybody. They would, they said, do an EIS when they got to the plaza and roads. That, opponents pointed out, locked everything down: once a bridge was up there would be no choice about the plaza and roads.

The City of Buffalo refused to give the PBA the easements it needed to begin construction. The City and the Episcopal Church Home both sued the PBA, saying that the bridge and plaza projects were not two separate activities, they were one big project and they should therefore be subject to New York environmental law. The PBA, in turn, sued the City for the easements. Judge Eugene Fahey threw out the PBA's suit and agreed totally with the City and Episcopal Church Home. He said the segmentation was a fiction, that the PBA had to obey New York environmental law, which meant they had to do an environmental impact study. Which meant the people had to be heard.

At first, the PBA board resolved to fight Judge Fahey's decision. Before and after his decision came down they spent a huge amount of money on a newspaper, radio and television disinformation campaign (similar to what Kaleida Health did before it came to the current cease-fire with the Children's Hospital physicians).They hired a new batch of lawyers from New York City, who told them they didn't stand a chance of getting it thrown out. PBA chairman Victor Martucci, who had been a fervent advocate of the steel twin span, convinced the rest of the board that there was no point continuing the fight, that the only rational course was to stop warring with the community and to do the EIS.

Vincent "Jake" Lamb, who had been a vice president of Parsons design, the PBA's principal consultant, was put in charge of the environmental impact study, which formally began its work on 1 January 2001. Things changed immediately. Instead of keeping everything secret from the public, the public was invited to take part in committees, panels and hearings. Lamb set up consultant groups with so many respectable and respected participants no one has suspected them of being bought off or fixed. I don't know anyone who has ever caught Lamb lying or distorting or misleading. Even some of the PBA's fiercest opponents trust Jake Lamb. I trust Jake Lamb.

WHAT'S GOING ON NOW

The consultant panels are meeting. Air quality, noise levels, and bird migration patterns are being studied and monitored. Traffic modeling is ready to make what-if analyses on some of the 50 bridge alternatives that have been suggested thus far. Things are moving more or less on schedule. Lamb's team came up with a draft scoping document – a statement of what the EIS is supposed to accomplish and what the alternatives might be. The City of Buffalo is scheduled to comment on that document on March 18 and Lamb's team promises a revised scoping document on March 28, followed by a Citizen's Advisory Council meeting on April 10 or 11, a public information meeting shortly after that, and then a meeting of the so-called "partnering group" – representatives of the

PBA, the City of Buffalo and the Town of Fort Erie. Then they'll begin screening site proposals. Once they settle on one or more preferred sites, they'll start talking about actual design.

If all goes according to plan, Lamb expects to have his new bridge system in operation at least by the time the PBA had planned to have its system in operation, about eight years from now.

Nothing of substance has happened yet: no design is on the table, no determination has been made where the new bridge will make land or what size it will be or what will happen to the old bridge.

Lamb says things like this: "I want to do this thoroughly, I want to do it the right way, I see the city as a partner, I've always seen the city as a partner, and if they're not in it, it ain't gonna work. Be patient. What's important is the substance and the process is as well. So you've got to work through things. You can't get emotional, upset about them." No senior representative of the Buffalo and Fort Erie Public Bridge Authority would have thought, let alone said, anything like that two years ago.

The amazing part of all this is that in so short a time an organization that hardly anyone trusted or liked has become an organization that isn't even on most people's radar scope any more. Jake Lamb is in charge of the project and Paul Koessler is current chairman of the PBA board. We can sleep well tonight right? Well.... More on that later.

REVERSE PROCESSING

One of the ideas that surfaced several times during the Public Consensus Review Panel hearings was moving all the customs and immigration activities to the Canadian side. There's a lot more room for it over there, and clearing that business out of the plaza would permit restoration of the large segment of Front Park the Public Bridge Authority has incorporated over the years.

At the time – two years ago – people said it was too complex, it would take consent of two federal governments, and the unions and professional organizations of the American customs and immigration workers didn't want to work across the river. Moreover, the American agents might not be able to carry guns in Canada and they didn't want to work without their guns.

But complex things sometimes simplify over time, and what starts out in one place often winds up in another. Because of the terrorist acts of September 11, it's far easier to talk with Washington about border issues now. And the public interest that started by focusing on design of the bridge has now expanded to include, and perhaps be dominated by, desire to restore Front Park, establish access to that part of the waterfront, and develop a proper gateway to the country and the city.

Moving the customs and immigration operations to Canada, while still a complex matter, is far more a possibility than it used to be.

But there's a new impediment to it: the notion of "reverse processing" currently being lobbied for very strongly by the IBTOA, the International Bridge and Tunnel Operators Association. The idea is to have all Canadian immigration and customs operations on the American side of the river and all American immigration and customs operations on the Canadian side of the river. That way, vehicles can be checked before they go on the bridge and there are no backups on the bridge itself because once a vehicle passes through the inspection point, it's got open road across the river and into the other country.

That's all it does: reduce congestion on the bridge. Reverse processing doesn't speed anything up, it doesn't reduce congestion. It just moves where that congestion occurs from the bridge and bridge plaza to the QEW on the Canadian side and residential areas of Buffalo on this side. It uses at least as much, and perhaps more, land for the processing operation.

I've heard that it's a reaction to 9/11: the IBTOA says this will keep terrorists from blowing up bridges and tunnels in the eastern part of the Canadian-American border. If so, it's the kind of knee--jerk substitute for thinking that says we need two bridges at our location in case a terrorist or old-fashioned mad bomber blows one of them up. Like Osama bin Laden is going to hit the Fort Erie and Buffalo connection rather than the Golden Gate Bridge, the George Washington Bridge, the Lincoln Tunnel, Holland Tunnel, Queen's Midtown Tunnel, or the Verrazano Narrows Bridge. Or that he or whoever has his job at that hypothetical point in time would be smart enough to blow up one bridge but not smart enough to blow up its sister fifty feet away.

Reverse processing is one of those ideas that, like the steel twin span, seems so stupid and dysfunctional on its face that you think you must be missing something. Then you look at it and think about it and it turns out you're not: it is stupid and dysfunctional. Maybe, as with the steel twin span, somebody somewhere will make money or get job security out of this, but for most of us it's worse than what we've got now.

In the twenty years I did prison research, time and again I'd see interesting projects killed and horrible procedures initiated by prison administrators in the name of "security". Security trumped everything: intelligence, experience, and common-sense. Raise the specter of security and all reasoned arguments are sidestepped on the spot. Almost always "security" was, so far as I could tell, a mask for laziness, incompetence, simple fear of anything new or different.

The IBTOA is lobbying for reverse processing to be the basic procedure everywhere. It might be a good idea in some locations. It might, Jake Lamb says, work perfectly well down on the Mexican border. But it's a bad idea here. Lamb says a more rational procedure would be for the whole thing to be flexible: find the method that fits the traffic and the place.

What's to be done? Lamb says if you think it's a lousy idea (it is) get on the phone and to write letters to your local, state and federal representatives (do it). He reminded the Common Council's Peace Bridge Task Force on March 14 that citizen involvement was a major reason he was standing there representing the Authority now.

SIX MILES UP, SIX MILES DOWN

He also told the Task Force that the Buffalo and Fort Erie Public Bridge Authority had what amounted to a franchise for bridge crossings on the Niagara River from where it begins at this end of Lake Erie up to the middle of Grand Island. This bit of information, buried deep in a 75-year-old piece of Canadian legislation, may dampen what ever dreams the Detroit International Bridge Company still has for setting up a money-making operation in Buffalo and Fort Erie.

DIBC is a private corporation owned by reclusive Grosse Point businessman and heavy-duty Republican party contributor Manuel Moroun. It operates the Ambassador Bridge between Detroit and Windsor, Ontario. DIBC has been making forays into this area at least since fall 1999 in the hopes of building and controlling a bridge that would replace the Peace Bridge entirely or in large part. They've made elaborate, albeit vague and airy, presentations to Buffalo citizens' groups and politicians. A year ago, they retained a local PR firm, set up a presentable Buffalo office, and hired Jim Kane (former Senator Pat Moynihan's main man in Buffalo) and Robert Knoer (a prominent Buffalo environmental attorney).

At first it seemed that they were planning on building their bridge next to the International Bridge – the railroad bridge that has its American-side landing not far from the foot of Amherst Street. They talked about how easy it would be to make connections on this side to I190 and, taking advantage of now-unused railbeds on the Canadian side, to the QEW. They had huge maps and elaborate PowerPoint programs showing all of this. The last time I saw Remo Mancini, their vice president handling the operation, he said the International Bridge site was just a for-instance, that they might put a bridge anywhere in the area, that this was all to be worked out.

They first came on the scene when the Buffalo and Fort Erie Public Bridge Authority was in their maximum stonewall mode, so several local politicians were nice to them even though there was little to suggest they were anything other than out-of-town businessmen smelling Buffalo and Fort Erie money. Senator Charles Schumer even emceed one of their presentations at the Adams Mark. I hadn't heard anything of them lately, so last week I asked Jim Kane if they were still in the game. He said they surely are.

For now, anyway. The Canadian legislation that may complicate, if not terminate, their involvement in this project goes back to this passage in the act of Parliament permitting the creation of the privately-funded Buffalo and Fort Erie Peace Bridge Company, the predecessor of the public benefit corporation called the Buffalo and Fort Erie Public Bridge Authority:

> The Company may construct, maintain and operate a bridge across the Niagara River for the passage of pedestrians, vehicles, carriages, electric cars or street cars and for any other like purpose, with all necessary approaches, from some point in Canada within the corporate limits of the village of Fort Erie at or near Walnut street in the said village to a point within the limits of the city of Buffalo, in the state of New York at or near Hampshire Street in said city, so as not to interfere with navigation, and may purchase, acquire and hold such real estates, including lands for sidings and other equipment required for the convenient working of traffic to, from and over the said bridge as the Company thinks necessary for any of the said purposes.... *Provided, always, that no other bridge for a like purpose shall be constructed or located at any point nearer than six miles from the location of the bridge of the Company, except with the consent of the Company or of the Governor in Council.*
>
> *An Act to incorporate Buffalo and Fort Erie Public Bridge Company*
> 13-14 George V. Chap. 74 (Royal Assent June 13, 1923), Chapter 74, §7, ¶ 1)

The italics are mine. Jake Lamb says the PBA's lawyers say that italicized section means the PBA has an unambiguous franchise: nobody can land a bridge on the Canadian side six miles north or south of the current bridge unless the cabinet of the Federal government of Canada decides to permit it, in which case the Buffalo and Fort Erie Public Bridge Authority would have to be fully compensated.

Why the franchise and why six miles? "Franchise protection for toll operators was common in Canadian and British law," Lamb said. "This protection in England was normally within one or two leagues of the bridge. We think this was the basis for Canadian Parliament choosing six miles, which is the modern equivalent of two leagues."

I assume that if Manuel Moroun and his staff intend to pursue their idea of setting up a private bridge operation here they've already got lawyers trying to figure out a way to crack the PBA's franchise. If they're not successful, the most they can hope for is to build a bridge halfway across the Niagara River, or out in Lake Erie, or the other side of the Grand Island bridge.

EMINENT DOMAIN

After all the hearings are done and all the consultant reports are submitted there will be a plan to expand capacity at the Peace Bridge. No one yet knows what that bridge will look like, where it will make land or what will happen to the current bridge. That means no one yet knows whether the current Public Bridge Authority footprint on the American side will stay exactly as it is, shrink (thereby permitting rehabilitation of Front Park), or move. If it moves, the PBA will have to acquire property.

Some speculators have bought property in the Peace Bridge plaza area, hoping that new construction will move their way and they'll make a good profit when the PBA has to buy them out. Some people who live there say they won't sell no matter what. They won't have any choice in the matter.

Privately owned real estate is nearly sacred in American law. Criminal lawyers base their work on the Fifth Amendment to the US Constitution, which entitles people to a grand jury, prohibits double jeopardy, says people can't forced to testify against themselves or be punished without due process. Commercial lawyers are more interested in the final 12 words of the Fifth Amendment: "nor shall private property be taken for public use, without just compensation."

Therein lies the rationale for the power of eminent domain, or condemnation. Private property can be taken for public use, and the state has to pay a fair price for any property so taken. So long as it pays just compensation, whatever that is, the government can take your house or factory or store when it wants to build a highway, railroad line, school, hospital, police station, or bridge plaza.

If someone wants to buy your house and offers you less than you think it's worth, you can hold out a while longer. If you just don't want to sell at all because you like it here and you don't need or want the money, that's it. Everything stops right there with your decision. But if an agency with the power of condemnation makes an offer you don't like and you say "No," they just take it anyway, and they send you a check for whatever they think the place is worth. You can sue to stop them, but it's very rare that government agencies lose eminent domain lawsuits.

The question isn't whether property can be taken for Peace Bridge plaza and roadway connections – it can – but who will control that taking, who will have the power to condemn.

Lately, Jake Lamb has been having conversations about eminent domain and the Peace Bridge expansion project. It's not a hot burner issue now, he says, but at some point down the line, whatever plan is finally adapted may very well require the acquisition of property from people who don't want to sell, or who don't want to sell at a reasonable price ("My price is $25 million, not a penny less. I'm sentimentally attached to this four-room wood frame house."). Some agency will have to exercise eminent domain, otherwise the whole project could stall while complex legal proceedings work themselves out in court.

Lamb says neither he nor his employers are pushing to have the PBA given condemnation power, they just want clarification as to who will do it and how it will be done. Most people who listen to Lamb talk about it conclude that the PBA really wouldn't mind at all if it were given the power, that they would, in fact, very much like to have it because if they don't have it they'll have to get the City of Buffalo to do it for them, and they worry about their condemnation requests getting stalled in city hall bureaucracy or because of competing political interests. They think everything would be faster, cleaner, and neater if the PBA could just do it themselves.

"We are preparing some draft legislation," Lamb said. "The Port Authority of New York has it." He met with the Western New York delegation to talk about it. Their reaction seems positive, though guarded. As soon as the subject of eminent domain went public a week or so ago, two of the PBA's former opponents, the New Millennium Group and the Episcopal Church Home – backed the idea.

The reason, they said, was because Jake Lamb and PBA board chairman Paul Koessler inspired confidence and trust.

"While NMG does support exploring the option to grant the PBA the power of eminent domain for this project," wrote Bill Banas, chair of the NMG Peace Bridge Action Group, after reading an earlier version of this article,

> we do NOT do so based solely on blind trust in any group or individual. We're grateful for this new process and that the PBA's relationship with the community seems to have improved , but we haven't donned the rose-colored glasses just yet. In the interest of efficiency and practicality, we would support such an option only with the proper controls in place. I am interested in getting this project done – but I am even more interested in getting it done RIGHT. As always, I'm not willing to sacrifice everything else on the altar of expediency. As I am not an expert in the legal nuances of eminent domain, I'm not exactly sure what form these controls might take – but it would be a knee-jerk reaction not to at least explore the option. If we can find a solution that buys us both expedience and integrity, I'd be stupid and hypocritical not to consider it. Lamb says they wouldn't ask for open season condemnation rights. It would be just for this specific project. Hoyt and most people who have come out in support of the idea say the same thing: if the PBA gets eminent domain power, it must be drafted in a way that cuts it off immediately once the plaza and connecting roadways are completed.

Some people continue to worry. "Even if it's limited to this project," one government attorney said, "you've established the precedent. Who to say they won't be back for more? It's the nose of the camel." I think he was referring to the Bedouin proverb: "Once the nose of the camel enters the tent can the rest of the camel be far behind?" Or maybe he was referring to the story that goes with the proverb: The camel sticks its nose in the tent because it's so cold outside, and then it sticks its shoulders in the tent, and then it is entirely in the tent, but the tent is small, so the Bedouin is pushed out onto the cold desert sand.

"I think eminent domain ought to remain with the city, where it is now," said former Erie County D.A. Edward Cosgrove. "It's checks and balances. That's one of the basic aspects of our system of government. Checks and balances." I said that the PBA thought it would be much slower and perhaps more complicated if these actions were routed through the city. "So what's wrong with that?"

Cosgrove said.

THE PRICE OF LIBERTY

The Buffalo and Fort Erie Public Bridge Authority is composed of five Americans, three of whom serve ex officio and two of whom are appointed by the governor, and five Canadians, all of whom are appointed by the Canadian minister of transport. There have been times when some of the American representatives on the PBA seemed to be dancing to Canadian rather than American tunes. Even if we can trust Jake Lamb and Paul Koessler as much as everyone thinks, what happens when they're gone? What guarantee have we that the people running the PBA then will have Lamb's and Koessler's sense of ethics and public service rather than the patronizing and contemptuous attitudes that seemed to dominate Peace Bridge Plaza only two years ago? The Canadians are going along with the EIS – they have to – but no Canadian member of the PBA board and no Canadian politician has yet expressed anything but annoyance or contempt for the environmental, social and aesthetic concerns that have been so important on this side of the border.

"The worst thing that could be done in a case like this," Sarah Kolberg, a member of the New Millennium Peace Bridge Action Group and a New York State Assembly staffer wrote in a recent email, is to make a decision of this magnitude based on "people" not "systems." If we're going to grant the PBA eminent domain we should do it because of any potential number of reasons (i.e.; valid need, parity with other public authorities, etc), not because we trust Paul Koessler. Keep in mind that the Canadian half of the PBA has railroaded us for years, and continues to do so. What happens next year when it's their side that has the Chair? Either way, just because we trust Paul is not a good enough reason. Even if the next five chairs are people that we trust, what happens when there's one that we don't? We lose the rest of Front Park to make a parking lot for trucks? When I read Sarah's comments I remembered something that happened in 1966 when I worked for the Cambridge, Massachusetts consulting firm Arthur D. Little on a narcotics policy study for LBJ's Commission on Law Enforcement and Administration of Justice. Senior officials at ADL made us take out of our report what we thought our most important recommendation: decriminalize, because law enforcement was doing more social and economic harm than the drugs were doing. The ADL vice president in charge said he would transmit the group's conclusions to the Treasury Department liaison guy directly, but if we had it in the report Treasury would get defensive and nothing would happen. (The Federal Bureau of Narcotics, then the primary Federal illegal drug law enforcement agency was located in Treasury because the legislation that made opiates controlled substances, the Harrison Narcotic Tax Act of 1914, was a tax law.) I don't know whether or not he did what he promised, but it didn't matter because very shortly thereafter the liaison guy quit Treasury to become general counsel for ComSat, the first commercial satellite operation, and even sooner after that he was killed in a small plane crash. All those recommendations, if he did know them, died with him.

We have a chance to get a decent bridge here, and to get back a beautiful Olmsted park that we lost because of inattention and commodity. It's important to work with the process, to contribute to it and, most important of all, to pay attention to it. Now is no time to relax, no matter how quiet things seem. The forces of greed, idiocy, carelessness and bad taste haven't left town. They never leave town.

July 1, 2002

58 WHAT 9/11 DID TO THE PEACE BRIDGE EXPANSION PROCESS

DOING THINGS TO OURSELVES

9/11 changed the political landscape. The realities central to – and the craziness at the periphery of – our lives have been redefined. Some people worry about what a second round of suicidal terrorists might do. Others worry about the damage we're doing to our own way of life because of that fear, or because of people capitalizing on it. Both groups are right to worry.

George Bush now refers to himself as a "wartime president." He says he is in a fight against Evil. How shall we ever know if he's won that battle? If can't tell when he's won, how can we ever abandon him without being unpatriotic? That's Vice President Cheney's mantra: any question about current policy is evidence for Cheney of a defect in the questioner's Americanism.

Attorney General John Ashcroft is the scariest of the lot. He's gone after the Bill of Rights like a Missouri hound dog after squirrels. As soon as he got to Washington, Ashcroft begin working to undo decades of federal progress on gun control. 9/11 and the egregiously misnamed PATRIOT Act licensed him to rachet that assault up by an order of magnitude.

By virtue of that act, Ashcroft's FBI agents now visit libraries in search of suspicious book-borrowing patterns, and other FBI agents are on the lookout for librarians who tell anyone about the FBI reading checks. Librarians who squeal about FBI scans are guilty of new major felonies. U.S. citizens merely suspected of being connected with unnamed terrorists can now be locked in jail without the right to see a lawyer or communicate with their families or the press. Trials can be held in secret and no one outside of Ashcroft's Justice Department knows how many of them are going on or are planned. People sucked up in that vacuum are, basically, convicted on the basis of suspicion. This guilt by accusation is like the Spanish Inquisition and Stalin's NKVD where the question wasn't "Are you guilty of this charge?" but rather "If you're not guilty then why have we locked you up?" For the first time, the United States has legislation permitting us have our own Disappeareds.

Scary times. And the scariness seems to license profligacy in a startling number of realms many of us had thought safe from domestic ecoterrorists. Bush's energy people use 9/11as an excuse to clearcut more first-growth national forest timber and engage in more oil exploration in the few unpoisoned natural habitats left. We can't trust that those Arabs will deliver the ever-increasing amounts of oil we need, they say, so let's drill, drill, drill. How does polluting a stream or river or sea or disturbing a herd of caribou compare to our (ostensible) increased need for domestic oil?

The same political economy lets them cut back on depoisoning major toxic waste sites. We need the money, they say, for armaments; there is no money to reduce domestic poisons dumped into the environment by major corporations. (see, for example, http://nytimes.com/2002/07/01/national/01SUPE.html for an discussion of how the Bush administration killed funding for investigation of 33 major pollution sites in 18 states).

We see the effects of 9/11 every time we fly out of town. Buffalo's airport used to be one of the most laid-back in the country. You could get there ten minutes before takeoff and, if they hadn't overbooked your flight, you could be in your seat when the plane pulled away from the gate. Buffalo is no longer a laid-back airport. There aren't any more laid-back airports. If you want a seat, you better get there early because you never know when they're going to get weird.

And then, there's the Peace Bridge. After years of increasing congestion, traffic last year started speeding up at the Peace Bridge because of operational improvements. But the increased inspections on both sides after 9/11 have cut into those improvements. Lately, Attorney General Ashcroft is pushing for vastly increased inspections and ID procedures coming and going. He wants more vehicles and more people inspected for more things by more agents. Crossing is going to get worse long before it gets better.

So how does this fear of further terrorism affect the Peace Bridge expansion project? Those folks were coming up with some nice ideas for opening the border, reducing the bureaucracy, cutting the red tape. Now it seems that Ashcroft & Co. are busily reversing all that progress.

The point man on all of this is Vincent "Jake" Lamb, who has been managing the environmental impact study and planning process for the Public Bridge Authority since Judge Eugene Fahey ordered the PBA to obey New York environmental law by incorporating community needs and opinions in its planning process. Lamb has amassed an impressive array of design, engineering, public health, environmental quality and legal consultants to work on the project, and he has set up a binational citizens' committee to consider every recommendation anyone has for expanding bridge capacity.

I asked Lamb to comment on the impact of 9/11on Peace Bridge design, thought, and operation. This is what he said:

WHAT JAKE LAMB SAID

> Certainly there's more focus now on what happens at the border by everybody that's involved, including people who cross the border. Everybody wants to know what's happening and what the changes are going to be to make the borders more secure toward the overall objective of fighting terrorism. It's affecting everybody that's planning, everybody that's using, everybody that may be building, everybody that's involved in the process.
>
> The most important thing at this point is the uncertainty that surrounds just how the border operation is going to change, because it really hasn't been fully defined yet by the federal authorities on both sides of the border. I know they're working hard on it. I know they're actively discussing it within the context of a broader security issue for both countries. So there's a lot of work that has to go into what they do at the border. And of course it's not just one location, it's not just Peace Bridge. It's the entire northern border, the entire southern border and our coastlines. It's going to take time for the federal authorities to get their ideas gelled and make the right decisions on what to do with the border.
>
> Obviously security is the main focal point at this time. We're definitely expecting changes in the way the border's operated. This definitely will impact the spatial requirements, layout, even the functioning at the border—how goods are transported, and people.
>
> It seems that there's going to be more checking and not less checking. There was a time when there was a lot of discussion about seamless borders, about going to a more open arrangement to facilitate movement of goods and people. I think that some of those ideas are still there, but there's probably a little more caution being taken with respect to innovative ideas or innovative technologies, to apply them in this environment, the terrorist-threat type of environment that seems to have enveloped us. I think there's more

caution about these systems and reliance on the new systems than might have been the case otherwise.

What we're hearing in the community is a lot of people hope to focus attention on this area as an attractive place to visit, to live, to work. That pushes the tourist part of it and the natural resources that we have here. If the border becomes the type of place that nobody wants to go to, it's certainly contrary to and would be interfering with these plans that are waiting for projects like the Peace Bridge to go forward. I think that's a major concern for the region.

We need to help the government as best as we can accomplish to their objectives, while at the same time not sacrificing the potential for these kinds of developments and the quality of life that everybody enjoys here and that we want to preserve for the future.

Some people want Fort Porter resurrected for historic significance, to make it an attractive place to visit. The resurrection of Fort Porter as a fortress is another matter, one that certain people would be concerned about.

I think that the major impact is what it's going to do to the area. How it's going to change our style of life and quality of life. How it's going to change the image. Everybody was for a positive change of the image to make it an attraction. People that want the signature bridge, people that want something better. It's consistent with "Give us back our waterfront, give us back our parks." All these things. Let's concentrate on hooking up our park systems with Canada, and our bike trails and pedestrian ways. Let's accent the historic sites, archeological sites, the history that's here. Let's generate that kind of enthusiasm.

What seems to be happening here is the development of a potential situation that's beyond our control.

The trick here is going to be preserve those goals, preserve those ideals and things that people want this place to become, and at the same time accommodate in a balanced, reasonable way the national interest for security and the rest.

Commerce has to be preserved. We have to preserve the relationship that we have with our neighbor, Canada. If that's going to be sacrificed in the name of public security and the greater interest of the country – it's going to be a heavy price that's going to have to be paid locally if we don't get this done right.

If that happens, then the bad guys win.

That's exactly right. If we strangle ourselves and we do things to diminish the potential that we have here to improve and better ourselves, it's just what they wanted to accomplish. We have to be smarter than that. I think we can be. We must be. We just need to work through these issues and accommodate in a realistic balanced way everybody's concerns without sacrificing that potential. That's extremely important. We have to be vigilant to make sure that happens. We have to really work at it.

You said a while ago that this was going to change the way we look at the space needs here. Could you say a little more about that?

There is the possibility of more checking and not less checking. We had hoped there''d be less checking, that there would be consolidated, integrated checking, meaning, maybe the Canadian and US authorities could do things together, maybe even on one side of the border, with cooperative information sharing, intelligence sharing. We'd all get on the same page with respect to the conventions that are used and make it more efficient. Hopefully we're still headed in that direction.

But there seems to be a tendency to have more control, more inspection, more checking. And there's going to be a reliance on the tried and true ways of doing that. That's what I meant before when I said I think there's going to be some reluctance to try innovate approaches.

So the infrastructure needs, the plazas, would tend to increase in size substantially to achieve all these objectives. To achieve the objective of maintaining commerce and to provide for the growth of commerce and the provide for convenient and attractive passage of people back and forth across the border. If you take that desire, together with the increased inspection or checking that may occur at the border by the authorities for security reasons, enforcement of the law, customs and immigration things, obviously you need more space.

If you didn't worry about making it an efficient operation, and if you didn't worry about choking commerce, or making it more difficult for people to cross the border, then you're less concerned about space. So we have to provide enough space to preserve that goal of maintaining and improving the attractiveness of the location, the attractiveness of the Niagara region, together with the security demands. You're looking at a hell of a lot more space to make it work.

So this implies that we may have to rethink the notion of all of the traffic crossing here, that it may be necessary to revise how we look at the whole crossing idea.

Yes, I think it complicates the crossing idea. We know from going through the scoping process and listening to the public that there are already ideas put on the table by the public. It's a good idea to think about approaching the movement of goods and people in a different way than we have been doing it traditionally.

What I'm referring to is the concern expressed by many of the citizens when we went and had our public meetings and we had scoping meetings and emails that we received, a lot of people want to see the truck traffic separated from the car traffic. A lot of people want to see the truck traffic separated from the car traffic from the standpoint of air quality, public health issues. This is what we were hearing from the standpoint of making the parkland attractive, making the passage attractive. It was in keeping with some of the initiatives that communities on both sides of the border have in mind with respect to attracting tourism and developing the tourist trade, the tourist industry.

We have the added element now that the complications of maintaining and providing facilities to adequately take care of security while taking care of our other issues, of moving people and goods efficiently. It adds another complication that I think leads us to examine the options in an even more thorough way than we would have otherwise.

THE LIKELY CONSEQUENCES OF ALL THAT

Jake Lamb is cautious when he makes public statements about the Peace Bridge expansion project, and properly so. He has defined his job as facilitating a complex information gathering and evaluating process that involves scores of technical experts, government representatives and ordinary citizens.

Before Judge Fahey's decision, the Buffalo and Fort Erie Public Bridge Authority met in secret, ignored community needs, and came up with a bridge design that was both ugly and expensive. In a few years it turned itself from an agency hardly anyone knew existed to an agency that thousands of citizens on this side of the border hated and wanted abolished.

Lamb changed all that. People trust him and he has become the PBA's point man on just about all major public issues. He has had the full backing of two widely-trusted and liked PBA chairs who were also willing to meet and talk with anyone – Victor Martucci and now Paul Koessler.

Until 9/11, the process Lamb developed seemed to be slowly moving toward an interesting design of a companion or replacement bridge, along with a reconfigured binational customs, immigration and tolling operation that would significantly reduce the PBA's land needs on the American side.

Which is to say, we stood an excellent chance of getting Front Park and Fort Porter back, thereby restoring a significant part of Frederick Law Olmsted's grand design.

But vastly increasing the number and complexity of border inspections of vehicles entering and leaving the country threatens to change all that. The forces of security would chew up more of Buffalo's land rather than giving us some of our land back, and would have those huge vehicles idling and pouring noxious fumes into the local air for more rather less time.

Lamb insists that everything is on the table, that all design and location options will be given equal consideration. The security consequences of 9/11 changed the economy of the border. Projects that seemed to make little economic sense previously – like Robin Schimminger's idea of moving the truck traffic up to the southern end of Grand Island – are a lot closer to feasibility than they were a year ago. Schimminger's plan would require a much longer, hence much more expensive, bridge but it would pull all the through trucks away from Buffalo's waterfront, which would mean the city would get nearly all of Front Park back, people could have direct access to the water, and we'd be rid of the health problems caused by trucks idling in a residential area.

Everything connected with crossing the border has gotten more complex, more cumbersome, thicker with agents and agencies, rules and regulations, procedures and processes. On the other hand, the range of feasible design options has significantly expanded.

Something good may come out of this yet.

59 HOW THE BUFFALO NEWS HYPED CHRISTIAN MENN'S BRIDGE DESIGN TO THE DISADVANTAGE OF ALL OTHERS

Patrick Lakamp's page-one article in the October 20 Buffalo News, "Striking new visions for the Peace Bridge," is 1661 words long, of which slightly over 1000 words are quote or paraphrase Christian Menn or are directly about Christian Menn.

The other three bridge architects and their designs combined get slightly over 300 words, or 18% of the text.

The continuation of Lakamp's article on page A6 has a two-column headline that reads: "Bridge: Construction could begin in 2006."

"Striking new visions for the Peace Bridge," including the four bridge images and the photograph of Lamb, is online in its entirety at the Buffalo News website, as is Lakamp's 761-word Q&A: "Anatomy of a design and its place in the selection process."

Sandra Tan's Buffalo News article, "West Side asthma rate is 35.7% of homes," is on line, but without the map showing the asthma rates in several areas of Buffalo. Tan's article on the lung disease research is 524 words long, of which 190 words, or 36%, consist of Lamb's responses on behalf of the Public Bridge Authority.

None of the three articles is linked to one any other and the Q&A was not listed in any of the Buffalo News web site tables of contents on Sunday. To get to it you had to run Search. By Monday night, the Q&A didn't even turn up on Search: you could get to Lakamp's page 1 article, but the Q&A had been disappeared.

If you any article on the City and Region page of the Buffalo News website, you get 15 categories of city and region news: Buffalo, Northern Suburbs, Eastern Suburbs, Southern Suburbs, Erie County, Western New York, Niagara County, Niagara Sunday, Ontario/Niagara, Around Town, Columns, Schools, Corrections, Sunday Neighborhoods, Peace Bridge Choices.

Click on any of the first 14 of those and you get all the articles about that subject for the past 10 days. Click on "Corrections," for example, and you get a day-by-day list of the errors the News wants to acknowledge.

Click on "Peace Bridge Choices" and you go directly to Lakamp's page-one article. His Q&A isn't listed and neither is Tan's article. Just Lakamp's page-one article with its four computer-generated images and the photograph of Jake Lamb.

It is the only one of the hundred or so categories in the Buffalo News contents frame that leads you to a single article.

60 HEADING FOR A WRECK AT PEACE BRIDGE PLAZA

WARNING SIGNS

In case you thought the Peace Bridge War was over, three related but widely-separated articles in the Buffalo News for Sunday October 20 might give you cause you to think again.

The three articles indicate that the long, costly and complex development process being directed by Vincent "Jake" Lamb on behalf of the Buffalo and Fort Erie Public Bridge Authority is a late phase in the Peace Bridge War, but it is not, as many people have been lulled into thinking, a benign exercise of the War's peace terms.

Which is to say, the bad guys are still primarily interested in moving as many trucks through this border crossing in as short a time as possible, and they don't give a damn about the health or environmental or aesthetic aspects of that crossing. If folks around here relax at this critical juncture in the process, those same bad guys who three years ago tried to ram through an anachronistic, expensive and environmentally noxious steel twin-span design are poised to move in and do what they've wanted to do all along, only with a design that is cosmetically less stupid and offensive.

The three articles remind us that the powerful interests that attempted to impose the steel twin span on us may have gone quiet these past several months, but they surely have not gone away.

Two of the articles deal with recently-released images of four bridge designs commissioned by the Public Bridge Authority. Two things are disturbing about those articles. The first is that Jake Lamb and the Public Bridge Authority (which the Buffalo News consistently refers to as the "Peace Bridge Authority," as if it didn't have any public responsibility and didn't exist only because it has a public charter) are publicizing these designs with particular stress on one of them before the public discussion of bridge options scheduled for 8:30 am-1:30 pm next Saturday at Crystal Ridge Community Centre in Crystal Beach. The second is that the Buffalo

News has chosen to hype one of those options to the almost total exclusion of all others.

The smallest of the three articles and the one physically separated from the other two is Sandra Tan's "West Side asthma rate is 35.7% of homes," which begins midway down the far right column of page B1, the first page of the newspaper's second section, then concludes on page B4, below the fold, with a map showing asthma rates in different parts of Buffalo.

This is material the Buffalo and Fort Erie Public Bridge Authority would prefer to ignore and which Lamb seems to dismiss out of hand: health problems on Buffalo's West Side caused and exacerbated by idling and slow-moving trucks being processed at Peace Bridge Plaza or getting on or off the bridge's I-190 ramps. If that's a valid issue, then all designs likely to increase idling and slow-moving trucks in that area – which all four of the bridge designs just released by Lamb and publicized by the Buffalo News seem likely to do – may very well be putting the wrong bridge in the wrong place.

Most of the article is based on research done by Dr. Jamson S. Lwebuga-Mukasa, whose work has been published in such peer-reviewed professional journals as *New England Journal of Medicine, Chest, American Journal of Epidemiology,* and *American Journal of Respiratory and Critical Care Medicine.* Much of his epidemiological research on Buffalo's West Side was done with federal funds obtained with the assistance of John Lafalce.

The most important point in Tan's article is in the headline itself: "West Side asthma rate is 35.7% of homes." That rate is double anywhere else in the city. The second important point is the reaction of the Public Bridge Authority: denial.

"Peace Bridge officials," Tan reports, "say it's too early to know whether Lwebuga-Mukasa's findings have merit, because his studies are not finished and they have not yet reviewed his underlying data. They also have their own data-gathering process to measure the environmental impact of the bridge. 'We're proceeding on our analysis,' said Jake Lamb, director of the Peace Bridge Expansion Project.'We're certainly going to consider any other information that is credible, based on scientific measurements and procedural criteria.'"

That response reminds me of the responses of tobacco companies to epidemiologists' reports that cigarette smoking causes cancer: it's too early to know, studies are still going on, we have to check their data, we have to do our own analysis, there are other explanations.

Before saying more about the way the PBA and Buffalo News seem to be pushing one design plan over any others in a way likely to foreclose any serious discussion of other designs, it might be useful to take a moment to remember how we got here, why Jake Lamb is conducting those huge public meetings on both sides of the border and why the Buffalo and Fort Erie Public Bridge Authority has hired more than a dozen high-priced technical and policy and public relations consultants for this project. (If you remember all this history well, just skip to the next section.)

It was supposed to be simple, like most other major public works construction projects around here: a small group of people meeting in private decided what they wanted to do, how they wanted to do it, who would profit from it, then the construction would go ahead and then everybody would move on to something else; somewhere along the line the public would be informed about what was being done to them, to their environment, to their community.

In this particular case, it was the Buffalo and Fort Erie Public Bridge Authority deciding long before going public with their plans that they would put up a steel bridge alongside the current 75-year-old bridge, using pretty much the same technology and adopting pretty much the same design as the older bridge. The Buffalo News and the Buffalo Niagara Partnership organized design charettes, after which the PBA selected the twin span design it had come up with years earlier. The Buffalo News/Buffalo Niagara Partnership design charettes turned out to have been design charades, frauds, nothing more than the PBA's way of fooling the public into thinking it was having some input into the process.

Buffalo architect Clint Brown and businessman Jack Cullen said with all the money this was going to cost, the PBA should take a deep breath and rethink the whole thing: this might be an opportunity to build something really beautiful. Bruno Freschi, then dean of UB's architecture school, worked with San Francisco bridge designer T.Y. Lin and came up with the sort of thing Brown and Cullen had been talking about: a beautiful, soaring, curving, cable-stayed, single-pylon bridge made of concrete, using the most modern materials and technologies.

The PBA said, Too late, you should have made your suggestions earlier. People said, How could we, when you"d made up your mind about what you were going to do before the first public meeting was held, and when you dismiss out of hand every suggestion not your own? The PBA said, Too late, too bad.

But something happened this time that was different from the other times the rich and powerful guys had their way with the city.

A huge groundswell of opposition developed, and it didn't go away.

The New Millennium Group took on the bridge as a major public interest project. The Buffalo News ran editorials saying that a signature bridge would be a better idea than what the PBA had in mind, but the PBA had the responsibility for all this so the New Millennium Group and all the other people asking for something better should just let this one go. Artvoice, a weekly newspaper, took exactly the opposite position: it published dozens of articles by a half-dozen authors challenging the PBA on aesthetic, economic, political, environmental and other grounds. Every time the Buffalo News editorial page said "roll over," Artvoice said, "stand up." Every time the Buffalo Niagara Partnership (the former Chamber of Commerce) urged immediate construction of the anachronistic steel bridge, the New Millennium Group's Bridge Action Committee urged people to look at the implications of hundreds of thousands more diesel trucks with their engines idling in a residential neighborhood, the foolishness of duplicating an ugly and anachronistic design, and the need for public participation. Andrew Rudnick, the Partnership's CEO, kept saying that citizens should stop interfering in things that were none of their business; Jeff Belt, spokesman for the New Millennium Group kept saying that if citizens didn't interfere the businessmen would romp over them one more time.

The Margaret L. Wendt Foundation, the Community Foundation, the City and the County, contributed money to set up the Public Consensus Review Panel. The PCRP hired consultants and had hearings about all the things the PBA was trying to avoid. The PBA refused to have anything to do with the PCRP until it seemed a legal judgment was going to go against it, at which point it seemed to join but in fact worked very hard to make a mess of the PCRP's work.

The PBA came up with a scam to make the legal part easier than it otherwise would be: they said that the bridge construction work and the consequent work on the US plaza and connecting roads were two separate projects. They got a finding of "no significant impact" from the US Coast Guard, which, they said, meant they did not have to do an environmental impact study [EIS] before starting construction on the bridge. That meant they did not have to talk or listen to any citizen concerns. They did not have to respond to anybody. They would, they said, do an EIS when they got to the plaza and roads. That, opponents pointed out, locked everything down: once a bridge was up there would be no choice about the plaza and roads.

Buffalo Common Council President James Pitts established the Peace Bridge Task Force to take a broad look at what was going on. Then the City of Buffalo refused to give the PBA the easements it needed to begin construction. The City and the Episcopal Church Home both sued the PBA, saying that the bridge and plaza projects were not two separate activities, they were one big project and they should therefore be subject to New York environmental law. The PBA, in turn, sued the City for the easements. Judge Eugene Fahey threw out the PBA's suit and agreed totally with the City and Episcopal Church Home. He said the segmentation was a fiction, that the PBA had to obey New York environmental law, which meant they had to do an environmental impact study. Which meant the people had to be heard.

At first, the PBA board resolved to fight Judge Fahey's decision. Before and after his decision came down they spent a huge amount of money on a newspaper, radio and television disinformation campaign (similar to what Kaleida Health did before it got a new boss and came to its cease-fire with the Children's Hospital physicians).They hired a new batch of lawyers from New York City, who told them they didn't stand a chance of getting it thrown out. PBA chairman Victor Martucci, who had been a fervent advocate of the steel twin span, convinced the rest of the board that there was no point continuing the fight, that the only rational course was to stop warring with the community and to do the EIS.

The Buffalo and Fort Erie Public Bridge Authority is composed of five Americans, three of whom serve ex officio and two of whom are appointed by the governor, and five Canadians, all of whom are appointed by the Canadian minister of transport. There have been times when some of the American representatives on the PBA seemed to be dancing to Canadian rather than American tunes. The Canadians are going along with the EIS — they have to — but no Canadian member of the PBA board and no Canadian politician has yet expressed anything but annoyance or contempt for the environmental, social and aesthetic concerns that have been so important on this side of the border.

Vincent ""Jake"" Lamb, who had been a vice president of Parsons design, the PBA"s principal consultant, was put in charge of the environmental impact study, which formally began its work on 1 January 2001.

The four designs publicized in Sunday's Buffalo News were produced by three of the consultants Lamb hired for that project.

THE DESIGN THE BUFFALO NEWS PREFERS

The largest and most prominently placed of the three articles is Patrick Lakamp's "Striking new visions for the Peace Bridge," which begins above the fold on page one with a three-column-wide color computer-generated image of a two-pylon suspension bridge over the Niagara River adjacent to and thirty feet closer to Lake Erie than the current Peace Bridge. This is the design provided by Swiss bridge engineer Christian Menn. It is a six-lane bridge with space for pedestrians and bikes. Should the current bridge ever have to be taken down, Menn says, the space for pedestrians and bikes can be given over to trucks and autos.

Below the full-color image of Menn's design is a two-tier 72-point headline and a good deal of text. The article concludes on page A6, where, under the headline "Bridge; Construction could begin in 2006," it occupies most of the page, and includes three more computer-generated images and a photograph.

Below that article and filling up the remainder of the page – five columns wide by 6" deep – is a 761-word Q&A postscript by Lakamp, "Anatomy of a design and its place in the selection process." The Q&A is with Lakamp himself, Christian Menn and with Jake Lamb. It includes a two-column wide photo of Menn with an out of-focus-Lamb to his left.

Lakamp's two articles occupy, for the Buffalo News, an enormous amount of space. Few breaking public events get three columns with a huge photograph on page one and an entire inside page. And this is not even a breaking event. Lakamp's two articles are occasioned by the PBA's release of four computer-generated drawings delivered by three of its consultants. That's all that's news here. Four consultants delivered drawings.

Nearly all of the page-one article and most of the Q&A deal with Christian Menn's design, opinions, predictions. The article begins with eleven consecutive paragraphs about Menn, the first of which describes him as "a visionary who could help resolve the Peace Bridge controversy."

Even if some of the other designers had more space it would be hard for them to catch up. Christian Menn is drenched in the light and the others are just shadowy peripheral players on Lakamp's stage.

WHOM DID MENN HEAR?

Not that Menn is unimportant. I think anyone who has studied 20th century bridge design must have high regard for Christian Menn's work. He has built some splendid bridges. His engineering colleagues praise him. He is impressive and likable as a person. He is thoughtful, forthright, a man of vision, experience and sensitivity. He listens to people.

Which may be part of the problem here. Not with Menn, but with his design and how it came about. When Menn visited this region he was treated to an earful of people in Fort Erie talking about how important the old bridge was to them, how much it meant to them, how they would lose their identity without it. He met with people at the Buffalo Niagara Partnership who told him the same thing. He told me how moved he was by those conversations and how he would have to take them into consideration when he set about creating a design. Almost from the beginning, before he considered any of the other needs or opportunities, he seemed to feel he had to accommodate the old bridge, he had to keep the old bridge up and running.

And he did. In his Q&A article, Lakamp quotes Menn saying how important it is to preserve the old bridge, especially its Parker Truss. (I can't tell from the Q&A if Lakamp actually spoke to Menn or if he was writing from a handout. There is no mention of the fact of a conversation or where or when one might have occurred.)

One reason Menn gave for preserving the Parker truss is its historical importance: it represents well a kind of engineering done in North America more than 80 years ago. I can see how that makes sense for him, a retired bridge designer at the end of a long and distinguished career, but how much sense does it make for us? Do we really want to make our waterfront into a museum of anachronistic bridge design? Wouldn't a bunch of photographs on the wall of the PBA office or in a folio at the Historical Society serve that end? Should the fact that the truss was interesting in the early 20th century drive the design we have to live and work with in the 21st century? How much of Menn's design for the new bridge was wedded to his conviction that local interests demanded preservation and continued use of the old bridge?

Moreover, Menn's knowledge of Peace Bridge design history is flawed. The Parker truss was *not* part of the design; the engineer who built the bridge hated it when it was imposed on the structure and, I've been told, continued hating it until his dying day. The original design had one final elegant arc bringing the bridge to the Buffalo shore, but the US Coast Guard stepped in and insisted that the canal crossing be one hundred feet above the waterline from side to side. So instead of that last segment being supported from below with a lovely arc, it was suspended from above by that ugly iron box. The Parker truss isn't a piece of bridge history; it's a piece of klutzy bureaucratic interference.

Did Menn start from an image of a river, two shores and a

bridge that could not leave the scene until it collapsed of its own advanced age? Everything I've heard him say, and much in Lakamp's articles suggests that. If that is the case, then didn't Christian Menn design his bridge with conclusions in mind about location and connection and function that are properly the business of the citizens' committees still at work? And if that's the foundation of his design, why should the Buffalo News and the Public Bridge Authority be putting it before all others at this point, or ever?

FINDING AIDS

If you think I'm perhaps overreading the significance of the huge amount of space given to Menn's design in Sunday's News, consider this anomaly in the Buffalo News web site.

On the left side of every page on its web site, the Buffalo News has a table of contents of articles by category: City & Region, Sports, Opinion, Deaths, etc. If you click on any of those categories, that line opens into a bunch of sub-categories. Deaths, for example, opens up four sub-categories: Death Notices, In Memoriam, Card of Thanks, Obituaries. That are perhaps 100 of these categories in all. Click on any of them and you get the past 10 days' stories about it: North Buffalo, Schools, Editorials, whatever.

In all of those hundred categories there is only one link that goes only to a single article: "Peace Bridge choices" in "City and Region," which takes you directly to Lakamp's article with the big color picture of Christian Menn's bridge.

Why is this article the only one singled out of all the articles published in the Buffalo News to get its own standalone entry and link in the primary table of contents frame at the left of the screen? Why is this the only article with its own entry in a list of about one hundred categories of articles in that column?

QUESTIONS

Why, in other words, did the editorial board and publisher of the Buffalo News go to such unusual pains to make sure you don't miss Lakamp's report on Christian Menn's bridge, with all its computer-generated images?

Things like that don't just happen any more than the design of newspaper front pages and full-page jumps just happen. They're the result of conscious choices. Page layout isn't part of the natural order of the universe; it's deliberate and strategic. Everybody who saw All the President's Men knows that. So why did the publisher and editors of the Buffalo News decided to hammer this one home?

Is this the Buffalo News simply reporting on a possible design, stressing one far more than the other three, or is it the Buffalo News counting on the impact of massive coverage to influence public opinion and especially next Saturday's meeting? Is it mere coincidence that the only one of the four designs that suggests a new bridge without the old ridge at its side is the last one pictured and the one least discussed in Lakamp's very long articles?

Lakamp quotes Lamb as saying, "How the public reacts to Menn's idea will be a 'heavy factor' in deciding whether the Peace Bridge Authority pursues it and whether municipal officials in Buffalo and Fort Erie allow it."

How much influence will the two articles concentrating on Menn's design have in that public reaction? Are Jake Lamb and the Buffalo News simply creating the public opinion they want in order to get a conclusion they desire?

Lakamp's articles put all of this on the order of Fred coming home from work deep in a frozen Buffalo winter, dropping a pair of airline tickets on the kitchen table, and saying, "Hon, the boss gave me a week off and is giving us a fully-paid trip to the Bahamas for our vacation: first class all the way, walking-around money, and we're comped for all we want at the casino. But if you've got something else you'd rather do next week, I'll tell him never mind."

How does someone at Saturday's meeting, or any other meeting, respond to this proposal? What is the point of all this public input if the location is pretty much determined by the present location and if the design is delivered before the concept is worked out?

I think questions like these would be unavoidable for anyone who remembers how the Buffalo News collaborated with the Buffalo Niagara Partnership to run a meaningless bridge design charette four years ago, or the way the News editorial page for a year told the public to stay out of this.

LAKAMP QUESTIONS LAKAMP, AND ANSWERS HIM

"What were the previous forums about?" Lakamp asks in the Q&A at the bottom of page A6. He answers his own question: "They were held to screen river crossing locations and to develop a weighting system of the project's goals and objective used to rate alternatives. The authority didn't show any bridges. By the final workshop next year, public input will have helped trim the alternatives to two or three that have the best chance of being built."

Why is Lakamp answering this rather than Lamb? Is he spokesman for Lamb or the PBA now? Perhaps more important:

why does he go into the passive voice for the action part? You always have to take special care when someone with an agenda slips into the passive voice.

"Best chance of being built?" This isn't a discussion at Casino Niagara about what number is about to come up. The choice of which bridge will be built will be active, not passive, and the choice and the bridge will both be made by the Buffalo and Fort Erie Public Bridge Authority.

Chance has nothing to do with it. Power has a great deal to do with it. Two years ago, the PBA told the public to go to hell, said it had the power, and that was that. New York State Supreme Court Judge Eugene Fahey changed all that with a single court order telling them they had to obey they law. So long as the bridge they want to built and the process by which they build it conform to New York and US environmental law, as the PBA's previous plan did not, they can and will build whatever bridge they damned well please.

With one possible exception.

THE EASEMENTS

The one thing that might force the PBA to play it straight is backbone in the Buffalo mayor's office. Last time, the Peace Bridge Task Force set up and kept going by Common Council President James Pitts was the only place in city, state or federal government that kept pouring daylight on what the PBA was doing. Some people doubt that Buffalo Mayor Masiello would have had the gumption to refuse to give the PBA the permits it needed to begin construction had it not been for that Common Council-based task force.

Whatever Masiello's reasons, that was then, this is now. Masiello is notorious for his flip-flops, especially when a lot of money for developers is involved. This bridge construction project will make a lot of money for a lot of people, whatever the design and whoever gets the contracts.

Who will Tony be listening to this time around?

INFORMED, NOT FICKLE

The Peace Bridge War began with a demand from Buffalo citizens that the people who control public space stop imposing ugly and dysfunctional monstrosities of public architecture on the city's residents.

That is still a concern, people on this side of the river still want a beautiful rather than an ugly bridge, but enough was learned in the course of the war for the quest to broaden. Now there is far more concern with the plaza, with the restoration of Front Park, with the access roads, with the chemical and noise pollution from the trucks and the likelihood that both will increase unless the design is driven by plaza considerations.

It's not a matter of inconsistency, or wanting one thing one time and another now. Once the big flat rock was lifted and people began looking seriously at the Peace Bridge mess, they realized for the first time what a malignant operation it had become. What had started early in the 20th century as a friendly project more for vacations and booze had become hard cold business, had become manufacturers and shippers on both sides of the border and the Buffalo Niagara Partnership saying to hell with Buffalo, all that matters is moving the trucks, all that matters is who makes how much money on bridge construction and maintenance.

Buffalo has suffered a lot of ugliness imposed by politicians and developers, many of whom got very rich while the city got very poor, and not just in financial terms. It is an *infamia* that we have virtually no access to our waterfront, that more children die of lung disease within a half mile of Peace Bridge plaza than any other part of the city, that no attempt was ever made to rehabilitate neighborhoods destroyed by the Kensington, that scarce federal discretionary funds are even now being turned over to kick successful small businesses out of a downtown office building so a private developer can turn it into upscale condos.

The Peace Bridge War began over design of the bridge, but it expanded to consideration of the plaza. How those trucks enter the US in this region, how quickly they're processed, where they go, what kind of idling and access roads they need. The War gave people in Buffalo a chance to look at something that had been lost: a fine public facility. Frederick Law Olmsted's Front Park, had been whittled away by the Buffalo and Fort Erie Public Bridge Authority until the only parts left were virtually unaccessible to the public. People around here want a decent bridge, a bridge that doesn't put more poison in our air, and we want our park back.

How can you design a bridge without knowing what the plaza configuration is going to be? Only one way: you pour it into the current plaza and change nothing, other than perhaps take more land from the people of Buffalo and have more idling trucks pouring noxious fumes into Buffalo's atmosphere. Christian Menn says his new bridge would be 30 feet from the old one. What can that do but increase the traffic through the current plaza and congestion everywhere around it?

Putting those flashy bridge designs on page one of the Buffalo News a week before a design meeting that is supposed to be considering not bridge design but system design, particularly after a meeting that left a lot of people feeling the PBA was pushing the whole process in a direction it has wanted all along stinks.

Which of the proposers of alternative sites and plaza configurations can have the kind of money Jake Lamb and his employers had to pay Christian Menn and the others? If those architects are coming up with their designs – with one of them given the instruction "come up with the best design you think of" and the others given the instruction "work within these strictures" – what chance does any other idea really have? What marketplace really exists?

Menn will be the only designer not at Saturday's public session, the only designer not subject to questions about why he did what he did. The only one of the four design consultants who was given a free reign is the only one not available for public conversation. Jake Lamb told Patrick Lakamp that he wanted the citizens attending Saturday's panel in Crystal Beach to tell him what they think of the four designs. Maybe those citizens should first tell him what they think about seeing the design process once again being taken over by give Public Bridge Authority and by asking him what is really going on.

61 DESIGNING THE PEACE BRIDGE: COLLABORATION OR PREDESTINATION?

WORKSHOP #3

The team handling the environmental impact study for the Peace Bridge expansion project held its third Collaborative Workshop Saturday morning the morning of October 26 at the Crystal Ridge Community Center in Crystal Beach. About 300 people turned up for it. The purpose of the meeting was for the Project's bridge consultants to show, to anyone who showed up, their design concepts for a bridge that would increase capacity, primarily for truck traffic, between Fort Erie and Buffalo.

The meeting had four parts. For about 90 minutes beginning at 9:00 a.m., project director Vincent "Jake" Lamb gave a PowerPoint summary of the entire process and where things are now. He always goes through the whole thing at length at the beginning of public meetings – I assume so anyone coming in for the first time won't be confused about what's going on – but this time his comments were far more defensive than usual. He several times indicated that he was responding to remarks by participants and the press about what seemed to be confusion, ambiguity and preconceived conclusions in the previous meeting, the one where they cut out of consideration all possible bridge locations other than the present site.

All the other suggested sites had failed to score 5 or higher and whoever took part in the meeting before that had agreed that no bridge proposal failing to score at least a 5 should be kept. The twin span bridge locations and designs proposed by the Buffalo and Fort Erie Public Bridge Authority didn't score 5 either, but they were grandfathered in, leaving some people to wonder if the voting structure mightn't have been defective or loaded, which I guess comes to the same thing. Lamb said that everyone had agreed on that plan so it was time to move on.

Lamb said that the usual process in these projects is to figure out the orientation first – where the bridge lands on both sides of whatever it's crossing – and then to design the bridge itself. But here, he said for a reason I couldn't quite follow, it was necessary to do both at once, which is what they're doing though it may seem that they have in fact come up with a bridge design before they've decided where the bridge will go and how it will articulate with roads and communities on either end.

After he was done, his assistant, Bruce Campbell, spent an hour or so describing and showing four slides of each of the designs, after which there was a break, then Lamb asked Buffalo Mayor Anthony Masiello to make a speech, which Tony did. Then Lamb asked the design consultants to talk about what they'd proposed and why, which they all did, except for the one who wasn't there but who sent a former student to represent him. Then Lamb opened the meeting, for what time was left, to questions and comments from the floor.

THE PAMPHLET THEY HANDED OUT

Everyone attending the meeting received an agenda, a report on the voting at the previous meeting, a booklet with all of the PowerPoint screens in Lamb's talk, and a full-color booklet with drawings, water colors, and computer-generated images of all the bridge designs presently under consideration.

The two color images that stand first are the Buffalo and Fort Erie Public Bridge Authority's two companion designs. Both have the same horizontal design and the five arches starting on the Canadian side coming over to the Black Rock Canal. One has a more modern steel arch that begins like the other five, but has a sixth arch coming up from the waterline over the bridge to support it from above rather than below. The other replaces the clumsy Parker truss with a duplicate modern steel arch. Both of these drawings are almost rustic: the water is calm, the roadways are computer-putty-grey, and there are a few widely-separated cars on the bridge but not a single truck in sight.

The next two pages have watercolors of the four replacement spans suggested by the engineering consultants for the Public Consensus Review panel two years ago. One mimics the George Washington Bridge, one turns the current bridge upside down, with all the under-bridge support arches turned into over-bridge cable arches; one looks like the legs of a squatting ballerina; and one looks like two huge upside-down slingshots. There's a bit more auto traffic in these watercolors than the PBA's two companion designs, but still not a truck in sight.

There is no drawing of the lovely, curving design by former UB School of Architecture dean Bruno Freschi and San Francisco bridge designed T.Y. Lin that helped initiate all the resentment toward the PBA's very ugly twin span design. I think that is because one of the Canadian engineers hired by the PBA to screw up the work of the Public Consensus Review Panel two years ago said that such a design wasn't feasible because it was more complicated than a straight bridge.

That's all historical. The bulk of the pamphlet is 25 computer generated graphics of designs produced by the project's three design consultants: Figg Engineering, Christian Menn, and a team composed of two engineering firms, Mojecki-Masters and Buckland-Taylor.

Diane Christian has already summarized these as well as anyone might in "Peace Bridge Models" (*Buffalo Report* 26 October 2002): "Overall, the designs are ugly. The images suggest concrete MacDonalds' arches, inverted goalposts, wishbones, tuning forks or aliens, or masses that are very bottom heavy."

The only one Christian found at all elegant was Christian Menn's companion bridge, but that, she said, is "flat and uninspiring – just artful in making the old bridge palatable. Like a wart, the Parker truss on the old bridge controls the face."

That wart seems to be the single most important factor driving all the designs. All of the designers began with that as a given. Even in their designs for replacement bridges, you sense its absence, like the place on a sheet of drawing paper where a line has been erased but, even so, you somehow know something used to be there.

CHRISTIAN MENN'S LOVE OF BRIDGE HISTORY

Menn wasn't there but Paul Gauvreau, a civil engineering professor at the University of Toronto and a former student of Menn, was introduced as his spokesman for the day. Gauvreau talked about how and why Menn thought the old bridge had to be preserved for historical reasons:

> He regards the Peace Bridge as a kind of state-of-the-art thing from the time that it was built with the graceful and very well-proportioned arches, the detail structure, and this sort of served as a basis for the concepts that has been developed.
>
> In other words, how can we maintain the existence of the current Peace Bridge, how can we put the new modern bridge next to it, how can we make these two complement each other and be respectful of each other's dignity? In

> developing his alternatives he looked at various widenings, twinning the bridge with a companion bridge composed of arches or girder spans, and all of these seemed to detract from the historical and cultural significance of that existing bridge. In the sense that girders would make it hard to see the existing bridge from various points of view. If you had arches next to it along the same span lengths, it was almost as if you were saying, 'what's new, what's old?' ...
>
> This led him to some of the cable-stay bridge alternatives....culminating in the bridge that was featured in the Buffalo News, which is basically a two-towered cable-stayed bridge, which is his favorite of all the alternatives because of the slender depth which basically allows you to see both the old and the new with a minimum of distraction in that regard. The new bridge represents the state of the art of 21st century engineering, yet it frames the existing bridge in a way that respects its original dignity. Those are just some very initial comments on the consideration that led to the design.

What's interesting about that is that Gauvreau felt no need at all to say *anything* about Menn's design for a substitute bridge. He spoke only of a companion bridge. Clearly, for him, the companion bridge was the desired outcome.

Menn did do a drawing of a substitute bridge, but it's just his companion bridge with the old bridge subtracted. The Buffalo News reported last week that Menn had said he'd compensate for the loss of the old bridge by giving trucks and cars the new bridge's pedestrian and bicycle lanes.

As Diane Christian wrote, standing alone, the design is spare and dull. It's a nice enough suspension bridge, it's big, but it does not, as Daniel Patrick Moynihan several times urged, fire the imagination. That's because the object firing Menn's imagination is the 1927 bridge. He's looking to the past, not the present, and certainly not the future.

What remains bizarre about that is that the 1927 bridge was not, as Menn insists it is, a state of the art design anyway. Menn must know better. The part from Fort Erie to the Black Rock Canal might have been state of the art, but that design was mutilated when, quite late in the process, the U.S. Coast Guard insisted on the 100-foot box clearance over the Black Rock Canal. That's why the bridge has its high point so close to the American side: it had to make that steep rise to achieve that boxy clearance. There wasn't space to do that with another under-bridge arch, so they had to support the last segment from above with the Parker truss.

There's nothing elegant or historically important about the Parker truss. It's an historical fact, but so was my grandfather's appendectomy. All historical facts do not deserve perpetual preservation.

Why should Christian Menn want to build a bridge that honors an 80-year-old bureaucratic impediment to decent design? I don't know because he was the only designer who didn't come to Saturday's session and his surrogate could do no more than speculate about the answer and defend his old teacher's choice.

It would have been a simple enough matter to get Menn to join the conversation by speakerphone, if not closed-circuit TV. Maybe they should have done that so someone could have asked him.

THE PROBLEMATIC ANTHONY MASIELLO

When Tony Masiello got up to speak I thought that we were just going to hear the usual political fluff, but he turned out to say one of the most important things anybody standing at the front of the room said in the entire 4 ½½ hour meeting.

Tony's in a creepy place these days. He's alienated many of his longtime supporters in the Black community with his vendetta against James Pitts and his intimate alliance with the group that wants to get rid of all the at-large members of the Common Council. He's alienated many of his longtime Democratic party friends with his endorsement of Republican George Pataki and his appearance of Joel Giambra's pocket on city-county issues. He's alienated almost anyone who's not a developer or liquor salesman with his advocacy of a downtown casino that is all-but-guaranteed to deal a major blow to Buffalo's leukemic tax base.

So one wonders whom he might be taking marching orders from on the Peace Bridge. Is he doing his own thinking or are Carl Paladino and Andrew Rudnick (or people like them) coming into his office telling him which way to turn? Can he hold firm and think about the city's needs in this major issue or is he ready to roll over for the Buffalo-Niagara Partnership and other organizations who have long pushed hard for a bridge of any design and any ecological consequences, just so long as it gets more trucks moving sooner rather than later?

When he got up to speak I remembered how, two years ago, nearly everyone in the Common Council's conference room for a meeting of the Peace Bridge Task Force began giggling when someone read a message he sent from where he was vacationing in Florida suggesting a solution to the twin span/new bridge problem: do both. Fix up the old bridge and put a signature bridge right next to it. People made jokes about Tony's preternatural ability to take both sides of any issue at once.

And now, there we were, with the expansion project obviously favoring exactly in deadly seriousness what Tony had suggested so long ago.

This is what he said Saturday morning:

> It wasn't but a short time ago that this entire issue was beset with significant frustration, disenchantment, and serious concern. And I think the process that has been put in place over the last year, year and a half, is beginning to produce results. Significant input, significant dialog, obviously some differences of opinion and that's good. But I think it's important for all of us that we continue to stay the course and work the process.
>
> While today we saw some impressive bridge designs, I can say to each and every one of you that we need to continue to think not only of a beautiful and grand bridge, but as the mayor of the city of Buffalo I'm going to continue to push the envelope for a grand and significant gateway. A gateway that gives us not only a doable and dynamic statement of the bridge, but also a neighborhood, a park, a plaza that all fits into that wonderful vision we have of a gateway in and out of our city of Buffalo.
>
> You know, often times as I drive around the city I see those nice big beautiful billboards, 'Think bridge, think bridge, think bridge.' Well, as this process is evolving I'm becoming more and more confident that soon we can change those billboards from thinking bridge to thinking gateway, and for me that's very very important. It's a wonderful opportunity for all of us. This exchange, while some people are in a hurry to get it done, I think is working out well. I think we need to stay the course, as I said.
>
> And I really appreciate your support, your input and your differences of opinion, and I think something really great is going to happen to Buffalo and Western New York and similarly for Fort Erie and Southern Ontario. Thank you very much and have a great day.

So, yes, there was the usual political puff and flutter, but there was also the heart of the matter: Tony Masiello was the single person at the front of the room to articulate one of the key concerns of the Buffalo community. It's not enough to design a bridge; you've got to think in terms of a system that works for us, a system that restores what's been taken away and makes sure that the Buffalo and Fort Erie Public Bridge Authority doesn't do us further harm.

QUESTIONS FROM THE FLOOR

The question and answer and comment period was supposedly for the panel to hear and respond to the audience, but Lamb himself fielded many of the question and responded to many of the comments.

Lynn Williams, the second person at the floor microphones, had two questions. She first asked what would be the implications for Front Park of the southern companion proposals. Her second question was:"I hear a lot of talk about preservation of the current bridge. And I'm wondering where the preservation of the truss comes in it, which was not part of the original design.

"I'll comment on your first question," Lamb said. "That has to do – let me make sure I heard it right: does the southern companion span mean that the Peace Bridge Plaza would basically stay where it's at with very little possibility of vacating land for park use? Is that it? Is that the question?" Williams nodded. "Not entirely," Lamb said. "We have alternatives that use a companion alignment with a north plaza locations. They become a little bit awkward depending upon how we lay out the plazas. These kinds of issues and the question that you raised are the very reason why we're doing this together, why we're talking alignment and bridge type together. Usually we do alignment first. Because where we position the bridge and what kind of bridge we have has a lot to do with the plaza. So during these coming workshops you will see more development of plaza alternatives with various configurations with north bridges, north companion bridge, south companion bridge, and replacement bridge to the north or replacement bridge to the south, though it works better to the north, based on the plaza.... We're trying to develop alternatives that give back as much land as possible, or that vacates as much land as possible. And the second question had to do with the Parker truss. I'll ask our experts to address that question."

Barney Martin of MMMT said, "All of our alternatives today have been developed with respect for the Parker truss in mind. None of the alternatives we have developed are built on the assumption that the Parker truss will come out. So we have tried to come up with alternatives that are complimentary and compatible with what's there. If by chance it was decided at some point that the Parker truss was to come out, that would remove some of the cluttered perspectives in some of the alternatives and may result in some minor adjustments. But the approach we've taken thus far is that we respect the Parker truss as being an element of the bridge that is key to the issue, and we've developed our concepts around that assumption. "

Paul Gauvreau, Christian Menn's student, said, "There is a definite historical value to that Parker truss and that has to be factored into the decision-making process."

Lamb said they had no preconceived notion whether the final choice would be for a companion or replacement bridge.

Pam Earl asked "I'd like to hear what you think. What are the pros and cons of keeping the old bridge or not keeping the old bridge."

Jake Lamb: Are you asking me or are you asking them?

Pam Earl: I'm asking you all.

Jake Lamb: I'm gonna pass. Frankly *[frankly, I always get itchy when people start dropping the word "frankly" a lot. Why would anyone think it was necessary to tell you they're speaking frankly?]*, I'm sure I'll have an opinion later, but I don't have enough information to have my own opinions about this. Because I want to see these plazas developed, I want more dialogue with the communities, both the governments, the city and the town of Fort Erie, and of course the citizens on both sides. As we progress, I'll get to the point where I'll be able to answer that. To the extent that the experts want to get into that question I'll let them take a crack at it. Anybody? *[Laughter, after which Lamb immediately starts speaking again.]* I think they probably have the same issue I have, that it's too early to really stand up and say what they favor. Now Christian has. Christian looked at it for a total solution ... He looked at it and a critical decision in his development was that the existing bridge should stay. Basically, that was his favorite and he said so and said why. That was critical to the development of his alternative. But if you look at his alternative, you could see that you could build it and have it stand alone. His concept. I don't know if you want to comment about this Paul, if any of you want to comment about your favorites at this point or not. I frankly think it's a little too early.

Lamb then immediately moved to another questioner. That is, he first said he'd pass, then he spoke about the difficulty of answering; then he asked the panel to talk but he didn't give them a chance to say anything before changing the subject. It was really well done.

The question is: why was that avoidance necessary? I don't

think Pam Earl's question was all that difficult to answer. She didn't ask them, "Which do you prefer: keep the old bridge or tear it down?" That might have put them at risk with their employers, the Buffalo and Fort Erie Public Bridge Authority. She simply asked what the pros and cons were and surely Jake Lamb, if not the design consultants, must be aware of many of them by now. Surely most of the people in that audience could have named several of them and they had none of Lamb's access to and experience with the relevant political, economic, engineering, environmental and other kinds of detail.

THE KILLER FLAW IN THE PROCESS

When another questioner asked about the continuing cost of maintaining an old bridge, Lamb said, "Life cycle costs of companion bridges, alternatives or concepts will include the existing bridge – what it costs to maintain it, what it costs to replace components of it, et cetera. Then you'll be able to make an evaluation along with others as to the practicality of that alternative. Today we don't have those numbers and we're not looking for a decision on that today."

That is pretty damned vague and deflecting, if you ask me.

So the two times that question came up, Lamb moved it off the table as quickly as he could.

I'm pretty sure I know why. Jake Lamb wasn't going to point out the one stumbling block to rational thinking in all of this: the almost religious devotion and attachment among Fort Erie's politicians to the old bridge. I don't know what is behind it, if they really believe that stuff or if there is some economic reason making the old bridge useful to them (a ton of maintenance jobs that would be unnecessary on a new bridge system, perhaps), but they have never budged an inch since this process began on this. Fort Erie Mayor Wayne Redekop has never altered his mantra about replacing the old bridge not being on the table, nor has any other Ontario politician.

Which is why I think that the problem given to these engineering consultants and community members is unsolvable because at this point there are too many unknowns. All anybody can do is speculate.

Anyone who has ever taken algebra knows that you cannot solve for more unknowns than you have equations. Figuring out the values for **x** and **y** when you have $x + y = 7$ and $x - y = 4$ *is* easy*, but figuring out the values for **x** or **y** or **z** is impossible when you have $x + y = 7$ and $x - z = 4$. Three unknowns and two equations give you nothing but three unknowns and two equations.

That is what is happening with the Peace Bridge process: three unknowns, two equations.

They have teams studying bridge design and teams studying plaza design, but nobody is addressing directly the third and defining question: companion or substitute? The presence or absence of the old bridge drives everything. How can they possibly design bridge approaches and plazas and connecting roads when they don't know if they're building one bridge or if they're building half a bridge system? How can they possibly talk about the plaza if discussion of replacing the old bridge is politically verboten? What possible use can the projections of life cycle costs be if the secret given is that the old bridge must be maintained, whether for self-interest or sentimental reasons? How can these public meetings be anything other than window dressing if the key question, the political question, has already been determined by players not taking part at all in any of these discussions, workshops or votes?

The fate of the old bridge isn't something that will willy-nilly work itself out as the project lopes along. The warty old bridge, as the designs displayed in Workshop #3 so clearly indicate, is driving all the thinking on this project. There will be no clear thinking and rational decision making until the fate of the old bridge is clarified.

Is the old bridge sacred space, inviolate until it tumbles into the river of its own decrepitude? If so, then all discussion of liberation of occupied parkland in Buffalo is wasted time and air. Has the Buffalo and Fort Erie Public Bridge Authority ceded that issue? Then the citizens on this side of the border have a right to know that. Is there any point to these workshops and meetings? With that all-trumping wildcard hidden in someone's pocket, how can any of us possibly know?

* First you isolate either x or y, then substitute. If
we isolate y, then the first equation becomes:
$y = 7-x$
plug that into the first equation
$x-(7-x) = 4$
get rid of the parenthesis:
$2x-7 = 4$
isolated the unknown by moving the 7 to the other
side,
$2x = 11$
solve for x by dividing both sides by 2

$x = 5.5$
Solve for y by plugging 5.5 into either equation;
$y = 1.5$

But when you add the z, the third unknown, forget it. You're doomed to endless equations.

62 THE PEACE BRIDGE EXPANSION PROJECT NOW. VINCENT "JAKE" LAMB: "I'M LISTENING TO PEOPLE"

HOW WE GOT HERE

Three years ago, the Buffalo and Fort Erie Public Bridge Authority (which almost always refers to itself by the alias "Peace Bridge Authority") set about to expand its truck-handling capacity by building a steel bridge adjacent to the 75-year-old steel bridge currently connecting the city of Buffalo, New York, to the town of Fort Erie, Ontario.

They decided to duplicate the design and technology of the old bridge, a puzzling decision, given the huge advances in bridge construction technology in the eight decades separating the two projects. Their plan, developed with no public or outside professional input, was also costly, ugly and environmentally defective.

The Authority ran into a firestorm of public opposition that culminated in an order from Judge Eugene Fahey forcing them to obey New York environmental law, which they had vigorously and creatively attempted to avoid. Fahey told them they had to mount a full environmental impact study. (For all the details, visit The Peace Bridge Chronicles.)

Up to that point, the primary spokespersons for the Authority were the two general managers or whoever was chairman of the board at the time (the chair bounces back and forth across the river/border annually). Sometimes Andrew Rudnick, president and CEO of the Buffalo-Niagara Partnership (née the chamber of commerce) served as the spokespersons' spokesperson. In general, their attitude was, "We'll do what we damned well please; we don't have to pay attention to what you think; we're not going to listen to what you have to say."

Judge Fahey's order changed that *de jure* and *de facto*.

WHAT JAKE LAMB IS UP TO

Vincent "Jake" Lamb, who has been directing the Peace Bridge Expansion Project for the past 18 months, is perhaps the most inexhaustible listener connected with any Niagara Frontier public works project in living memory. He's hired an army of outside technical and design consultants; he's been meeting with citizens, public officials and groups on both sides of the border; and he's been conducting a series of large public meetings in Buffalo and Fort Erie. He's put a good deal of technical material on line at the project's web site, http://www.peacebridgex.com.

The Peace Bridge expansion project is now in its scoping phase, which means they are deciding where they want the new bridge to go and figuring out what kinds of landings the various options might require on either side of the Niagara River. When they've finished scoping, they'll move into the environmental impact study (EIS) itself. The EIS examines everything connected with a project: what it does to air, water, human communities.

Before construction can begin, Canadian and American federal, state, provincial, city and town agencies must all agree that the EIS has shown the project isn't harmful. The Canadians are expected to sign off on anything, as they did last time, since almost all of the potential the social and other environmental damage from the bridge project will happen on the Buffalo side. Last time the U.S. Coast Guard was the lead agency for permitting the project on this side and it totally ignored any concerns except those of the Public Bridge Authority. This time the agencies will be paying very close attention to local concerns and needs, in large part because this time the process is open to the public.

WORKSHOPS

At a public workshop at the WNED studios the morning of December 7th, the Peace Bridge Expansion Project's technical consultants discussed their short list of recommendations of which alternative sites should be retained for further study – four different alignments at the present site and an entirely new site near Grand Island. (Their report, Alternative Screening Process Technical Analysis Report: Technical Recommendations 04 December 2002, is online at http://www.peacebridgex.com/studies_reports.asp), after which the public was invited to comment on and then vote on those recommendations and on other sites it thought ought to be included. Several Buffalo Report correspondents discuss on that public meeting in "Peace Bridge Expansion: It's NIMBY Time," Buffalo Report 16 December 2002.

POLITICIANS AND THEIR CONTROLLERS

Several of the key politicians involved in Peace Bridge affairs have moved elsewhere since all this started:

– **Senator Daniel Patrick Moynihan,** who argued passionately for a glorious piece of public architecture, has been replaced by **Senator Charles Schumer**, who argues with less passion and who seems willing to settle for something that works pretty well.

– **Senator Alfonse D'Amato**, who wasn't particularly interested in the Peace Bridge, was replaced by **Senator Hilary Clinton**, who isn't particularly interested in the Peace Bridge either. She occasionally comes out in favor of something generically good being done, rather on the order of being in favor of highway beautification and well-maintained parks, and recently she came out for tightened security along the entire Canadian border. Clinton isn't interested in city life. She's been no help in this process and isn't likely to be.

– **Congressman John LaFalce**, who energetically backed the twin span, has been gerrymandered out of office.

– **Buffalo Mayor Anthony Masiello** remains as passive on the whole issue as he's been all along. Masiello does what Mark Hamister, Andrew Rudnick and the rest of the Buffalo-Niagara Partnership tell him to do, unless something else comes along to startle him into transient local responsibility. His continuing failure to demand excellence in this major public works project continues to astonish.

–The board members of the **Buffalo and Niagara Public Bridge Authority** have changed a little bit since this all began, but not much. The Canadian members are the same. The few changes are on the American side. All the Board members seem anxious to get the trucks moving. At the most recent meeting of the Binational Peace Bridge Task Force, PBA Chairman Paul Koessler argued eloquently for swift acceptance of the Memorandum of Understanding between the PBA and the City. Afterwards, I asked him why he was so pushy about that, since that seemed contrary to his usual laid-back style. The Canadi-

> ans, he said, were in despair that anything would ever happen on this side and the MOU was of both practical and symbolic importance. The MOU was passed and signed, and the process has moved along.

The day before the December 7 workshop, Lamb and I talked about the present state of the project. Two weeks later, after he had met in Washington with Homeland Security and other agency officials, we spoke again.
– B.J.

SCOPING

Lamb: We're taking the results from Workshop 2 and using them. We used those results to focus what we've done since Workshop 2. But we have more information now in certain areas. Especially on those locations that we've worked on. We've got new iterations of them. We had a traffic diversion analysis done. We have more information on the air inventory that we were doing. And when we focused on these alternatives, for instance, at the existing location, we also started developing information about the number of parcels that might be affected and the consist of those parcels, meaning how many are commercial, how many are residential.

We have the GIS mapping that shows land use and we're able to get more information about it. We've been carrying on more meetings with the city representatives. That has developed into a cooperative working relationship. We have meetings that take place whenever we have more information to share.

We have our consultants coming in and developing plaza alternatives and talking about how they're going to function with the street systems, with I-190. We even have estimated how much park land would be returned, or how much land would be vacated for park land use under each one of those scenarios, each of those iterations that we've developed at the existing locations. All of that information will be shared with the public on Saturday.

We had an initial screening at Workshop 2. We had always intended to have a second shot at that. I was surprised that using the criteria that I set up that none of them technically met that criteria. A lot of people said to me, "Then, Jake, knock them off the table." I said "No, we really can't do that. We really need to have a second and maybe more, better understood process for actually narrowing down and finishing the scoping process before we go into more detailed examination."

Scoping is to identify those alternatives that we want to consider in further detail and analyze in the preliminary design and draft EIS phase. So in the scoping phase we do only enough work to identify issues, environmental issues, public concerns, and enough information about formatting the alignments, configurations, the tentative location of these plazas, to say that we know enough by intuition, common sense and the information we have gathered to say "We don't have enough information to knock it out. We want more information to compare it with the other alternatives."

TENTATIVE RECOMMENDATIONS AND CHANGING MINDS

We're recommending what we should carry forward. These recommendations don't have anything to do yet with whether it's a replacement span, a companion bridge or what type of bridge it is – that is yet to be decided, and any one of those options is available to every one of these recommendations. That needs to be clear.

We're recommending, for instance, that we retain for further consideration and evaluation shifting the US plaza to the north to basically vacate as much as possible of the existing footprint. It's important to understand that. Now that includes what we've formally described as EBNP1, EBNP2 and the iterations of that that we've developed since Workshop 2. So it isn't like here is an alternative that's what we're going to look at and we're not going to wiggle it around or change the configuration. As a matter of fact, if we end up with an alternative to the north, if that's chosen ultimately, it will probably be not be exactly as anything that you see Saturday. It will be something that's evolved from that. Because as we are focusing on that location and that particular alternative, we try and manipulate it.

We're trying to make connections to 190 different than they are now. Not only are we working on vacating the existing plaza, we're also eliminating some of those high ramps there that would interfere with the vision or perspective from the park area. But as soon as we do this, it means that incoming traffic is on the residential side. So we said, "Let's try to develop something to put it further away from the residential side."

So you see how these things evolve: through development, development, development, development.

People say to me, "Jake, you changed your mind." And you know what this process is all about? It's all *about* changing your mind. We changed our mind to do this process, didn't we?

When you go through a process to come to a solution, whether your really recognize it or not, you are going through a series of tentative 'what ifs' in your brain, or sometimes you write them down. And you're weighing them. And sometimes you say, "This looks good." But then later on it doesn't look as good as it did. That's what this process is about.

SHARED BORDER MANAGEMENT

One of our recommendations is for what we call shared border management. The idea is to move some or all of what I'll call our border barrier operations, to Canada: all the customs, immigration and security issues. There's some headway being made in that regard.

The Commercial Vehicle Processing Center in Fort Erie was always designed to handle some form of US primary inspection. There are already some activities going on over there that help move operations on the US side. But the Authority could accommodate a measured – I say measured because I can't broadly predict anything that the feds would want to do – moving of primary US inspections for commercial now, or within some reasonable time. Which means what you would have on the US is secondary inspection. That, hopefully, would meant that we would need less land. That would affect every one of these alternatives, right? So when we show these and we talk about them, an overriding factor is shared border management.

We haven't focused on trying to work this out in detail yet. If we knew what the probability of that or the certainty of it, we could come up with a number of alternatives. For instance, there's no duty free in these drawings, there's no welcome center. We just haven't thought through any of this yet.

Maybe the security forces will want some kind of Checkpoint Charlie on our side. We'd have to work that through and that would affect all of our thinking.

BUFFER ZONES

We've talked about buffer zones for a long time. We can't size them yet. But we recognize there's an interface with residential or commercial properties and there certainly is justification to consider some kind of transitional land use. You've heard me say this at meetings. As we focus more on these plans we have to be thinking about that. We have to be thinking about the existing land use and how what we do affects it, and what the city may have in mind for changing land use.

For instance, they may want to change land use to commercial in some areas that are now not commercial and in areas that are considered to be adversely affected by our construction. We need to work together with the city and the public to develop that kind of thing.

OTHER SITES

Let's take a look at this from a common sense standpoint. We've got thirty alternatives at six locations. We know a Niagara Falls location doesn't work. Forget it. We know that Lake Erie bridge doesn't work. Forget it. We know that Niagara International Railroad Bridge is not an alternative to look at.

If we put a bridge anywhere close to where 198 hits 190, we're forced into connecting to 198 to make an interchange with 190, and of course rebuilding that whole thing. It's not an interstate system. We are on an interstate system now. We're trying to fix the congestion problem on a US interstate that's the equivalent of a Canadian interstate, the QEW. Even if you said, "No trucks on 198," which a lot of people want to have happen, you're still going to generate more traffic on 198 because of the autos. More traffic means queues on 198 because the intersections between 190 and Delaware Avenue and Elmwood Avenue, and all the others, were not designed for interstate traffic.

I know a lot of work and a lot of effort went into these ideas but they all fall on this one point.

Let me say this: I'm not interested in this really getting into a full-blown public discussion because it really gets us off-track. Theoretically the time may come someday that somebody wants to get rid of 198. Will that ever happen? I don't know. You don't have an east-west road. It would seem logical to replace it with something somewhere. I don't know where that is. You could think, "I'll dead-end the 198 somewhere." Well, that's almost the same thing as saying you're getting rid of it. It would be an access road to the park. That would completely change the nature of the traffic patterns and you'd have a new demand for something to replace it. But that's way beyond where we're at.

So I say, they just don't make sense from a transportation standpoint.

Tonawanda is a potential connection between the QEW and the Interstate system's 290 and 190. We know that about 80% of the trucks on the border crossing – that's Lewiston-Queenston and the Peace Bridge – are through truck traffic. They're destined for areas beyond our immediate region.

The other 18-20% of the truck traffic is local. When I say local I don't mean just Buffalo and Fort Erie. It could be Tonawanda. Some of the Tonawanda local would come over the Lewiston-Queenston. I'm talking about the local communities that are within reasonable commercial service distance of the existing bridge.

That's an important factor. When we talk about economic impact and people say "It doesn't have any economic impact on us," certainly it's hard to argue that the 80% is good for you, except for the general welfare of the two countries. But the 18-20%, that's another matter. I think that's important from the standpoint of being essential for what happens here.

Jackson: *Nobody's ever talked about that, that I know of. For two years we tried to get the PBA and the Partnership say what the truck traffic did to us other than harm and they never came up with anything.*

Lamb: Maybe there's a lot of other things that should be talked about.

We did a traffic diversion analysis. We created a model for the entire Niagara River. Our traffic engineers worked with the Ministry of Transport Ontario and New York State Department of Transportation. They have transportation models for planning their transportation networks. They're cranking in population information, demographics; they crank in economic activities, projections, land use, and they update it with as much reliable information as possible so that they can predict in some reasonable fashion impacts of changes in traffic patterns or even developments that are considered. It also helps them when they do their air quality containment analyses. It's damned good information. But it ends at the shoreline. What we did was work to connect them.

So we've connected them. And we've said, "Take the traffic projections in the future and do a what-if analysis on how much traffic is coming across one of the bridges." We have existing traffic at each one of the bridges – Rainbow, Whirlpool, Lewiston-Queenston for trucks and autos. We have scenario 1, where we expand at the existing Peace Bridge, and it shows expansions at the other bridges. And we have scenario 2 and 3, which are basically what happens if we did make a connection up at Tonawanda or even going across Grand Island. It's the same effect; they're so close to one another. We know Grand Island hates it, they don't want anything on Grand Island. We know that.

CROSSING AT TONAWANDA

Scenario 3 says we don't let any trucks at the Peace Bridge. Say they all have to go up to Tonawanda. Well, we need at least six lanes up there then. And at the existing Peace Bridge we need three lanes. Possibly you'd need a fourth one for functional reasons, but in terms of traffic, and this is strictly by the numbers, the traffic projections would mean that if you built this other bridge and took all the trucks up there, and cars would be allow to go up there as well, you would have no more traffic 30 years out at the Peace Bridge than you have now.

Well it's a compelling reason to say "that's an alternative that has potential." It has a lot, a lot of drawbacks. Bigtime drawbacks. There are serious issues. But you have redundancy from the standpoint of national security. You've got more flexibility on the border. You've got more alternatives for the convenience of more people, like in the northtowns and the upper edges of Buffalo. It's a potential catalyst for economic development for the state and province because you're facilitating the movement of that 80% to other areas. It's an alternative for future traffic growth. We need to do more so we can compare it to these other alternatives.

One of the major, major issues has to do with jurisdiction. It's an issue that we have to examine in detail to determine. We know that we have to get legislation from the United State Congress, Canadian Parliament, as well as Provincial and State legislation to authorize the Peace Bridge to build up there.

You've heard about the franchise. The franchise doesn't give authority to build. It prevents somebody else from building. And it's only on the Canadian side. That's a protective shield, if you will, from competition, but it doesn't give the Peace Bridge Authority any entitlement.

There are other things. It's not on the long-range master plan. I think if the need and warrants were demonstrated it would be reasonable to expect it would overcome. And it's a high cost and there's a good chance that because of the environmental considerations a tunnel might be the only viable option. Because you have Niagara Park on the Canadian side, a very sensitive area. You have sensitive ecological issues in the river itself, and if you put any piers out there it's going to be a major problem. Even the type of tunnel you would build, you would have to bore from the shore, not dig a trench and sink something because that would be even worse than a bridge. So there are lots of problems associated with it.

Nonetheless, I think it's an essential aspect of our process that we take a look at it as an alternative to answer the questions that are asked about the existing location. It's the only one of all the offsite alternatives – when I say the only one, the idea of connecting QEW with the interstate system – that makes transportation sense. The rest of them really don't.

We're recommending we retain it for further consideration. Does that mean we're recommending that you build it? No. We're not recommending any of them at this point.

EMINENT DOMAIN

A lot of questions have been asked, by the way, about how eminent domain works. People have to understand that if we do need to take your property, we would first come to you with independent appraisals of what the market value of your sale. You have a right of refusal, in which case we would go perhaps into eminent domain as a last resort. Eminent domain process guarantees you, within our structure and its capabilities, which I recognize has limitations and there are debates about, a fair market price through the court system. It also protects the public from the point of not paying too much for your property, which I think is an important element of eminent domain that I think some people haven't focused on here.

The idea of a project of this nature with the amount of property that may be required demands that justification in terms of public need to spend that money, and justification that the money spent is sufficiently and fairly spent. And that goes to avoiding any sweetheart deals on any land acquisitions or indeed even configuring a solution in anyway to sweetheart somebody. That just ain't gonna happen, at least as long as I'm involved in this thing. It's not gonna happen.

So it's important for the public to realize it isn't just a one-sided thing. I think that eminent domain is always thought of as, "My god, you don't want to give that power to the Peace Bridge Authority." But we've put all kinds of restrictions and limitations on it that we recognize are needed to build this project, and at the same time there's protection built in the for the public good. As you know, the Council approved the MOA. They had some changes that they wanted to make. The Peace Bridge Authority on the 3rd of December accepted and endorsed all those changes, so the MOA is getting ready to be signed.

One thing was, they wanted some calendar time limit when eminent domain would

Jackson: *expire?*

Lamb: Not necessarily expire, but if we didn't initiate eminent domain within 5 years of the ROD it would expire. I think the logic in back of that is "Look, indeed it's for this project. Indeed, if we are going to move forward with the project after the record of decision we don't want this sitting around. Because if nothing happens within 5 years there's probably something wrong." This is acceptable, we understand that. We have no problem with that.

If we have to do eminent domain, we have to initiate that within 5 years. That doesn't mean that it's all finished then.

There was a neighborhood meeting at which a young boy up in the community center had one of the most intelligent questions. I'm not saying there weren't other intelligent questions asked, but he said, "Why are you going to take my house? And my friends. I have friends. And there are old people who live there in the neighborhood. And what are we going to do? Why are you taking my property?" I feel a responsibility to answer that question.

First of all, I told him, that we have to consider alternatives that don't take your property. That's why we're looking at all of these. We have to take a look at alternatives that don't through recycling continue to do this. We have some alternatives that don't take his house.

It's a legitimate question. We have a responsibility to look at alternatives and options so that in the end if that boy's home has to be taken, we need to be able to go in that room where I went and look at him and his family and others and say, "Listen. We've looked at all of these alternatives. This is what we have found. We've done this with public participation and all this technical help and here are the reasons why for the public need we have to take your property."

We're not anywhere near there now. We have to go through the process to be there.

THE AIR WE BREATHE

We've got a draft air inventory report. It's going through a review by state agencies, and that's why I haven't released it yet. But I am talking to some of the findings of that inventory because of the importance it has to the public discussion.

The other night I said I was going to show it on Saturday but I'm not. That's because I thought we were close to the end of this review, but I was informed yesterday we were not. But I'll just briefly go through what I did show the other night.

We're going to meet all the guidelines, the study requirements. Actually, we're going beyond the study requirements. That's a sensitive issue in some areas. We felt it was important for the community to make an inventory of existing conditions with respect to air quality. And we went out and we put monitors in upwind and downwind. What we found was there's variations. That's obvious because the wind blows in all directions. And we used the closed meteorological station available, which is right down near the rowing club instead of the airport, which we could have relied on. We were trying to get local study information. And we did.

It's reflected especially on this date that happens to be very close to 9/11. You see it ticked up a lot at the crossing. So it's good stuff. Sometimes a certain type of process puts out a chemical that's nowhere else that's a fingerprint, so you can tell where it's coming from. We didn't find any fingerprints but we did find elemental carbon slightly higher downwind from where the diesel exhausts are.

We happen to have taken measurements before 9/11. The estimated average plaza contribution to downwind particulate matter in relation to the total quantity downwind represented about 6% of what was in it, what was at Busti avenue or downwind of Busti avenue. Before 9/11 the density of particulate matter we were getting was 2.2.

Then it went to 12.7. It went to 12.7 in one day, even with far less traffic. Why? [*I made a gesture of holding up my hand like a cop stopping traffic.*] Yes.

That, in and of itself, confirms that when you have efficient transportation it's good for a lot of things. It's good for moving people and goods more efficiently. But it's also good for our environment from a standpoint of pollutants that are emitted in the air.

So a crucial part of what we're doing is asking , "What's that operation going to be and how fast are we going to be able to move traffic across the border?

We continued these measurements and since then, even with the drop in traffic, we've had an increase in the total concentrations of particulate matter blown into the neighborhood. That raises additional concerns about making sure that we end up doing more to make our operations more efficient at the border, so that we can get back down to a more reasonable number. And it also raises the issue of we better look at that other site alternative, just to make sure.

In Canada it went from 4.4 to 10.1.

Over the distance of a block there was a substantial reduction in the PM. Dissipation, and more importantly, buildings knock down PM. Especially diesel PM for some reason. Buildings, trees, birms, walls.

People don't like walls. I mentioned it the other night and somebody said, "Take walls out of there. We don't want any walls." Well, I don't want any walls either. But walls are often used as a buffer instead of taking property. I said, "We're going to keep it on the table so we can show you what the impacts would be of this if that were the case."

At least we recognize the sensitivity for Buffalo. This helps us understand that this material does dissipate. We haven't done any analysis yet. We will do it for every option we get down to considering seriously. And most importantly is the throughput time. This traffic's coming through here: is it going three miles an hour or is it going 40 miles an hour? They're the variables that go into the model. And in discussions with our science guys.

WHAT NEXT?

Jackson: *What happen after tomorrow [the December 7th public meeting]?*

Lamb: We're hoping that the results of tomorrow's meeting, where we ask the public "Give us your five." We've given our five recommendations, in addition to the shared border management. We're hoping that they kind of confirm our recommendations. There may be exceptions, and there will be naturally. We will finalize our scoping report. We'll bring it back out to the public. Distribute it. And we'll have another meeting to say okay, we've got all this information, this is where we're going with it now. We have defined the alternatives to take further on.

There is always the possibility that we can resurrect something that was discarded before as we develop new information. People have to understand that. And they have to understand that when I say at the end of the meeting, "Here are the things we're going to do," and you look at it that day and you don't come back for three or four months and you can say "Hey, what happened?" It may look different. You have to understand that it's a dynamic process. We'll show how we've gotten from there to here. And we'd like you to be there as we're doing it. But if you're not, don't be surprised if it looks different.

We will go to the Partnering Group *[City of Buffalo, Town of Fort Erie, Buffalo and Fort Erie Public Bridge Authority]* for ratification. In the case of the Tonawandas, we will ask the town of Tonawanda to ratify considering that. We've got indications already that they support the idea. Grand Island absolutely is against it. They're going to bring an army down, they say, on Saturday. When I went

up there, they kind of beat me up. I see it as part of the process. It's necessary for people to get up and say, "Hey, you're out of your mind. Don't do this."
Then we go into the preliminary design and draft environmental impact study phase.

Jackson: *When do you settle on one location, one alignment?*

Lamb: We go through iterations of examining these and the things that we developed from them. We're going to meet with the public all the time. I want as many public meetings as people want. I'll go to them and explain as best as I can what we're doing.

Then let's say we have these five locations, these five that I'm talking about. Maybe about March or April we'll say, "Hey, look: we've got more information now to lead us in a direction that says this one and this one." Maybe have this discussion again, but now about three. Maybe we take three right to the end.

And at the end, we say, "We got one that we recommend. That's our preferred alternative." We've got input from the agencies, the permitting agency, approval agency. There's a reasonable potential of being able to build it from a jurisdictional standpoint, money standpoint. Community input is there and to the extent that there is consensus and at least acknowledgment, acceptance of that alternative – you'll never get 100% – being the alternative that we go forward with. That's it.

I want to get there.

Jackson: *And that's when you go to design?*

Lamb: No. We would have a public hearing, a formal public hearing. All these meetings I'm having now are not formally required. But there's a requirement for a formal public hearing. We would have a formal public hearing and then start the paper process of getting a record of decision that normally takes something like six months, and that's hurried. What we would try to do is expedite that, make it sooner, and when the record of decision comes, you're safe in going ahead with the design. It's a judgment call on whether or not you go ahead with the design before that. You could. And in many cases agencies do go ahead, especially if they've worked enough with the permitting agencies and there's like,"Hey, everybody's kind of satisfied, we don't have any wrinkles on this thing."

And if there was a repeat of what happened last time with one of the major senators, like Moynihan, saying, "Hey, look: don't do this. Coast Guard, don't give the permit"? We don't want that situation to develop. We're going to do everything we can by engaging the senators and congressional delegation and the rest, and the public, so hopefully that won't happen. That gives you less uncertainty. So you say, "My judgment call is I can go ahead with the design." But I need that ROD to really trigger certain things – serious discussion on right of way acquisitions, for instance. Because you need some lead time to get that done, and a lot of work would have to be done on the first part of the design to finalize exactly the properties that were needed.

I want the public input. I like those neighborhood meetings, like that kid getting up, people who live there being able to get up and ask questions. It's like everybody sitting in a living room and talking, talking about the subject. That to me is vital to the process. It's more important than meeting with 400 people or 300 people. Not that that isn't important. That *is* important. But it's trying to inform them. Trying to let them understand number one how the process works and number two that they do matter, their points of view do matter.

Why am I doing what I'm doing on Saturday? Because I'm listening to people.

(We continued our conversation on December 19th))

HOMELAND SECURITY

Jackson: *How does the Homeland Security affect us?*

Lamb: I think the formation of Homeland Security is a positive set of circumstances. I think it's a positive thing for all of us, and for the project itself, because it's bringing together and integrating Customs Service along with INS and Transportation Security Administration all under one roof so the work will be integrated and coordinated. I think that's a very positive sign that we can expect better definition and probably more efficient plans of action being produced because they're organized instead of having three independent bodies.

Jackson: *This will save you time?*

Lamb: I think it will save us time. I have very positive reaction to meetings that we had in Washington last week. [Peace Bridge General Manager] Steve Mayer and [BFEPBA board member] John Lopinski and I met with officials of the INS, Customs and Transportation Security Administration. I got the distinct impression that they've got their act together and they're developing what looks to be a plan of action.

They're focusing on action plans that I think are going to make the border operations more efficient. That obviously is a goal of theirs. They want to move people and commerce – that's what FAST is all about, you know, this new program FAST: Free and Secure Trade. The idea is to make trade easy to get across the border, commerce easy to get across the border and people easy to get across the border. So they're on the right track, I think.

Jackson: *So you think this might accrue to our benefit rather than screw things up further?*

Lamb: Yes. I think it's the precursor of things to come and those things to come I think are going to manifest themselves in a more efficient border operation. It's obviously due to the cooperation that's going on between the Canadian and US border officials.

BRUNO FRESCHI

Jackson: *When you were in Washington you also saw Bruno Freschi. Can you say a little about that?*

Lamb: It was my first meeting with Bruno. He's been out of Buffalo for some time. We did talk about the Freschi plan. He's fleshing it out for me with details about it that either I hadn't heard or had forgotten about when I first came to Buffalo. He's going to send me some more information to clarify and elaborate why he and T.Y. Lin believe that their plan represents a preferable solution for us at the Peace Bridge. When I say "plan" I'm talking about the bridge, I'm not talking about the plaza locations.

Jackson: *You're talking about their curved single-pylon –*

Lamb: – Yes, their curved cable-stayed bridge. Their scheme had a plaza that we questioned the practicality of, but when I talk about them demonstrating or clarifying their plan it would be the bridge itself, not so much the relationship that they had developed between the bridge and the plaza.

Jackson: *Things have moved a long way in plaza thinking since Bruno was involved in this.*

Lamb: Sure. So we don't tie that early concept of plaza to their total concept. We divorce the two of them. I think it was a positive meeting for me. I'm going to have further discussions with Bruno. He's interested in giving me more and better information and I'm interested in receiving it.

MAKING THE FINAL DECISION

Jackson: *Finally, some questions about the December 7 public meeting. One of the questions that a lot of people had after it was over was, how do all these differing opinions impact the final decision?*

Lamb: They all impact the final decision in one way or another. The whole idea of public involvement is to consult with the public. Let them know what's going on and let them weigh in with their opinions, suggestions, objections. And I think we're accomplishing that. What you saw on Saturday was a large outpouring of feelings, especially from Grand Island folks that got together and they wanted to let us all know, let you know, let everybody in Buffalo and the region know, and certainly let the Peace Bridge authority know, how they feel about certain alternatives that may impact directly the citizens there. That's what this is all about.

That's input for the process. These meetings are not decision meetings. They're input meetings. Meetings that give the people the opportunity to find out what's going on and weigh in and give their viewpoints. We take seriously what their viewpoints are.

Jackson: *How do you correct for people loading up a meeting?*

Lamb: We don't. There's no way you can correct and I think it's foolish to try and correct. What we look at are examples. I don't discount anything. What we had on December 7 was an expression of viewpoints about certain alternatives. When it comes to how we react to that, we look for issues that have been raised of substance. It's one thing to say, "I don't like this plan, get it out of my back yard." It's another to discuss issues, concerns of substance that drive that feeling. And questions that are raised on technical issues on any one of the alternatives, alternatives for example that were suggested by the group. The group came down basically to suggesting put it somewhere else, put it at the International Railroad site. There was not a lot of discussion about what the merits would be of putting it International Railroad. It was more, "We don't want it where we are. Let's put it somewhere else."

We appreciate that, we certainly have to be cognizant of viewpoints and opinions, but in terms of technical evaluation we look for issues of substance that we would have to address. Such as environmental issues – noise, air, the public health issues, the archaeological issues. And some of them were brought up by individuals. They're important even though we're aware of most, if not all, of the issues that were raised by people who got up and talked about the alternative.

So it's input, it's important input, and what I'm doing is, I'm reviewing, it all. I took notes on every one of the persons that spoke in the public meeting as well as I could. And I'm reviewing the video of the public input portion to make sure I haven't missed any points of substance that were raised.

But what I think is demonstrated here is that you cannot rely on these meetings to be representative of a decision by the public or even a comprehensive direction, let's say, that represents the public in the region. Because you could take any one of these workshops and mobilize or organize a large group to push for a certain issue or concern. That's very important and we need to listen to that, and I'm not saying that it shouldn't be done. We recognize it and want those things to come out earlier in the process instead of later.

But it's clear that it may not represent the public, the full spectrum of the region. We hope to get more diversity in these meetings. There were not as many Canadians as we would have liked. There were not as many people from Tonawanda as we would have liked.

We're interested in viewpoints from all the communities, and when they're all taken together and the public and ourselves can focus on what's best for the common good, we're going to end up with a better decision. That's kind of how I feel about it. I thought it was a great workshop and I thought it demonstrated the very point that I'm making now. As I said before, these are input, not decision meetings.

Jackson: *That's an important distinction.*

Lamb: It really is. And it certainly does not minimize the input from the public or minimize the impact on us and on the Peace Bridge Authority of that input. It's all important. But it's quite clear that the decision process will be exactly what we said it would be. We've shown that we need technical input, we need input from the public, we need input from the agencies, we need the recommendations and analyses that the consultant team makes, and we need all of that combined to give to the decision-makers.

Jackson: *And the decision makers are...*

Lamb: The decision makers are the partnering group. The Peace Bridge Authority is one of the partners. The others are the city of Buffalo and Fort Erie. In the case of an alternative like Tonawanda, we would go to the town of Tonawanda or any other town – in this case, that alternative is in the town of Tonawanda and town of Fort Erie – and we would ask the town of Tonawanda to weigh in officially with their viewpoints in respect to retaining that alternative in the next phase.

63 PUBLIC PRESSURE AND THE INTERNATIONAL RAILROAD BRIDGE OPTION

TRUCKS AT THE INTERNATIONAL RAILROAD BRIDGE

Bruce Campbell, deputy project manager of the Peace Bridge Expansion Project, said at the February 13 meeting of the Buffalo Common Council's Binational Peace Bridge Task Force that the PBA is "taking a hard look" at a $500-million dollar, four-lane, high-level bridge running roughly along the route of the International Railway Bridge, which hits Buffalo's west side close to where the route 198 meets the downtown section of the Thruway (I190). Such a project would require major highway construction to provide links to the QEW and I190, which means it would need a great deal of new Canadian and American cash.

Thus far, Canada has evinced little interest in putting much or even any money into the expansion project. It wants the extra lanes for trucks, but all of its design pressure has been for functional minimalism. The Canadian representatives on the Buffalo and Fort Erie Public Bridge Authority have continually voted for quick and ugly. Their overarching concern is faster truck processing. All the pressure for an environmentally responsible and aesthetically interesting crossing have come from the American side. Canada may come up with QEW connection money if that's the only way it can increase the truck traffic, but it is unlikely it will allow any of its resources to be utilized for anything else.

A crossing adjacent to the International Railroad Bridge had been all-but-dismissed by the Peace Bridge Expansion Project earlier, primarily because of the difficulty and great cost of connecting a landing at that site to the Thruway. The project's director Vincent Lamb pointed out that NY 198 and Delaware Park and its accompanying residential neighborhoods couldn't handle the huge amount of truck traffic that would result from a crossing at that site. Furthermore, the Coast Guard requirement that any non-drawbridge crossing over the Black Rock Canal have 100' water clearance meant it would have to be a high bridge, which in turn would mean a large landing and ramp area on the Buffalo side of the Canal.

But then a lot of noise interfered with that line of thought and things got turned around. At least for now.

GRAND ISLAND'S PRE-EMPTIVE ATTACK

The Expansion Project's fourth "Collaborative Workshop" on December 17 was swamped by a huge Grand Island contingent that voted en masse against a crossing at or near Grand Island and in favor of a crossing at the International Railroad Bridge. Suddenly the IRR option was back on the table. It's moved from 18th ranked position at Workshop #2 to 2nd rank at Workshop #4, with 347 votes, second only to the current Peace Bridge site with the plaza shifted north, which had 314. Slight variations of the IRR description also moved it into ranks 4, 6, 7 and 8. (The full rankings report is online at http://www.peacebridgex.com/files/ws4summary07 jan-03_sm.pdf).

More than two-thirds of all votes cast for IRR site were by Grand Islanders. If they hadn't come and voted as a block for that site and against all other sites, IRR would still be at the bottom of the list. I'd interpret their vote as a straightforward Not In My Backyard to a Grand Island site rather than a Yes, that's the best place for the IRR site.

No reason they shouldn't do that. I wouldn't want all those fume-spewing loudmouth trucks in my neighborhood either.

WHAT THEY WON

I asked Vincent Lamb, project director of the Peace Bridge Expansion Project, how seriously they were considering the IRR site. Lamb replied:

> The Consultant's recommendations going into Workshop #4 was to retain alternatives at the existing Peace Bridge site and at Tonawanda. After Workshop #4 the PBA directed us to include the IRR site for further consideration. We took another look at the IRR corridor and concluded that none of the alternatives suggested to date were without the fatal flaw of adversely impacting NY 198 and Delaware Park. So we (I am sure others have thought of this before) defined a alternative at IRR based on trucks only (autos only and no trucks at the Peace Bridge) and connections directly with I-190 designed in such a way to pre-

> clude bridge truck traffic from using NY 198. There would be no connections between this IRR truck bridge and the local street system. Our first thoughts are that this would probably be a 4-lane bridge, but this is not settled, nor is the cost estimate.
>
> Is this serious? We will consider the IRR concept as well as the Tonawanda concept to determine whether one or both are reasonable and doable or not as alternatives to expansion at the Peace Bridge. The reasonableness and doability of all retained alternatives needs to be determined and documented in the Draft Environmental Impact Statement. Is this likely? If you mean that we take a hard look, then the answer is yes. If you mean is the IRR a likely solution that will be selected. This can only be determined by the Process.

I take that to mean that the smart politicking by Grand Islanders put the Public Bridge Authority in the embarrassing position of having to honor their own process, even though the vote at Workshop #4 was obviously loaded. Like all other public works projects, part of what's going on is grounded in public relations project. The PBA got in so much trouble with the courts two years ago because it ignored public opinion in favor of private interests. Now it's bending over backwards in the opposite direction – spending a lot of money on options it already decided were fatally defective.

ADDITIONAL NIMBY IN TONAWANDA

The Tonawanda town board considered locating a bridge there at its meeting Monday night. There was a large and vigorous public turnout in opposition. Because of the great width of the river at that location and complex environmental considerations, any bridge in that area would be several times more expensive than adding lanes or building a new bridge at or near the present site. It's possible that a new bridge could go there, but given the economy it's not likely that funds will be available to do it. Unless the PBA can convince the Bush administration that national security requires two bridges some distance apart to reduce the likelihood of terrorists closing down this part of the border. Anti-terrorist craziness and spending are both so high that such a proposal might indeed find extraordinary federal support. But it still would have a difficult time getting past the local residents, who are more concerned with the more demonstrable threat of noise and air pollution from the trucks.

TIME FOR SOME BUFFALO SELF-DEFENSE?

Is the International Railroad Bridge a viable site for new truck lanes or is the current study just a move to satisfy public demand for current study? Is this look at the IRR site merely a truck stop on the way to expansion at the present site, which is clearly the favorite choice of the PBA's consultants, or is cars at the old bridge and trucks at a new IRR bridge a real possibility? Who knows what really moves those folks down in the conference room at Peace Bridge Plaza?

One thing does seem clear: if folks on Buffalo's West Side don't want this project totally muddled by negative interests, they better get out there for Collaborative Workshop #5 and vote defensively.

64 PEACE BRIDGE *V.* WTC

THE PEACE BRIDGE QUESTION ASKED MORE THAN ANY OTHER

"I have a question for you," my friend Ron said at dinner the other night. "I read in the Times that they'd settled on a design for the World Trade Center site." He paused. I knew what he was going to say next and so did everyone else at the table.

"So what's with the Peace Bridge?" I said.

"Yes. What's with the Peace Bridge? Do they have anything? Why is it taking so long?"

Those are the questions – one question really – I have heard more often than all others combined in conversations about the Peace Bridge Expansion Project. It is the question asked more than all others in emails to Buffalo Report about the Peace Bridge Expansion Project. I suspect it is the question heard more than any other by Vincent "Jake" Lamb, who is heading up the Project, and Common Council James Pitts and Niagara District Common Council representative Dominic Bonifacio, the only elected officials in Buffalo city government who seem to have a major continuing interest in it.

NYC HAS AN IDEA, NOT A DESIGN

The first thing to be said is that the folks in New York City don't exactly have a design. What they have is a concept for a design. No individual building has been proposed, let alone planned. Daniel Libeskind came up with a vision, the shape of an idea. The only specific in it is the height of the memorial tower – 1776 feet – and that specific number of feet is not a function of any sophisticated or even superficially considered engineering or design factor. It's a political number: Libeskind's tower commemorates the year the Declaration of Independence was signed. It's a good thing for Libeskind's concept that the New York isn't as old as, say Baghdad, in which case his memorial tower would rise only 762 feet above the ground and would be obscured by buildings already in place.

The WTC project is a long way from construction workers, and longer still from anyone sitting in an office and doing business. Every structure in Libeskind's design concept has to be proposed, designed, submitted for environmental approval, and then built.

FOUR REASONS WHY BUFFALO AND FT. ERIE LACK EVEN AN IDEA, LET ALONE A DESIGN, LET ALONE A PLAN

That said, they are still ahead of us because they do have that concept for a design and we haven't even decided where the new bridge lanes will strike land.

I can think of four major reasons for the difference. There are no doubt others, and I would welcome hearing from anyone about them, but these four seem to me primary:

1. The price of inaction

Perhaps the key reason New York moved so quickly on the design concept, which must precede any specific design, is the vast amount of money involved.

The space previously occupied by the World Trade Center's offices generated billions of dollars a year in rent, salaries, taxes, and local multipliers. Since September 11, 2001, the World Trade Center site has been a big aching hole in the ground in lower Manhattan. And a big aching hole in the economy of the entire region. When the new buildings go up, they will comprise a business complex producing a huge amount of private and public money, as well as a memorial to the people murdered there on September 11, 2001. Almost everyone involved, from the people needing a symbol to the people needing to make money to the people needing to spend the money generated there, feel the urgency of that.

The Peace Bridge, on the other hand, is a working bridge, doing its job pretty well. Traffic is slowed by processing, mainly on the American side, but increasing lanes won't necessarily solve that problem; it will just provide more space on the bridge for trucks to park while they're waiting to be processed. The Peace Bridge is not the unique engine moving that traffic, as the buildings occupying the WTC site were and will be the unique engine developing whatever economic good or loss accrues there. Other bridges in this region handle Canadian-American truck traffic, and if the shippers on either sides of the border decide they don't like them they can move to other crossings to the east or west of here.

So there is no real urgency about the Peace Bridge Expansion Project and there never has been. Some folks anxious to make a lot of money sooner beat the drums saying there was, but it was nonsense. An expanded bridge will permit trucks to move across the bridge more quickly, but the situation is hardly critical. Those lines of trucks on the bridge are an annoyance for the drivers, they raise the cost of shipping goods slightly, but the flow is not cut, nor even curtailed. The huge pressure from Andrew Rudnick and the Buffalo Niagara Partnership a few years to build anything, no matter how ugly and environmentally destructive, was based entirely on increasing profits for a small number of manufacturers, shippers and builders.

2. Buffalo learns things

From the beginning of the World Trade Center project there was agreement that whatever went in there should put that space back into the economy and it should also memorialize the atrocity of September 11, 2001. The Peace Bridge project has rather been one of expanding aspirations. As the community learned more about what it had lost to the Buffalo and Fort Erie Public Bridge Authority over the past half century it began to demand more in any new construction. That's not delay. That's education and progress.

Opposition to the anachronistic, ugly and expensive twin span advocated by the Buffalo and Fort Erie Public Bridge Authority, the town of Fort Erie, and the Buffalo Niagara Partnership, developed only after people found out about Bruno Freschi's concept of a single-pylon curving cable-stayed bridge across the Niagara River. The New Millennium Group was inspired by Freschi's idea and poured a huge amount of energy into forcing the PBA to behave more responsibly. Buffalo Mayor Anthony Masiello never saw the potential of the project to have a major influence on Buffalo's future so he never got involved in it until a great amount of public pressure forced him to. A court order stopped the PBA from building its ugly bridge and much of the time since then has been spent redoing work that should have been done in the first place. Proponents of build-anything-now see a calendar that started running when the PBA and the Partnership came up with that stupid and irresponsible idea; proponents of responsible civic planning see the process as having started only when the PBA decided to stop fighting Judge Fahey's court order and to engage in the required environmental impact study. If you start counting from there, the process hasn't been going on very long at all.

3. A mess at the border

The only border the WTC planners had to deal with was the Hudson River. That is because the site is owned by the Port of New York Authority, a bistate agency. We've got a whole country to negotiate with, plus the petty politics of Fort Erie, Ontario, plus a new assumption of function by U.S. Customs that threatens to impose a whole new layer of bureaucratic clutter on everything going on here.

The Buffalo and Fort Erie Public Bridge Authority is a balanced agency representing an unbalanced situation. It has five members from the US, five from Canada. Simple enough, right? No, not at all simple. The tiny town of Fort Erie, Ontario, has a far greater voice in its affairs than does Buffalo, New York State's second largest city. Fort Erie's concerns are local, parochial, and stubborn. They are willing to consider problems of native artifacts on their side, but nothing about environment or quality of life on the American side. Four of the US board members are controlled by the governor of New York; all the Canadian board members are appointed by a member of Canada's government, the minister of Transport.

The single member of the American delegation on the PBA not appointed by the governor is the representative of the state attorney general. For the past four years that has been Barbra Kavenaugh. Since she announced that she's a candidate for Buffalo's comptroller earlier this month, there's been a good deal of noise saying she was the one defender of the city's interests against the forces of Canadian self-interest and American corporate greed on the PBA board. I don't know anybody who offered any evidence of that. Many of the PBA's most atrocious votes were unanimous, and Kavanaugh refused to return calls from reporters asking about any of them afterwards.

September 11, 2001 complicated anything having to do with the border. Everyone on this side of the border seemed to be moving toward a good solution to the Front Park, West Side air pollution, and bridge congestion problems: place all the Customs and Immigration facilities for both nations on the Canadian side of the river, where there is far more room to maneuver. It seemed, recently, as if that was going to happen. It would take an act of Congress and a similar act on the Canadian side, but Senator Charles Schumer has been aggressive on the issue and seemed to be making progress. It is difficult to get Hillary Clinton's attention for anything this local, but she seemed willing to go along with Schumer on this. In the House, Louise Slaughter has none of the twin span baggage that weighed down John LaFalce in these discussions.

But now comes U.S. Customs claiming a new portfolio: they no longer have to focus on smugglers of contraband drugs (their primary mission before 9/11) or on foreign malfeasants (their primary mission after 9/11) but say they now are responsible for infrastructure. Not only do they have to watch out for what bridge users are carrying and who those users might be, but they've got to bodyguard the bridge itself. To make that job easier, they want full inspection facilities on *both* sides of the river to check vehicles coming and going. That is a nightmare for everyone involved and right now everything is on hold while Schumer and Lamb try to get Customs to settle back down to earth.

4. Leadership Buffalo

Everyone says that is an oxymoron, but it isn't. There is leadership in Buffalo. Unfortunately, it rarely serves the public; it serves narrow corporate and private interests. Is that short term? Yes, but the corporations, as we've recently seen, will just move elsewhere when the rents and power and tax deals are cheaper in some new place or when corporate politics favors such a move. Remember Lucas-Verity and Adelphia? Corporations are not interested in long-term benefit to the community and it is naive of us to expect that of them. Corporations don't think in those terms, save when they serve immediate self-interest. The cowboy moneymakers, the hotshot developers coming in for the kill and making it because they've got this or that politician in their pocket – they think of themselves, not of community. You can't blame the corporations and developers for being like that. That's what they are. They're doing what they do.

We've had leadership – but it's been lousy leadership, or leadership in the service of private interests rather than the public good. The business community has continued to seek profit, but it has poured its money into making city government obedient to its will rather than making government better or more efficient. What else is new?

With the Peace Bridge, no one in Buffalo city government tried to get ahead of that mess when it was happening except maybe Jim Pitts, and it if weren't for Pitts's Binational Bridge Task Force city hall wouldn't be doing anything about it now.

WHAT WILL HAPPEN WITH THE BRIDGE?

the Peace Bridge Expansion Project, directed by Jake Lamb, will plod on. Accommodation will be made to Fort Erie's disproportionate voice in the decision. Barbra Kavanaugh will run her comptroller campaign claiming she was the people's advocate in the PBA boardroom and, since the key meetings were all *in camera*, no one will ever be able to prove her wrong. Customs will continue expanding its mission. Chuck Schumer will continue trying to make Washington behave reasonably. Hillary will continue assiduously avoiding this issue except for utterances about national security, when she can wave the flag.

The design process will plod along. If the ground isn't cut from beneath his feet by wheelers and dealers on the PBA board or by bureaucratic work-expanders at Customs or bottom-feeders in the Buffalo development community, Jake Lamb's consultants will come up with a decent design. Whether or not that decent design gets built will be up to the PBA board and the politicians we elect to represent our interests.

My friend Ron who asked me that question about the state of the Peace Bridge design was also asking a second question: what can we do to help make sure that we're not screwed in the process, just like all the other times.

Thinking about that brings to mind a one-hour radio documentary called "Who Killed Michael Farmer?" done years ago by the great Edward R. Murrow. It was about a boy killed by a gang in a New York City park. He had nothing to do with the gang; he was just walking through a park on his way home from school and they killed him. The first half of the program was about what a good kid Michael Farmer was and what a bad bunch the gang that killed him – the Egyptian Kings and Dragons, as I recall – were. By the end of that half, the world was simple.

Then Murrow spent the next 30 minutes showing how the Egyptian Kings and Dragons had sought help and how the city had pulled it away, how they had depended on a detached street worker and how the city had abolished the street worker's job. He ended his program with a line I think of just about every time I hear someone complain about how we've been sold out, betrayed, let down or just received lousy service from this or that councilperson or the mayor:

"Why do these things happen?" Murrow said. He paused. "Because we let them."

It's easy to blame Tony Masiello for doing whatever this or that developer tells him to do or to blame the Common Council for behaving like a bunch of squabbling fools. Those people are what they are. You can't fault them for that. It's we who should be faulted, those of us who voted for them after they disappointed us last time, or who didn't push the political parties to give us candidates worth a damn, or who didn't write a check or carry a petition for a new face in the game who wasn't burnt out or worn down or paid off.

A lot of key choices have yet to be made in the Peace Bridge Expansion Project and some of them will be influenced, if not decided, by people we put in office in the next election. One more reason to take that election seriously, even at the most local level.

INDEX OF NAMES

Individuals who appear under several names are listed under their largest name: "Doug Turner" is listed under **"Douglas Turner"**; "Clint Brown" under **"Clinton Brown"**, "Pam Earl" under **"Pamela Earl"**. Acronyms that appear only as acronyms are listed that way with the full name in parentheses, as **"ADL *(Arthur D. Little, Co.)*"**. Organizations and laws that appear as full names and acronyms are listed under the full name, with a reference to the full name at the acronym location. The name of the city Buffalo, which occurs on nearly every page, is not indexed. The entry for "Buffalo and Fort Erie Public Bridge Authority" includes "Bridge Authority,""Public Bridge Authority," "PBA" and "BFEPBA."